# FACING HAZARDS AND DISASTERS
## UNDERSTANDING HUMAN DIMENSIONS

Committee on Disaster Research in the Social Sciences:
Future Challenges and Opportunities

Division on Earth and Life Studies

NATIONAL RESEARCH COUNCIL
*OF THE NATIONAL ACADEMIES*

THE NATIONAL ACADEMIES PRESS
Washington, D.C.
**www.nap.edu**

THE NATIONAL ACADEMIES PRESS    500 Fifth Street, N.W.    Washington, DC 20001

NOTICE: The project that is the subject of this report was approved by the Governing Board of the National Research Council, whose members are drawn from the councils of the National Academy of Sciences, the National Academy of Engineering, and the Institute of Medicine. The members of the committee responsible for the report were chosen for their special competences and with regard for appropriate balance.

This study was supported by Contract No. CMS-0342225 between the National Research Council and the National Science Foundation. Any opinions, findings, conclusions, or recommendations expressed in this publication are those of the author(s) and do not necessarily reflect the views of the organizations or agencies that provided support for the project.

International Standard Book Number 0-309-10178-6 (Book)
International Standard Book Number 0-309-65985-X (PDF)
Library of Congress Catalog Card Number 2006931516

Copies of this report are available upon request from Byron Mason, the National Academies, Division on Earth and Life Studies, 500 Fifth Street, N.W., Keck 610, Washington, DC 20001; (202) 334-3511.

Additional copies of this report are available from the National Academies Press, 500 Fifth Street, N.W., Lockbox 285, Washington, DC 20055; (800) 624-6242 or (202) 334-3313 (in the Washington metropolitan area); Internet, http://www.nap.edu.

# THE NATIONAL ACADEMIES
*Advisers to the Nation on Science, Engineering, and Medicine*

The **National Academy of Sciences** is a private, nonprofit, self-perpetuating society of distinguished scholars engaged in scientific and engineering research, dedicated to the furtherance of science and technology and to their use for the general welfare. Upon the authority of the charter granted to it by the Congress in 1863, the Academy has a mandate that requires it to advise the federal government on scientific and technical matters. Dr. Ralph J. Cicerone is president of the National Academy of Sciences.

The **National Academy of Engineering** was established in 1964, under the charter of the National Academy of Sciences, as a parallel organization of outstanding engineers. It is autonomous in its administration and in the selection of its members, sharing with the National Academy of Sciences the responsibility for advising the federal government. The National Academy of Engineering also sponsors engineering programs aimed at meeting national needs, encourages education and research, and recognizes the superior achievements of engineers. Dr. Wm. A. Wulf is president of the National Academy of Engineering.

The **Institute of Medicine** was established in 1970 by the National Academy of Sciences to secure the services of eminent members of appropriate professions in the examination of policy matters pertaining to the health of the public. The Institute acts under the responsibility given to the National Academy of Sciences by its congressional charter to be an adviser to the federal government and, upon its own initiative, to identify issues of medical care, research, and education. Dr. Harvey V. Fineberg is president of the Institute of Medicine.

The **National Research Council** was organized by the National Academy of Sciences in 1916 to associate the broad community of science and technology with the Academy's purposes of furthering knowledge and advising the federal government. Functioning in accordance with general policies determined by the Academy, the Council has become the principal operating agency of both the National Academy of Sciences and the National Academy of Engineering in providing services to the government, the public, and the scientific and engineering communities. The Council is administered jointly by both Academies and the Institute of Medicine. Dr. Ralph J. Cicerone and Dr. Wm. A. Wulf are chair and vice chair, respectively, of the National Research Council.

**www.national-academies.org**

*v*

# Preface

The United States and many other countries throughout the world are vulnerable to a wide variety of natural, technological, and willful hazards and disasters. In this nation, while local decision makers and other stakeholders have the final responsibility for coping with disaster threats, federal agencies have developed science-based activities, including research and applications programs that are intended to further the understanding of such threats and provide a basis for more effective risk reduction efforts in vulnerable communities throughout the country. The National Science Foundation (NSF), sponsor of this study, has been in the forefront in providing support for social science hazards and disaster research, including research carried out through the National Earthquake Hazards Reduction Program (NEHRP), which was established in 1977. Since the creation of the Department of Homeland Security (DHS) in response to the September 11, 2001 terrorist attacks, that agency also has emerged as a potential major sponsor of social science hazards and disaster research.

Given the changing hazards and disasters landscape in recent years, brought on by such factors as new demographic trends and settlement patterns and the emergence of new kinds of disaster threats discussed in this report, NSF requested that the National Research Council (NRC) conduct an analysis of hazards and disaster research in the social sciences, a research community that is vital to understanding societal responses to natural, technological, and willful threats. In particular, NSF asked the NRC to provide the agency and other stakeholders with an appraisal of the social science contributions to knowledge on hazards and disasters, especially as a result of NEHRP funding; the challenges facing the social science hazards and disaster research community; and opportunities for advancing knowledge in the field and its application for the benefit of society. The study is expected to provide a basis for planning future social science disciplinary, multidisciplinary, and interdisciplinary research and application activities related to the threat of natural, technological, and willful disasters.

In response to this charge, the NRC established the Committee on Disaster Research in the Social Sciences, an ad hoc committee under the Division on Earth and Life Studies. The committee was comprised of experts from various social science disciplines, public health, and emergency management. The committee met six times during the course of the study. As part of the input to the study, the committee reviewed in detail the scientific literature in the field. The committee also benefited from presentations and discussions that took place during two workshops held in conjunction with committee meetings, one in Washington, D.C., at the National Academies' Keck Center and the other in Irvine, California, at the National Academies' Beckman Conference Center. Participants in the first workshop included researchers from the multidisciplinary hazards and disaster research community, practitioners, and representatives from various agencies. All participants in the second workshop were practitioners.

The many people who provided input to the committee through oral presentations or in writing are listed in the acknowledgments. On behalf of the committee, I extend appreciation and thanks to all of these individuals for contributing to the study. The committee also extends special appreciation to William A. Anderson, study director for the project, whose substantive knowledge and experience in hazards and disaster research are enormous and whose contributions to the study were essential to its successful completion. Thanks also to Patricia Jones Kershaw, who was senior program associate during part of the study, and especially to Byron Mason, program associate, who provided very effective substantive and logistical support for all phases of the committee's work. Finally, I wish to thank the members of the committee for devoting substantial time and effort to the project. Their commitment to the field has been matched by their hard work on this committee.

Gary A. Kreps
*Chair*

# Acknowledgments

This report was greatly enhanced by the participants of the three public meetings, including two workshops, held as part of this study. The committee would like to acknowledge the efforts of those who gave presentations at the meetings: James Ament, Michel Bruneau, Caroline Clark, Joseph Coughlin, Penny Culbreth-Graft, Frances Edwards, Joshua M. Epstein, Steven French, Gerard Hoetmer, Eric Holdeman, Howard Kunreuther, Rocky Lopes, Larry Mintier, Jack Moehle, Poki Namkung, Robert O'Connor, Anthony Oliver-Smith, Laura Petonito, Ralph B. Swisher, Roger Tourangeau, Larry Weber, Dennis Wenger, Thomas Wilbanks, and Rae Zimmerman. The committee would also like to acknowledge the written contribution of Thomas E. Drabek.

This report has been reviewed in draft form by individuals chosen for their diverse perspectives and technical expertise, in accordance with procedures approved by the National Research Council's Report Review Committee. The purpose of this independent review is to provide candid and critical comments that will assist the institution in making its published report as sound as possible and to ensure that the report meets institutional standards for objectivity, evidence, and responsiveness to the study charge. The review comments and draft manuscript remain confidential to protect the integrity of the deliberative process. We wish to thank the following individuals for their review of this report:

Ruzena K. Bajcsy, University of California, Berkeley
Eve Gruntfest, University of Colorado at Colorado Springs

Peter J. May, University of Washington, Seattle
Dennis S. Mileti, University of Colorado at Boulder
Robert B. Olshansky, University of Illinois at Urbana-Champaign
Adam Z. Rose, Pennsylvania State University, University Park
David M. Simpson, University of Louisville, Kentucky
Neil J. Smelser, University of California, Berkeley
Seth A. Stein, Northwestern University, Evanston, Illinois
Susan Tubbesing, Earthquake Engineering Research Institute, Oakland,
    California

Although the reviewers listed above provided many constructive comments and suggestions, they were not asked to endorse the conclusions or
recommendations nor did they see the final draft of the report before its
release. The review of this report was overseen by Enrico L. (Henry)
Quarantelli, Disaster Research Center, University of Delaware, and Carl
Wunsch, Massachussetts Institute of Technology. Appointed by the
National Research Council, they were responsible for making certain that
an independent examination of this report was carried out in accordance
with institutional procedures and that all review comments were carefully
considered. Responsibility for the final content of the report rests entirely
with the authoring committee and the institution.

# Contents

# Summary

Recent catastrophic events—in 2005, the earthquake at the borders of Pakistan, India, and Afghanistan as well as Hurricane Katrina along the United States Gulf Coast; in 2004, the Indian Ocean tsunami, and in 2001, the terrorist attacks on New York City and Washington, D.C.—are stark reminders of the global importance and implications of natural, technological, and willful disasters. Response to such events before, when, and after they occur are matters of both hazards and disaster management practice and public policy at national and international levels. Responses to the September 11, 2001 terrorist attacks has led to a wide range of policy changes that may affect all phases of emergency management, including the newly created U.S. Department of Homeland Security (DHS), the U.S. Patriot Act, and the Aviation and Transportation Security Act. The inclusion of the Federal Emergency Management Agency (FEMA) within the DHS may have important implications for U.S. response to major natural disasters such as Hurricane Katrina.

Studies of hazards and disasters by social scientists is the primary focus of this report, particularly research undertaken during the past three decades with support provided by the National Science Foundation through the National Earthquake Hazards Reduction Program (NEHRP). Since the establishment of NEHRP in 1977, a cadre of social science researchers—from such disciplines as geography, sociology, political science, psychology, economics, decision science, regional science and planning, public health, and anthropology—has made continuing contributions to the development of knowledge about societal response to hazards and disasters. Among other advances, these contributions have helped to dispel myths about crisis related behaviors, led to improvements in early warning and evacuation systems, and facilitated the ways communities and regions prepare for disasters.

Disaster research, which has focused historically on emergency response and recovery, is incomplete without the simultaneous study of the societal hazards and risks associated with disasters, which includes data on the vulnerability of people living in hazard-prone areas. Historically, hazards and disaster research have evolved in parallel, with the former focusing primarily on hazards vulnerability and mitigation, the latter primarily on disaster response and recovery, and the two veins intersecting most directly with common concerns about disaster preparedness. It is vital, however, that future social science research treat hazards and disaster research interchangeably and view the above five core topics of hazards and disaster research within a single overarching framework (see Figure S.1). Such integration also provides the foundation for increased collaborative work by social scientists with natural scientists and engineers.

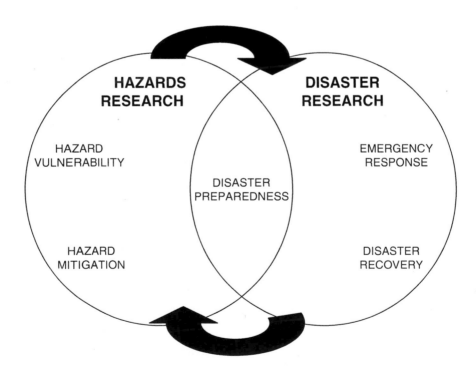

**FIGURE S-1** Core topics of hazards and disaster research.

This report, conducted with support from the National Science Foundation, assesses the current state of social science hazards and disaster research and provides a set of recommendations that reflect opportunities and challenges in the field. Although research to date has revealed much about how societies respond to natural and technological disasters of various types, it is clear from the following report that we need to learn more. Among the most needed types of research are studies that compare systematically the unique circumstances of catastrophic events such as major earthquakes, hurricanes, and acts of terrorism. Such comparative studies will allow researchers to examine societal response in relation to variables such as the amount of advanced warning, the magnitude, scope, and duration of impacts, and the special requirements for dealing with chemical, biological, and radiological agents. Among the report's other recommendations is the need for systematic studies of how societies complement expected and sometimes planned responses with improvised activities. In the September 11, 2001 terrorist attacks, for example, first responders had to work around the loss of New York City's Emergency Operations Center, which was located in one of the towers.

## CHARGE TO THE COMMITTEE

The committee's primary mission is to provide NSF and other stakeholders with a detailed appraisal of the short- and long-term challenges facing social science hazards and disaster research, and also new and emerging opportunities for advancing knowledge within the social sciences and through interdisciplinary collaborations with the natural sciences and engineering. Of central importance to its statement of task, the committee is charged with examining the contributions and accomplishments of the social sciences since the establishment of NEHRP in 1977, the program that through NSF has provided much of the support for social science research on hazards and disasters for more than 25 years. The committee is also charged with assessing the impact of key societal changes on the way social science hazards and disaster research will be carried out in the future and what should be studied nationally and internationally. Finally, in the context of these societal changes, the committee is charged with considering the special challenges of post-disaster investigations, advancing the application of research findings, and meeting future social science workforce needs in this field. In completing the above mission and tasks, the committee has drawn on the experience and expertise of its 13 members, the voluminous social science research literature on hazards and disasters, and information and insights from two workshops that were held during the course of the study.

## STUDY CONCLUSIONS

The committee's assessment of the current state of social science research can be summarized succinctly in the following conclusions:

**Social science hazards and disaster research has advanced in the United States and internationally.** Under NEHRP social science knowledge has expanded greatly with respect to exposure and vulnerability (physical and social) to natural hazards in the United States, such that the foundation has been established for developing more precise loss estimation models and related decision support tools for hazards and disasters generally. The contribution of NEHRP to social science knowledge on natural hazards is less developed internationally as is its contribution nationally and internationally on exposure and vulnerability to technological and willful threats.

**Social science knowledge about the responses of U.S. households to natural hazards and disasters is well developed.** There is a solid knowledge base at the household level of analysis on vulnerability assessment, risk communication, evacuation and other forms of protective action, and expedient disaster mitigation activities—for example, how people in earthquake or flood prone regions communicate about risks and warning messages, and how they respond to warning messages. The knowledge base and related explanatory modeling under NEHRP are skewed toward natural hazards (most notably earthquakes) as opposed to technological and willful hazards, and so far they have been confined primarily to national rather than international contexts.

**Far less is known about how the characteristics of different types of hazards affect disaster preparedness and response.** There has been little systematic comparative work on the special characteristics of natural, technological, and willful disasters (e.g., predictability and controllability; length of forewarning, magnitude, scope, and duration of impact) and their relationships with physical and social impacts. For example, how does the variation in warning time—little or no warning for an earthquake, short-term warning for tornados, longer-term warnings for hurricanes, and indeterminate warnings for terrorist attacks—affect preparedness and response? Greater understanding of event/impact relationships would directly facilitate the adoption of more effective disaster preparedness and mitigation practices.

More is known about immediate post-disaster responses of groups, organizations, and social networks than about mitigation or disaster recovery policies and practices. While less so than the post-World War II studies that preceded NEHRP's establishment in 1977, NEHRP-sponsored social science research has still tended to focus more on the immediate aftermath of disasters (post-disaster responses) and related emergency preparedness practices than on the affects of pre-disaster mitigation policies and practices, disaster recovery preparedness or longer term recovery from specific events. Research over several decades has contradicted myths that during disasters panic will be widespread, that large percentages of those who are expected to respond will simply abandon disaster roles, that local institutions will break down, that crime and other forms of antisocial behavior will be rampant, and that psychological impairment of victims and first responders will be a major problem. The more interesting and important research questions have become how and why communities, regions, and societies leverage expected and improvised post-impact responses in coping with the circumstances of disasters. While much of organizational response to disaster is expected and sometimes planned, improvisation is an absolutely essential complement of predetermined activities.

**The circumstances of terrorist threats could alter societal response to disasters.** The possibility exists that some future homeland security emergencies could engender responses that are different from those observed in previous post-disaster investigations of natural and technological disasters. Particular attention is being given post-September 11, 2001 to vulnerability assessment of national energy, transportation, and information systems, terrorist threat detection and interdiction, the special requirements of nuclear, biological, and chemical agents, and the organizational requirements of developing multigovernmental preparedness and response systems. Fortunately these concerns are readily subsumed within the historically mainstream topics of hazards and disaster research depicted in Figure S.1 above.

**NEHRP has made important contributions to understanding longer-term disaster recovery.** Prior to NEHRP relatively little was known about disaster recovery processes and outcomes at different levels of analysis (e.g., households, neighborhoods, firms, communities, and regions). While research on disaster recovery remains somewhat underdeveloped, NEHRP funded projects have refined general conceptions of disaster recovery, made important contributions in understanding the recovery of households (primarily) and firms (more recently), and contributed to the development of statistically based community and regional models of post-disaster losses and recovery processes. Moreover, interest in the relationship between disaster recovery and sustainable development has become sufficiently pronounced in this field that the committee has allocated an entire chapter of the report to its consideration.

**The management and accessibility of data needs immediate attention.** Thus far social scientists have not confronted systematically issues related to the management and accessibility of data—from its original collection and analysis, to its longer-term storage and maintenance, and to ensuring its accessibility over time to multiple users. What the committee has termed the "*hazards and disaster research informatics problem*" is not unique to this research specialty, or to the social sciences, natural sciences, and engineering generally. But the informatics problem demands immediate attention and resolution as a foundation for future research and application of findings.

**How research is communicated and applied is not well understood.** More systematic research is needed on how hazards and disaster information generated by the social sciences and other disciplines is disseminated and applied. Such research will provide clearer understanding of what can be done within hazards and disaster research to further the dissemination of knowledge, thereby advancing sound mitigation, preparedness, response, and recovery practices.

**A more diverse, interdisciplinary, and technologically sophisticated social science workforce is needed in the future.** Given the national and international importance of natural, technological, and willful disasters, the next generation of social scientists studying these events should become larger, more diverse, and more conversant with interdisciplinary perspectives and state-of-the-art research methods and technologies than the previous generation.

## RECOMMENDATIONS

Grounded in the above conclusions of its assessment, the committee has offered 38 separate recommendations in Chapters 3 through 9 of the report, with the majority relating to the need for comparative studies of societal responses to natural, technological and willful hazards and disasters. No explicit priorities among these recommendations have been established by the committee, primarily because traditional topics within, respectively, hazards and disaster research necessarily are interrelated. The committee also wishes to ensure that NSF and other stakeholders have considerable flexibility in addressing the broad range of research and application issues included in its statement of task from NSF. For purposes of this report summary, the 38 separate recommendations are encapsulated within three global recommendations. In discussing each one, the committee offers guidance to NSF and other stakeholders for their future consideration.

**Summary Recommendation 1:** *Comparative research should be conducted to refine and measure core components of societal vulnerability and resilience to hazards of all types, to address the special requirements of confronting disasters caused by terrorist acts, and to advancing knowledge about miti-*

*gation, preparedness, response, and recovery related to disasters having
catastrophic physical and social impacts.*

The recommended comparative research is essential for isolating common
from unique aspects of societal response to natural, technological, and
willful hazards and disasters. A key contribution of NSF through NEHRP
over the years has been that, while necessarily emphasizing earthquakes,
since its inception the program has encouraged and supported comparisons
of societal responses to earthquakes with other natural as well as technological hazards and even with terrorist-induced events, though less so. This
historical emphasis within NEHRP dictates that a rigorous approach should
prevail in making generalizations to terrorism and that there is a continuing
need for systematic comparisons of all societal hazards and disasters using
the conceptual and methodological tools summarized in this report. A comparative perspective should be sustained within NSF and also prevail in the
new DHS.

The five core topics of hazards and disaster research depicted in Figure S.1
are referenced explicitly in both the summary recommendation for comparative research as well as the more detailed lists of research recommendations
found in the report. These five core topics are deemed by the committee to
be equally important to the development and application of social science
knowledge. Thus, the committee sees no useful purpose for establishing
priorities among what have traditionally been termed disaster research
topics, on the one hand, and hazards research topics on the other. On the
contrary, a major priority demanded by the conceptual approach adopted
by the committee is to capture to every extent possible within specific
studies the essential relatedness of these core research topics. Accomplishing this research goal will require research designs that are both comparative and longitudinal.

**Summary Recommendation 2:** *Strategic planning and institution building
are needed to address issues related to the management and sharing of data
on hazards and disasters (hazards and disaster informatics), sustain the
momentum of interdisciplinary research, advance the utilization of social
science findings, and sustain the hazards and disaster research workforce.*

Of particular importance because of its direct relationship to Summary
Recommendation 1 is the call for strategic planning to address issues of
data management and data sharing. A Panel on Hazards and Disaster
Informatics should be created to guide these efforts. The Panel should be
interdisciplinary and include social scientists and engineers from hazards
and disaster research as well as experts on informatics issues from cognitive
science, computational science, and applied science. The Panel's mission
should be, first, to assess problems of data standardization, data management and archiving, and data sharing as they relate to natural, technological,

and willful hazards and disasters, and second, to develop a formal plan for resolving these problems to every extent possible within the next five years.

Post-disaster investigations inherently have an ad hoc quality because the occurrence and locations of specific events are uncertain. That is why special institutional and often funding arrangements have been made for rapid response field studies and the collection of perishable data. But the ad hoc quality of post-impact investigations does not mean that their research designs must be unstructured or that the data ultimately produced from these investigations cannot become more standardized, machine readable, and stored within data archives. Having learned what to look for after decades of post-disaster investigation by social scientists, the potential for highly structured research designs and replicable data sets across multiple disaster types and events can now be realized. Pre-impact investigations of hazards and their associated risks are no less important than post-impact investigations of disasters, less subject to the uncertainties of specific events, arguably more amenable to highly structured and replicable data sets, and no less in need of data archives that are readily accessible to both researchers and practitioners.

Addressing hazards and disaster informatics issues within the next five years requires interdisciplinary collaboration. This collaboration can build on the momentum of interdisciplinary research that has been achieved at NSF's three earthquake engineering centers during the past decade and advance the sharing of more highly structured data and findings within the entire hazards and disaster research community. Resolving informatics issues within this community will then lead to greater accessibility of hazards and disaster research to policy makers and practitioners at national and international levels. The assessment of knowledge utilization in this field calls for the continuing role of social scientists because of their special expertise in evaluation research.

The committee's call for strategic planning on interrelated informatics, interdisciplinary research, and knowledge dissemination logically precedes specific recommendations in the report for interdisciplinary centers and workforce development. One recommended interdisciplinary center could serve as a natural site for implementing a strategic plan on hazards and disaster informatics. Among other functions, such a center could serve as a distributed social science data archives that would be accessible to the entire research community. A second recommended center would promote, also on a distributed basis, the application of state-of-the-art modeling, simulation, and visualization techniques to terrorist events as well as natural and technological disasters.

Workforce development is a continuing issue for social science hazards and disaster research, and an integrated strategy to replenish and expand the current research workforce is needed. The workforce problem will be difficult to resolve in the short term, and it requires more careful assessment than the resources of the committee have allowed. As an interim step, the committee recommends that a workshop be held to facilitate communication, coordination, and planning among stakeholders from governmental, academic, and professional constituencies. Representatives from NSF and DHS should play key roles in the workshop because of their historical (NSF) and more recent (DHS) shared commitment to foster the next generation of hazards and disaster researchers.

**Summary Recommendation 3:** *NSF and DHS should jointly support the comparative research, strategic planning, and institution building called for in Summary Recommendations 1 and 2.*

The proposed leveraging of NSF with DHS support is critical because these two agencies are focal points of federal funding for research on all types of extreme events. The two agencies should take advantage of opportunities to leverage their resources by jointly funding social science hazards and disaster research whenever possible. This could lead to a better understanding of the similarities and differences between natural, technological, and human-induced hazards and disasters. It could also provide the foundation for sound science-based decision making by policy makers and practitioners, whether they are developing measures to counter a major natural disaster like Hurricane Katrina or a terrorist-induced event like the September 11th attacks on the World Trade Center and Pentagon. Social science research on the September 11, 2001 terrorist attacks as well as more limited observations that have been made thus far on Hurricane Katrina indicate, first, that many previous findings about societal response to hazards and disasters remain valid, and second, that there is still much to be learned about responses to truly catastrophic events.

## A VISION OF SOCIAL SCIENCE CONTRIBUTIONS TO KNOWLEDGE AND A SAFER WORLD

While NSF social science studies supported through NEHRP are summarized in some detail in the report that follows, the committee's overall vision of future hazards and disaster research underlies the summary recommendations that have been developed. The committee envisions a future:

- where the origins, dynamics, and impacts of hazards and disasters become much more prominent mainstream as well as specialty research interests throughout the social sciences;
- where traditional social science investigations of post-disaster responses become more integrated with no less essential studies of hazard vulnerability, hazard mitigation, disaster preparedness, and post-disaster recovery;

- where disciplinary studies of the five core topics of hazards and disaster research within the social sciences increasingly become complemented by interdisciplinary collaborations among social scientists themselves and between social scientists and their colleagues in the natural sciences and engineering;
- where there is continuing attention throughout the hazards and disaster research community on resolving interdisciplinary issues of data standardization, data management and archiving, and data sharing;
- where there is continuing attention throughout hazards and disaster research on the dissemination of research findings and assessments by social scientists of their impacts on hazards and disaster management practices at local, regional, and national levels;
- where each generation of hazards and disaster researchers makes every effort to recruit and train the next generation; and
- where the funding of hazards and disaster research by social scientists, natural scientists, and engineers is a cooperative effort involving the NSF, its partner agencies within NEHRP, the Department of Homeland Security, and other government stakeholders.

With the foundation established by previous basic and applied studies of hazards and disasters, and guided by the committee's recommendations, the above vision is attainable. Describing and explaining societal response to hazards and disasters is both a continuing challenge and major opportunity for the social sciences. Natural, technological, and willful hazards and disasters faced by humankind are continuous, global in nature, and increasing with demographic expansion, technological change, economic development, and related social and political dynamics of enormous complexity. Considerable progress has been made during the past several decades by social scientists studying different types of hazards and disasters, sometimes working collaboratively with investigators from other disciplines. But the continuing challenge for the social sciences centers on unraveling the complexity of individual and collective action before, during, and after disasters occur, on providing research findings that improve loss reduction decision making, and on assessing hazards and disaster related policies and programs. The major opportunity for the social sciences is to employ state-of-the-art theories, methods, and supporting technologies to further this type of knowledge development, which can in turn further science-based decision making by policy makers and practitioners. The responsibility for attaining the committee's vision is in no sense the sole responsibility of NSF. That responsibility can and should be shared with the entire hazards and disaster research community, with those who fund hazards and disaster studies, and certainly with those who stand to learn from these studies.

# 1

# Introduction

Τhis opening chapter is organized as follows: The charge and major tasks of the Committee on Disaster Research in the Social Sciences (DRSS) are summarized initially. An orienting definition of disasters (and hazards and risks as key related concepts) is then offered along with an explicit framework that addresses central conceptual and measurement issues in hazards and disaster research. An historical overview of social science research within the National Science Foundation's (NSF's) National Earthquake Hazards Reduction Program (NEHRP) is then presented, and this is followed by a summary of key issues that inform the committee's charge and tasks. The introduction concludes with a brief characterization of the remaining chapters of the report.

DRSS is an ad hoc committee under the Division on Earth and Life Studies. The study project was initiated in February 2004 with funding from NSF. The charge to the committee for the 18-month study is stated in Box 1.1.

In carrying out its charge, the committee has drawn on the experience and expertise of 13 members of the hazards and disaster research community from the disciplines of psychology, geography, political science, sociology, economics, decision science, regional science and planning, public health, and emergency management. In preparing its report, the committee has drawn on the literature in the field as well as information and insight from two workshops that were held during the course of the study.

As noted in Figure 1.1, adapted from Tierney et al. (2001), components of hazards and disaster research have evolved historically with different

---

**BOX 1.1**
**Statement of Task**

The objective of the study is to provide the National Science Foundation and other stakeholders with a detailed appraisal of the short- and long-term challenges facing the social science disaster research community and new and emerging opportunities for advancing knowledge in the field and its application for the benefit of society. The study should provide a basis for planning future social science and multidisciplinary research related to natural, technological, and willful disasters in response to challenges and opportunities presented by a changing nation and world.

In order to put future projections into context, the study will initially examine the contributions and accomplishments of the social sciences in the field starting with the creation of the National Earthquake Hazards Reduction Program (NEHRP), the program that through NSF has provided much of the support for the social science effort to date. Attention will be given to the contributions of the social sciences to understanding the full range of natural, technological and human-induced disasters that social scientists have studied during the past 25 years since NEHRP was established.

Overall the study will examine the following areas:

— Social science contributions under NEHRP, both in terms of knowledge creation and utilization.
— Contributions of the social sciences since the creation of NEHRP to the understanding of natural, technological and human-induced hazards faced by communities in the nation.
— Challenges posed for the social science disaster research community due to the expectation that, like other relevant disciplines, it become a major partner in integrated hazard and disaster research.
— Opportunities for bridging the gap between social scientists that study natural disasters and those that investigate technological risks.
— Likely impact of key societal changes—such as the emergence of new technologies, emphasis on new hazards, and a changing emergency management profession—on how disaster research is done by social scientists in the future, as well as what is studied.
— Challenges of post-disaster investigations and opportunities to increase their value.
— Future opportunities for collaborative international research.
— Opportunities for meeting the challenge of furthering the application of research results.
— Future workforce needs and opportunities to meet them.

---

emphases, depending on the types of hazards and disasters studied and research topics related to them. Given the above charge and tasks of the committee, further integration of hazards and disaster research, as depicted by the overlapping circles and two-directional arrows in Figure 1.1, is a fundamental future requirement for the social sciences. Such integration

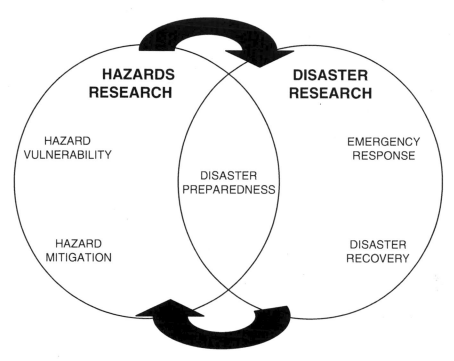

**FIGURE 1.1** Core topics of hazards and disaster research.

within the social sciences also can provide the foundation for increased collaborative work by social scientists with natural scientists and engineers.

## THE DISASTER CONSTRUCT

*Disasters are non-routine events in societies or their larger subsystems (e.g., regions and communities) that involve conjunctions of physical conditions with social definitions of human harm and social disruption.* (Kreps, 2001:3718)

This entry, from the latest edition of the *International Encyclopedia of the Social and Behavioral Sciences*, draws on the historically rich tradition of hazards and disaster studies within the social sciences, most notably since the post-World World II era (for earlier to more recent statements see Fritz, 1961; Barton, 1969; Dynes, 1970; White and Haas, 1975; Quarantelli and Dynes, 1977; Kreps, 1984; Burton et al., 1993; Kreps and Drabek, 1996; Kunreuther and Roth, 1998; Mileti, 1999b; Tierney et al., 2001;

Cutter, 2001; Montz et al., 2003a,b). So defined, disasters are both physical events and public policy issues with distinctive qualities. As further clarified in the above encyclopedia entry:

> *The phrase "nonroutine events" distinguishes disasters as unusual and dramatic happenings from everyday issues and concerns. The dual reference to "physical conditions" and social definitions means that each is individually necessary and both are collectively sufficient for disasters to occur in social time and space. The designation "societies or their larger subsystems" means that human harm and social disruption must have relevance for larger social systems. . . . Poverty, hunger, disease, and social conflict are chronic societal concerns. Economic depressions, famines, epidemics, and wars are disasters as defined above. Global warming and ozone depletion have become defined objectively and subjectively as environmental hazards or risks. The possible disastrous consequences of these hazards . . . remain matters of scientific and public debate. . . .* (Kreps, 2001:3718)

While the term *disaster* is part of popular parlance, it also has important bureaucratic meaning (e.g., disaster declarations). Potential disasters are associated with *hazards* of various types and the *risks* (i.e., probabilities) of specific events occurring. Distinctions among these three terms are useful and important. As Cutter (2001:3) notes:

> the distinction between hazard, risk, and disaster is important because it illustrates the diversity of perspectives on how we recognize and assess environmental threats (risks), what we do about them (hazards), and how we respond to them after they occur (disasters). The emphasis on hazard, risk, and disaster is also reflective of different disciplinary orientations of researchers and practitioners. . . . However, as the nature of hazards, risks, and disasters became more complex and intertwined and the field of hazards research and management more integrated, these distinctions became blurred as did the differentiation between origins as "natural," "technological," or "environmental."

The blurred distinctions highlighted by Cutter, a geographer whose research focuses more heavily on the left-hand side of Figure 1.1, have contributed greatly to breaking down historical barriers between hazards and disaster research. This positive development has been affirmed by two sociologists (Tierney and Perry) and a social psychologist (Lindell), whose interests focus more heavily on the right-hand side of Figure 1.1 (Tierney et al., 2001:22):

> . . . more comprehensive perspectives are needed that consider both disaster events and the broader structural and contextual factors that contribute to disaster victimization and loss. While the functionalist approach that characterized classical disaster research mainly addressed the fact of disaster, not the sources of disaster vulnerability, other work has sought to better

understand the societal processes that create vulnerability, how vulnerability is distributed unequally across societies, communities, and social groups, how vulnerability changes over time, and how and why these changes come about.

Definitions of core subject matter necessarily are matters of intellectual discussion and debate within any science. Studies of hazards and disasters are no different. During the past decade, for example, there have been two books (Quarantelli, 1998; Perry and Quarantelli, 2005) wherein authors from several social science fields have grappled with the question: What is a disaster? A diversity of perspectives on the meaning of disasters, hazards, and risks is to be expected (1) because the social sciences are not homogeneous disciplines either theoretically or empirically, and (2) because these constructs are of interest to scholars nationally and internationally. In reviewing this continuing dialogue about core subject matter, the committee agrees with Perry's conclusions in both of the above volumes that there is more agreement than disagreement on the definitional fundamentals (Perry, 1998:197-217, 2005). With respect to its mission and tasks, the committee makes the following assumptions about disasters and the hazards to which they relate.

First, while all concepts in science are nominal, consensus about objects of inquiry is essential to developing and applying knowledge about them. Second, disasters have physical impacts and involve subjective definitions formulated by individuals and social entities. Third, disasters are disruptive of social systems at small to more inclusive levels and are intertwined with broader dynamics of change. Fourth, the characteristics of disasters themselves must be distinguished from their antecedents and consequences. Capturing these antecedents and consequences is part and parcel of constructing descriptive and explanatory models of hazards and disasters. Fifth, given the broad range of hazards and disasters that can be studied, developing typologies and taxonomies is an essential component of theory building. As discussed in this report, classification schemes have frequently been based on defining characteristics of disasters such as their length of forewarning; detectability; speed of onset; and magnitude, scope, and duration of impact. Such dimensions allow for comparisons of multiple disasters, thus bridging the gap among social scientists studying hazards that are natural, technological, or willful in origin. Sixth, research on hazards and disasters requires an appeal to the scientific logic of discovery and explanation, regardless of substantive topic and regardless of whether the research is discipline based, multidisciplinary, or interdisciplinary. Finally, before, when, and after they occur, disasters are physical and social catalysts of collective action.

This last observation merits a further comment. Some years ago an influential social science meta-theorist made the following point (Dubin,

1978:115-116) about the theoretical importance of social catalysts. As quoted below, his basic argument remains fundamental to the study of risks, hazards, and disasters:

> There does not seem to be any theoretical reason why we may not think of social catalysts and use them in theoretical models. . . . For example, in the study of behavior of populations under conditions of disaster . . . disaster is the catalytic unit whose presence [actual or potential] is necessary for the interaction of psychological and social units that are studied by disaster [and hazard and risk] specialists. It makes no difference whether the event studied is a flood, an earthquake, an explosion, or whatnot.

The phrase "or whatnot" is important and resonates nicely with the inclusive range of natural, technological, and willful events that are under consideration by this committee. By intent and design, a multihazard approach has been adopted by the committee in responding to its mission and tasks.

The previously referenced encyclopedia entry derives from the above assumptions and serves as a starting point for the committee. The definition of disaster adopted by the committee will not, of course, end debates about the theoretical and practical implications of achieving clarity about the meaning of risks, hazards, and disasters (e.g., Dynes, 2004; Perry and Quarantelli, 2005). However, this definition does provide a heuristic tool for examining a broad range of environmental, technological, and willful events on their own terms and for comparing systemic adjustments to actual or potential events with societal responses to other social problems and public policy issues (Barton, 1989).

## THRESHOLDS OF DISASTERS

Defining disasters raises fundamental questions about how they should be demarcated. Although thresholds of disasters have been debated, researchers, practitioners, and policy makers affirm clearly that such thresholds exist (see Wright and Rossi, 1981; Kreps and Drabek, 1996; Kreps, 1998). There is no argument about whether the three hurricanes on the United States Gulf Coast in 2005, the earthquake on the borders of Pakistan, India, and Afghanistan in 2005, the Indian Occan tsunami in 2004, the September 11, 2001 terrorist attacks on New York and Washington, and a host of other natural, technological, and sociopolitical events that have occurred in the recent or more distinct past were disasters. Potential disasters, such as the current threat of an avian flu pandemic and other environmental hazards of various types, are just as important to consider as those that have actually occurred, and this is the essential preoccupation of what is now termed vulnerability science (Cutter, 2003a).

When assessing the actual or potential severity of human harm and social disruption, the level of society that is being analyzed must be focused on—the entire society or subunits within it, such as communities, neighborhoods, and households (the relevant literature includes recent work by Cutter et al., 2003). Thus, natural disasters are relatively frequent at the societal level, and the absorptive capacities of large, technologically advanced societies are considerable. However, not all societies are large and technologically advanced (e.g., Bates and Peacock, 1993), and even when they are, disasters become less common and their impact ratios change as the unit of analysis moves from the societal to the regional, community, and household levels. Hazard vulnerability and mitigation, disaster preparedness, emergency response, and disaster recovery take on different meanings depending on which systemic level is being considered. The focus of the above encyclopedia entry is on the societal level and its major subsystems.

There are at least two key questions: What are the thresholds of actual or potential disasters, below which events do not score high enough to be included analytically and above which it is possible to distinguish smaller- from larger-scale events? How can these thresholds capture both physical conditions and social definitions of human harm and social disruption? The requirement is clear: To be useful the committee's definition must drive more precise specification of disasters as objects of inquiry (Dynes, 1998).

The committee's approach to specifying disasters empirically is as follows. Disaster metrics must capture the magnitude and scope of physical impact and social disruption at the community, regional, or societal level *and* the social significance attached to these effects on human populations. Physical impact and social disruption are tied to loss of life, injuries, structural and property damage, economic losses, and a variety of other measures of human harm. Social significance is a function of past experience with and future expectations of these effects. Comparatively speaking, for example, a 100-year flood potentially has much greater social significance than a 10-year flood. Oklahoma City and 9/11 are certainly benchmarks of social significance for terrorist attacks, Chernobyl for nuclear power plant accidents, Bhopal for toxic chemical releases, and numerous historical events cross-nationally serve the same purpose for wars, earthquakes, hurricanes, floods, droughts, famines, and other hazards.

The precise determination of physical impacts and social disruption is highly complex because disasters produce a host of primary, secondary, and indirect effects. As Tierney et al. (2001:6) note:

> Direct effects include the deaths, injuries, and physical damage and destruction that are caused by the impact of the disaster agent itself. Research has recently begun to emphasize the importance of secondary disaster impacts, such as fires or hazardous materials releases that are triggered by earthquakes and environmental pollution resulting from flooding. These kinds

of occurrences can produce significant impacts and losses over and above those caused by the primary disaster agent. . . . A distinction can also be made between direct and secondary impacts and the indirect losses resulting from disasters. Those losses include "ripple effects" resulting from disruption in the flow of goods and services, unemployment, business interruption, and declines in levels of economic activity and productivity.

A key intervening factor in assessing primary, secondary, and indirect effects can be termed "information effects," which are those resulting from revised expectations of losses in the future (Yezer, 2002). Information effects are of central importance to the social significance of disasters. Willful disasters such as the September 11, 2001 terrorist attacks have negative information effects, and indeed, that is what they are designed by terrorists to accomplish (see NRC, 2002a:267-313). However, all disasters have information effects, and these effects can lead positively to increased vulnerability assessment, hazard mitigation, and emergency preparedness as well as more efficient and effective emergency response and disaster recovery. Whether negative or positive, information effects are important catalysts for increasing or decreasing uncertainty about hazardous conditions before, during, and after disasters.

Despite the complexity of measuring primary, secondary, indirect, and information effects, disasters can be distinguished conceptually from non-disasters by keeping the following definitional points in mind (see Barton, 1989, in press; Dynes, 1998). First, disasters are a subset of societal problems, and the committee does not attempt in this report to equate them with all other forms of trouble in the world. Second, regardless of their origins, disasters are acute events that involve a conjunction of physical conditions and social definitions at systemic as opposed to individual levels. Third, historical circumstances are not disasters until they are defined as such. Although who is doing the defining (e.g., the general population, professional experts and practitioners, institutional elites, or the mass media) is an important research issue (Stallings, 1995), once made, social definitions of disasters are consequential (May, 1985; Birkland, 1997). The research problem then becomes one of comparing events in terms of levels of physical and social impact. These levels increase as the magnitude of the effects are evidenced at community, regional, societal, and cross-societal levels. This is why development of databases on hazards and disasters and maintaining central data repositories are so important to future social science research. Fourth, while no less complex to measure than physical impacts, social impacts are a function of the proportions of populations and organizations involved at various systemic levels, the duration of individual and organizational involvement, the uncertainty of impact conditions, and the probability of disaster recurrence. Finally, the social significance of disasters reflects the difference between physical impacts and social disruption on the one hand

and expectations about their severity on the other. Logically speaking, a fully anticipated event would not be defined as a disaster (Turner, 1978; Perrow, 1984; Clarke, 1989; Weick, 1993; Cutter, 2003b).

## MAINSTREAM TOPICS OF HAZARDS AND DISASTER RESEARCH

Note that Figure 1.1 includes five topics of mainstream research within this field: hazard vulnerability, mitigation, disaster preparedness, emergency response, and disaster recovery.

*Hazard vulnerability* is the potential for physical harm and social disruption to societies and their larger subsystems associated with hazards and disasters. There are two general types of vulnerability. Physical vulnerability represents threats to physical structures and infrastructures, the natural environment, and related economic losses. Social vulnerability represents threats to the well-being of human populations (e.g., deaths, injuries, other medical impacts, disruptions of behavior and system functioning) and related economic losses. Social vulnerability also includes the relative potential for physical harm and social disruption to subpopulations of societies and their larger subsystems based on socioeconomic status, age, gender, race and ethnicity, family structure, residential location, and other demographic variables (for recent discussions of social vulnerability and its measurement see Cutter et al., 2003; Buckle, 2004).

*Hazard mitigation* includes interventions made in advance of disasters to prevent or reduce the potential for physical harm and social disruption. There are two major types of hazard mitigation. *Structural mitigation* involves designing, constructing, maintaining, and renovating physical structures and infrastructures to resist the physical forces of disaster impacts. *Nonstructural mitigation* involves efforts to decrease the exposure of human populations, physical structures, and infrastructures to hazardous conditions. Nonstructural mitigation approaches include enacting land-use measures that take into account potential disaster impacts; regulating development in high-hazard zones such as hillsides that are prone to landslides and coastal zones subject to storm surge; and even in some cases buying out and relocating communities or parts of communities, a measure that is now used for areas that have experienced repetitive flood losses.

*Disaster preparedness* includes actions taken in advance of disasters to deal with anticipated problems of emergency response *and* disaster recovery. These actions include the development of formal disaster plans; the training of first responders; the maintenance of standby human, material, and financial resources; and the establishment of public education and information programs for individual citizens, households, firms, and public agencies. Of particular importance to disaster recovery preparedness, hazard insurance is designed to provide financial protection from economic losses caused by disaster events, the purchasing costs of which are based on actuarial risk.

*Emergency response* includes activities related to the issuance and dissemination of predictions and warnings; evacuation and other forms of protective action; mobilization and organization of emergency personnel, volunteers, and material resources; search and rescue; care of casualties and survivors; damage and needs assessment; damage control, restoration of essential public services; public information; and maintenance of political and legal systems.

*Disaster recovery* includes activities related to the reestablishment of predisaster social and economic routines (education, cultural activities, production, distribution, and consumption); the provision of financial assistance and other services (e.g., mental health care) to victim populations; replacement and repair of damaged and destroyed housing and business properties (sometimes a long-term process); and in some cases, determination of responsibility and legal liability for the event. The concept of recovery encompasses both objective measures, such as reconstruction and assistance efforts, and the subjective experiences of disaster victims and processes of psychological and social recovery.

The above core topics and their definitions apply generally to the broad range of hazards and disasters of interest to the committee. With respect to willful events such as terrorist attacks, particular attention is being given post-September 11, 2001 to vulnerability assessment (e.g., of societal energy, transportation, and information systems), disaster prevention (i.e., detection and interdiction), special requirements associated with nuclear, biological, and chemical agents, and the organizational requirements of developing multigovernmental preparedness and response systems (NRC, 2002b). These highlighted concerns are central to the five mainstream topics of hazards and disaster research depicted in Figure 1.1 and the conceptual model developed in this chapter. As highlighted throughout the report, social science knowledge about natural and technological hazards and disasters can and should be applied rigorously and systematically to willful events, which have been studied by social scientists funded through NEHRP, but less frequently so. While findings from social science research on natural and technological disasters are clearly relevant to willful events, it is clear that much more needs to be learned through comparisons across these different risks. For example, does the fact that willful incidents occur without warning—a trait they share with earthquakes—and are induced by human adversaries who can alter their strategies, tactics, and targets have a different impact on mitigation, preparedness, and response when compared with natural and technological disasters?

The application of social science knowledge by hazards and disaster management practitioners is an important issue for the committee. The reorganization during the mid-1970s that led to the creation of the Federal Emergency Management Administration (FEMA) was based on the principle that federal mitigation, preparedness, emergency response, and recovery programs related to peacetime and wartime disasters should be integrated. A major rationale underlying this principle was that multigovernmental responses to more frequent peacetime disasters provide an essential experience base for dealing with lower-probability, albeit enormously important, wartime events. The integration principle has remained sound for decades, central to FEMA's cross-hazards approach, and consistent with support for social science hazards and disaster research within NEHRP. The recent inclusion of FEMA in the new Department of Homeland Security (DHS) appears to be based on the same principle and rationale. This means that FEMA's continuing and highly visible role in peacetime disasters serves as a potential resource for societal response to terrorist events. The extent to which that potential will be realized in the future is an empirical question.

Figure 1.1 is useful for highlighting substantive and overlapping foci of hazards and disaster research. Through overlapping circles and two-directional arrows the figure directs attention to essential interactions among these topics and the simultaneity of collective actions related to them. For example, vulnerability assessment informs mitigation and disaster preparedness activities. These relate to each other and, in turn, influence conditions of vulnerability. Insurance programs can further disaster mitigation as well as preparedness, and under certain circumstances, disaster recovery influences insurance policy and actuarial rates. Disaster preparedness affects emergency response and recovery, and the experience of disasters has important (short- and longer-term) consequences for the level of preparedness, the conditions of vulnerability, and mitigation adjustments, and so on. The interactions among these topics are numerous and varied, as are systemic adjustments related to them, which require analysis for both theoretical and practical reasons (Bankoff, 2004).

## A CONCEPTUAL MODEL OF SOCIETAL RESPONSE TO DISASTER

Figure 1.2 adapted from Kreps (1985), Cutter (1996), Lindell and Prater (2003), has been constructed to represent a more refined conceptual model developed by the committee to complete its charge from the NSF. The mainstream research topics depicted in Figure 1.1 appropriately remain central to Figure 1.2, thus again capturing the primary research interests of hazards and disaster research. However, what is now represented, in effect,

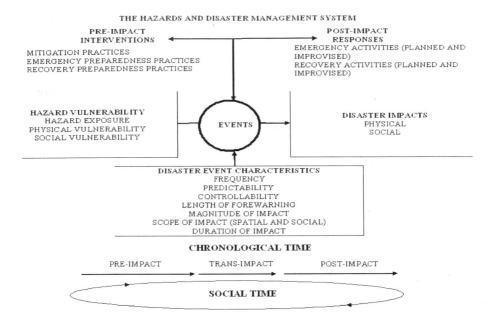

**FIGURE 1.2** Societal response to disaster.


is a process model of societal response to disaster within which the physical and social impacts of catalytic events are a function of conditions of systemic vulnerability, disaster event characteristics, and what has been termed the hazards and disaster management system. As represented in Figure 1.2, specific disaster events (whether environmental, technological, or willful) are placed in the center circle as social catalysts of collective action before, when, and after they occur. Represented to the left, the events circle is the causal importance of antecedent conditions of hazard vulnerability (hazard exposure, physical vulnerability, social vulnerability). Represented below, the events circle is the causal role of key defining features of disasters (frequency, predictability, controllability, length of forewarning, and magnitude, scope, and duration of impact) that allow for comparisons of environmental, technological, and willful events of various types. Represented above, the events circle is the causal relevance of the hazards and disaster management system. That system is represented as the intersection of pre-impact interventions (disaster mitigation and preparedness practices) and post-impact responses (planned and improvised emergency and recovery activities).

Viewing Figure 1.2 in its totality, the hazards and disaster management system interacts with hazard vulnerability and disaster event characteristics in determining levels of disaster impacts as outcomes of the model. The unity of hazards and disaster research that the committee considers essential is thereby revealed. The interactions among the five core topics of hazards and disaster research—introduced in Figure 1.1 and depicted more pointedly in Figure 1.2's process model—are important on both theoretical and practical grounds. Both theoretically and empirically, hazard vulnerability, hazard mitigation, disaster preparedness, emergency response, and disaster recovery are mutually related. Indeed, they are components of a highly complex but comprehendible response structure. Practically, collective actions related to these constructs and their interactions increase or decrease the human harm and social disruption of disaster as the committee has defined that term. Thus, research on hazards and disasters has important implications for both basic science and public policy.

### Chronological and Social Time

Both chronological time and social time are essential constructs in hazards and disaster research. As depicted in Figure 1.2, chronological time is linear, unidirectional, and readily calibrated using standard physical measurements. Chronological time allows for the partitioning of collective actions by time phases of disaster events (pre-impact, trans-impact, post-impact) and the examination of their interactions. In chronological time, pre-disaster vulnerability assessments influence hazard mitigation and disaster preparedness decisions under more routine, pre-impact circumstances. The trans-impact period constitutes the time immediately prior to and during an actual event when specific hazard mitigation and preparedness interventions are set in motion. Such planned interventions intersect with improvised emergency response and recovery activities during and after the event has occurred. Chronological time is also an essential tool for making comparisons between disasters in terms of such characteristics as frequency, predictability, length of forewarning, and duration of impact.

The scientific value of chronological time is unquestionable and taken for granted. Yet its value for analytical purposes is not unlimited; and thus, Figure 1.2 calls for a complementary treatment of social time. Social time is more complex than chronological time, but the concept is very useful for expressing the singularity of hazards and disaster research. The distinction between chronological and social time has heretofore rarely been mentioned by the hazard and disaster research community (see Forrest, 1993; Quarantelli, 1998:255-256), let alone seriously examined (for a notable exception, see Bankoff, 2004). The committee thinks that the distinction has scientific value and directly informs its work (Zerubavel, 1981, 1997, 2003).

Social time is nonlinear and multidirectional and may be experienced differentially by individuals and social entities of various types. Within social time, *the past may be reconstructed from the present*. History itself has been variously reconstructed by individuals (citizens, professionals and practitioners, public officials, journalists, and scholars) and by what Zerubavel refers to as "mnemonic communities" (see Zerubavel, 2003, especially Chapters 1, 2, 4 and related literature referenced in that volume). Mnemonic communities are small to more inclusive social systems (families, ethnic groups, organizations, communities, and societies) whose memories of the past are collectively shared and often commemorated in various ways. The reconstruction of history is, indeed, a complex process. Long expanses of chronological time may be cut up and compressed into historical eras by mnemonic communities, and substantial "mental bridging" is required to maintain a sense of continuity across or even within these discrete periods. Particularly helpful in maintaining this sense of continuity are catalytic (watershed, benchmark) events such as the founding of new nations or religions, wars, the development of new technologies and inventions, and the creation of new modes of artistic expression (Zerubavel, 2003:12, 85-88, 97-100). Disasters, as defined above, provide important additional examples of catalytic events in social as well as chronological time.

Some catalytic events are only defined retrospectively. This is the case, for example, in what historically have been characterized in hazards and disaster research as "chronic" or "creeping" disasters (e.g., Fritz, 1961; Barton, 1969, 2005; Turner, 1978). For example, a 30-year drought-induced famine ultimately becomes defined as a multiple disaster. This disaster exists in social time only when changing historical conditions over decades have been collectively reconstructed to define them as acute. Yet how acute are these conditions? In chronological time, famines and droughts are physically characterized as slower-onset disasters with considerable forewarning in comparison to disasters such as earthquakes, tornadoes, hurricanes, and explosions of conventional, biological, or chemical weapons (Kreps, 1998:34). Chronological time is arguably central for comparative studies of the above disasters in terms of hazard vulnerability, hazard mitigation, and disaster preparedness. It is also a resource for taking preventive steps. In social time, however, the temporal uniqueness of droughts and famines is far less important. Once a disaster has been socially constructed, the "luxury" of time no longer exists. A previously unidentified disaster has now been located in social time and space. Chronological time and social time have become coterminous, as have collective actions related to hazard vulnerability, hazard mitigation, emergency preparedness, emergency response, and disaster recovery. Simultaneous activities are directed to meeting demands that are defined objectively and subjectively as acute in all of these areas.

It is also the case that in social time *the present may be reconstructed from the past.* As opposed to the previous example in which certitude ultimately exists in both social and chronological time, here there is openended uncertainty about whether a set of historical conditions constitutes a disaster. A useful example is global climate change. In chronological time, global climate change draws primary attention to hazard vulnerability and mitigation activities to reduce its effects before they become disastrous. However within social time, equal attention is warranted to sustainability and perhaps survivability of the planet (i.e., to disaster response and recovery activities). Thus, whether global climate change is a potential or actual disaster is a non-issue from the standpoint of social science research. Just as with droughts, famines, earthquakes, tornadoes, explosions, and other hazards, the research interests of hazards and disaster researchers can and, in the committee's opinion, must be seen as coterminous.

Finally, *the future is inextricably linked to the present and past in social time.* For example, decisions to build alternative types of physical structures and infrastructures in floodplains, in coastal zones, along fault lines, and in highly vulnerable urban areas are based on prior disaster experiences *and* future disaster expectations as both relate to assessments of hazard vulnerability. Moreover, decisions to make development investments necessarily involve decisions about disaster mitigation and preparedness measures, and these decisions are based on prior disaster experiences and future disaster expectations, including those related to emergency response and disaster recovery. Decisions about development, hazard mitigation, and emergency preparedness give rise to one of the most important economic issues in this field: Do increased levels of hazard mitigation and disaster preparedness increase risk taking by individuals and social systems? Thus, from an economic perspective, there is an implicit component of hazard exposure in Figure 1.2 that reflects decisions by individuals and social systems to locate in harms way.

The committee concludes that the past, present, and future of chronological time are interchangeable features of social time. In effect, chronological time compresses and expands within social time as individuals and social systems *create, define, and adapt to environmental hazards, the risks associated with them, and the disasters that occur from them.* The interests of those studying environmental hazards, risks, and disasters are coterminous, and equally important, and they must be captured within a common framework. Thus, the committee has had very specific objectives in mind for Figure 1.2: first, to further elaborate conceptual issues attending the above encyclopedia entry; second, to identify the common interests of hazard and disaster researchers; and third, to capture graphically both the interactions among central research topics in this field and their simultaneity. The individual and collective decisions and actions subsumed within these research topics demand the kind of causal framework depicted in Figure 1.2. So also do the needs of policy makers, practitioners, and other stakeholders. This framework has been used by the committee to meet its charge and prepare this report.

## SOCIAL SCIENCE AND THE EMERGENCE OF NEHRP

Created in 1977, the National Earthquake Hazards Reduction Program (NEHRP) was mandated to include the social sciences within a broader program of research in the earth sciences and engineering. This original mandate has been sustained in the latest NEHRP strategic plan (FEMA, 2003a). The inclusion of the social sciences in NEHRP was facilitated in the mid-1970s by the fact that hazards and disaster research had become an established, although relatively young, area of inquiry in the social sciences. It was therefore thought by champions of NEHRP in government and academia that the social sciences could contribute to the goals of the program.

Social science hazards and disaster research in North America is usually traced to Samuel Prince's research on the 1917 Halifax, Nova Scotia, ship explosion, considered the first empirical social science disaster study in the region (Prince, 1920). Another important line of early work can be traced to studies of human adjustments to natural hazards under the direction of Gilbert White at the University of Chicago (began in the 1940s). A crucial growth period in the field occurred during the 1950s when multihazard and disaster research programs were established at the University of Chicago's National Opinion Research Center, the University of Oklahoma, the University of Maryland, and the National Academy of Sciences. These programs were succeeded in the 1960s and 1970s by other multihazard and disaster research programs established at institutions such as the Ohio State University (where the Disaster Research Center was located from 1963 to 1985 before it moved to its present location at the University of Delaware), the University of Colorado (which became the home of the Natural Hazards Research and Applications Information Center in 1976), and Clark University (where the Center for Technology, Environment, and Development was established in 1978).

With respect to earthquake research, social scientists became a part of a multidisciplinary effort to understand major events that occurred during the 1960s and early 1970s. In fact, findings from studies of these events provided part of the rationale for the creation of NEHRP (Anderson, 1998). The first earthquake to receive serious attention during this period was the 1964 Alaska earthquake, which at the time was arguably the most studied seismic event in U.S. history (NRC, 1970). The second was the 1971 San Fernando earthquake, which clearly demonstrated the vulnerability of the nation to this hazard. These two disasters served as catalysts for the creation of a national program of earthquake research and application. The program's supporters included academics at institutions conducting earthquake research, officials at federal agencies such as the U.S. Geological Survey (USGS) and NSF, and a few members of Congress. Their goal was to use findings from studies of these events to convince federal decision makers and other stakeholders of the need for a national program (Hamilton, 2003).

Further support for inclusion of the social sciences in NEHRP was the timely publication in 1975 of two highly relevant reports stemming from studies led by social scientists. These reports appeared just a few years before the program was finally authorized by Congress, at the point when discussions were at a critical juncture. One of the reports was *Earthquake Prediction and Public Policy* (NRC, 1975), produced by an NRC panel led by sociologist Ralph Turner from the University of California Los Angeles (UCLA). This report provided an assessment of possible socioeconomic consequences of earthquake predictions. Its recommendations were considered very germane to the future NEHRP. One of its key arguments was that such a program would facilitate the development of earthquake prediction science and engineering, and that social scientists could play an important role by conducting complementary research and analyses related to the timely and effective issuance of earthquake predictions to the public. The other report, *Assessment of Research on Natural Hazards* (White and Haas, 1975), analyzed the state of the art of hazards and disaster research and offered recommendation on future research and application needs related to earthquakes and other hazards. The White and Haas report attracted the attention of the earthquake community not only because of its reference to earthquakes and other hazards, but also because of the authors' advocacy of multidisciplinary research (Hamilton, 2003). The report also provided impetus for establishing the Natural Hazards Research and Applications Information Center at the University of Colorado at Boulder.

This emerging awareness of the relevance of the social sciences within the earthquake research community was reinforced by a highly influential 1976 report entitled *Earthquake Prediction and Hazard Mitigation: Options for USGS and NSF Programs* (NSF and Department of the Interior, 1976), more popularly known as the Newmark-Stever report after its two lead authors. Also important was a report published in 1978, one year after the establishment of NEHRP, entitled *Earthquake Hazards Reduction: Issues for an Implementation Plan* (Working Group on Earthquake Hazards Reduction, 1978). The Newmark-Stever report provided a research plan that included major social science research tasks under the rubric of "research for utilization." This rubric was later reflected in the Earthquake Hazards Reduction Act that established NEHRP. J. Eugene Haas, a co-founder of the Disaster Research Center at Ohio State University, was an important contributor to the Newmark-Stever report. The Working Group on Earthquake Hazards Reduction was established by the Office of Science and Technology Policy (OSTP) to prepare the second report. The working group reported to the prominent seismologist Frank Press, then director of OSTP and science advisor to President Jimmy Carter. The working group included two social scientists from federal agencies and representatives from the engineering and earth science communities. Its external advisory committee included two prominent social scientists, Charles Fritz of the National Research Council (NRC) and Ralph Turner from UCLA. The report of this working group addressed implementation issues that NEHRP and the nation faced, including those that could best be understood from a social science perspective (e.g., emergency preparedness, disaster warning, risk communication).

Participating agencies in NEHRP include USGS, NSF, the National Institute of Standards and Technology (NIST), and FEMA, with FEMA serving as lead agency during most of NEHRP's existence. When NEHRP was established in 1977, NSF was already the focal point for federal government funding of social science hazards and disaster research, principally through what was later to become the Engineering Directorate. This social science element became a part of NSF's continuing contribution to NEHRP, and funding is now provided through the directorate's Division on Civil and Mechanical Systems (CMS). Over the years, this social science component has been variously named Societal Response to Natural Hazards, Earthquake Systems Integration, and more recently Infrastructure Management and Hazard Response. Some social science hazards and disaster research is also funded within programs of the Social, Behavioral, and Economic Sciences Directorate, including the Decision, Risk, and Management Sciences Program. As is the case with support received by other disciplines from NSF, most social science funding for earthquakes and other hazards goes to academic institutions. A relatively modest amount of funding has also been made available by FEMA and USGS to the social science research community under the auspices of NEHRP.

## KEY ISSUES THAT ARE RELATED TO AND INFORM THE COMMITTEE'S CHARGE AND TASKS

*Social Science, Disasters, and Public Policy.* This report summarizes a body of social science research that informs and can influence public policy. A classic definition of public policy is the "things that government chooses to do or not to do" (Dye, 1992). This definition encompasses the idea that governments—and, more to the point, the people that work in government—make choices about what government should do (the policy goals) and what government does to achieve these goals (the policy tools). These decisions are in turn influenced by basic and applied scientific research.

Because the findings of this report inform and are influenced by the political process and by political institutions, it summarizes past research and recommends future studies that have the potential to influence public policy. Most recently, after Hurricanes Katrina and Rita, policy makers have sought ideas to improve the nation's preparedness for and response to natural and other types of disasters. Key ideas for addressing these problems are being developed by members of the research community described in this report. While the committee does not make specific policy recommendations, the research recommendations in this report can influence public policy in ways that can reduce vulnerability and promote hazard mitigation and preparedness. Further, the committee acknowledges the influence of public policy on social science research on hazards and disasters. For example, the Earthquake Hazards Reduction Act created the National Earthquake Hazards Reduction Program, which has supported a considerable amount of basic and applied social science research.

Research supported under NEHRP suggests that there is much variation in the nature of government policies intended to address natural hazards. Policies vary by level of government, by physical location, and by hazard. This report therefore summarizes past research and calls for additional research that could be useful to policy makers. At the same time, the policies adopted by government at all levels to mitigate, prepare for, and respond to disasters have inspired research about the nature and effectiveness of these policies. The ultimate translation of scientific knowledge into policy is subject to the usual economic, social, and political factors that can either further or impede policy changes in political systems.

*Societal Change and Social Science Hazards and Disaster Research.* Societies worldwide are undergoing significant changes that will require major adjustments on the part of social science hazards and disaster research in terms of what is studied and how. For example, as discussed in Chapter 2, the populations in the United States and elsewhere are undergoing significant change, affecting the vulnerability of various groups that social scientists study. The emergency management profession, seen by social scientists as a major user of the knowledge the field generates, has new responsibilities (including those brought on by the increased threat of terrorism) and new institutional arrangements to meet these responsibilities. Social scientists must address both of these changes within an inclusive framework of hazards and disaster research, as depicted in Figure 1.2. Moreover, technologies now available to the general public, (e.g., the Internet, cell phones, geographic information systems (GIS), remote sensing) and formal and informal groups and organizations using these technologies will influence all aspects of behavior and decision making related to hazards and disasters studied by social scientists (Cutter, 2001). Also, many of these same technologies are likely to have a profound impact on the way disaster researchers carry out their investigations and disseminate their results, which already appears to be the case for the Internet and GIS.

*Social Science Contributions Under NEHRP.* In Congress's definition of NEHRP's roles, NSF is responsible for activities such as funding research, especially at universities, on problems that can be addressed by the disciplines of earthquake engineering and the earth and social sciences. Thus, as part of its NEHRP role, NSF has provided much of the external support for social science hazards and disaster research conducted by U.S. investigators during the past 25 years. This work has included studies carried out by researchers in disciplines represented by experienced specialists on the committee, and their expertise has been supplemented as necessary by workshop presentations from other experts. NSF support has been critical for enabling the social science research community to pursue a long term program of research on hazards and disasters, to train succeeding cohorts of graduate students, and to pursue new strategies to disseminate knowledge. Much of this research has focused on the United States; however, significant international work has also been carried out by U.S. investigators, often in collaboration with international colleagues.

The social and behavioral science research funded by NSF has included both individual investigator awards (i.e., projects involving a single researcher and perhaps one or more graduate students) and team awards with multiple investigators. Examples of the latter include projects that cut across social science disciplines as well as the even more challenging multidisciplinary research in which social scientists collaborate under the auspices of the three NSF-supported earthquake engineering research centers: the Multidisciplinary Center for Earthquake Engineering Research (MCEER), administered through the State University of New York at Buffalo (which succeeded the National Center for Earthquake Engineering Research); the Mid-America Earthquake Center, administered through the University of Illinois at Urbana-Champaign; and the Pacific Earthquake Engineering Research Center, administered through the University of California at Berkeley. Additionally, over the years social scientists have participated on multidisciplinary post-earthquake reconnaissance teams organized by the Earthquake Engineering Research Institute (EERI) as part of its Learning from Earthquakes program (also funded by NSF).

In the broadest terms, the research that social scientists have carried out during NEHRP's 25-plus years has focused on activities related to pre-, trans-, and post-disaster time periods, as depicted in Figure 1.2. Appropriately enough, a large portion of this research has targeted earthquake hazards. However, over the years, NSF has been quite flexible about the type of social science research it was willing to fund under NEHRP. Thus, NSF has permitted social science researchers to study other types of hazards and disasters as surrogates for earthquakes and has concurred with the importance of carrying out research projects that included other types of hazards for comparative purposes. This concurrence is important because most social scientists have a preference for engaging in cross-hazards research, rather than specializing in specific hazards as earthquake engineers, atmospheric scientists, and earth scientists tend to do. NSF's flexibility has set the stage for significant leveraging of knowledge across hazards and disasters, making it more likely that a holistic knowledge base can be generated. Additionally, this more comprehensive approach is valuable to emergency managers, urban and regional planners, and other practitioners who face the reality of confronting multiple hazards.

A key task for this committee, then, is to document succinctly the key contributions that social scientists have made under NEHRP in developing knowledge of earthquakes and other hazards and disasters, and also advancing appropriate collaborative research activities subsumed by the research topics represented in Figure 1.2. Recent discussions have been suggestive, including a workshop (National Earthquake Hazards Reduction Program at Twenty-Five Years: Accomplishments and Challenges) held in Washington, D.C., on February 20, 2003. The workshop was organized by the National Academies' Disasters Roundtable at the request of the four participating NEHRP agencies. At this workshop, it was argued by some participants that social science research supported through NEHRP has resulted in a greater understanding of the social and economic consequences of earthquakes, including the effects on regional and national economies, the economic impacts on individual firms, and the effects on individuals, families, and communities. It was also suggested that much has been learned about the way individuals, organizations, and government entities respond to earthquake threats and seismic events, about how to communicate risk more effectively, and about how to design and implement mitigation policies and programs. Finally, participants recognized that the social science research conducted under NEHRP is relevant across all of the various types of hazards and disasters studied. Clearly, much of what has been learned by social scientists through the study of earthquakes is applicable to other natural, technological, and human-induced disasters, and vice versa.

A similar theme was struck during another workshop (Contributions of Earthquake Engineering, Seismology, and Social Science, held in San Francisco on June 18-19, 2003) that was organized by EERI. Like their counterparts in earthquake engineering and earth science at the workshop, participating social scientists outlined what they considered to be some of the major contributions their disciplines have made that apply to earthquakes as well as other types of hazards. Among the contributions noted was the creation of a knowledge base on factors that facilitate and hinder mitigation and preparedness efforts. This knowledge base was seen as providing insights about the degrees of vulnerability that characterize various segments of society, specifying major principles of emergency preparedness and management, and documenting challenges and opportunities presented during disaster recovery. On the implementation side, it was noted that one of the major contributions of the social sciences during the past few decades has been increasing the availability of their research results to emergency managers and other practitioners, thereby contributing to the latter's ability to better cope with today's array of hazards. Various strategies have been employed, including the establishment of college- and university-based emergency management courses and programs at the undergraduate and graduate levels.

Finally, the First Assessment of Research on Natural Hazards conducted by Gilbert F. White and his collaborators at the University of Colorado at Boulder called attention to the relevance of the social sciences to a future NEHRP. A second assessment which was initiated in the 1990s under the leadership of Dennis Mileti (and also at the University of Colorado), resulted in a number of important publications, including the summary volume *Disasters by Design: A Reassessment of Natural Hazards in the United States* (Mileti 1999b). The Second Assessment produced four additional books, which focused, respectively, on hazard insurance (Kunreuther and Roth 1998); land-use planning for disaster reduction (Burby, 1998); disaster preparedness and response (Tierney, Lindell, and Perry, 2001); and the risks and vulnerabilities associated with different geographic locations in the United States (Cutter, 2001). The Second Assessment took a comprehensive look at advances in hazards and disaster research since the results of the First Assessment were published in 1975. The committee has drawn on these and other publications in documenting social science contributions to hazards and disaster research and, more importantly, in identifying gaps in and future opportunities for the development and application of knowledge.

It is important to determine, for example, gaps in knowledge about natural as opposed to technological and other types of hazards and disasters. Historically some social science researchers have shown a preference for studying one type over another, perhaps interacting primarily with like-minded researchers, thereby reducing opportunities for sharing research results and theoretical insights. As previously noted, however, many social scientists investigate a variety of hazards and disasters, including terrorist incidents. This tendency seems to be especially true of researchers affiliated with social science centers that have sustained programs of research, such as the Disaster Research Center at the University of Delaware, the Hazard Reduction and Recovery Center at Texas A&M University, the G.P. Marsh Institute at Clark University, and the Hazards Research Lab at the University of South Carolina. There also are some specific topics, such as risk perception and communication, about which specialists devote most of their time to either natural or technological disasters, but tend to meet fairly frequently (e.g., the Society for Risk Analysis) and share insights in specialty journals (e.g., *Risk Analysis*). Building more and better networks among specialists in respective areas seems an important requirement for the future. For example, forums for social scientists studying hazards and disasters have become institutionalized to varying degrees in professional associations and meetings. One example is the Regional Science Association International (RSAI), an interdisciplinary association that brings together economists, geographers, sociologists, planners, and engineers, as well as some public officials. Sessions on disasters have been organized at RSAI's annual North American meetings since the early 1990s and have helped build up interest, community, and an established literature on disaster research in the regional science community. Similar specialty groups have been established by the Association of American Geographers and the American Sociological Association.

*Interdisciplinary Research: Challenges and Opportunities.* Figure 1.2 provides the framework used in subsequent chapters to document what is known and not known about hazards and disasters and the opportunities for future research. While there is a compelling need for disciplinary research within the social sciences, physical sciences, and engineering, there is a similar need for collaborative research across disciplines. Simply put, hazards and disasters pose problems that require multidisciplinary and interdisciplinary solutions. The challenges are major and the opportunities to meet them merit careful consideration.

One of the key justifications for the creation of earthquake engineering research centers was that they would provide a platform for significant interdisciplinary research involving engineers, earth scientists, and social scientists. As noted, NSF currently supports three such centers, and all are expected to promote an integrated research program that includes the social sciences. In addition, over the years NSF has supported other interdisciplinary activities, and this type of research is receiving increasing emphasis. In 2003, the Engineering Directorate and the Social, Behavioral, and Economic Sciences Directorate launched a joint program to support collaborative engineering and social science research that include hazards and disasters. If successful, it is expected that NSF will make a long-term commitment to this program. The committee has therefore examined the experience with interdisciplinary research on hazards and disasters and identified the challenges faced both within the social sciences and between the social sciences and natural science and engineering fields.

*Opportunities for Collaborative International Research.* The United States is viewed as a world leader in the field of hazards and disaster research.

Because of their education, resources, and experience, disaster experts in this country are highly sought after by stakeholders looking for research partners. A significant amount of the international collaborative research on hazards and disasters funded by NSF has been related to earthquakes. U.S. collaboration with China and Japan has been particularly strong over the years and in the case of Japan has included at least modest social science participation. Because damaging earthquakes are rare in this country, U.S. investigators have generally been keen to undertake studies of events in foreign locales. These studies have often involved scientific collaboration with researchers in affected societies, as was the case following the 1985 Mexico; 1995 Kobe, Japan; 1999 Kocaeli, Turkey; 1999 Chi Chi, Taiwan; 2001 Gujarat, India; and 2003 Bam, Iran earthquakes.

Although earthquake hazards have been a favorite subject of U.S. researchers involved in collaborative international research, NSF has funded collaborative research on other types of hazards as well (e.g., research on Hurricane George and Hurricane Mitch, which struck the Caribbean and Latin America in 1998). During the course of its work, the committee has therefore developed ideas to facilitate opportunities for collaborative international research involving the social sciences.

*Role of New Technologies and Methodologies for Enhancing Studies of Disasters Before, During, and After Their Occurrence.* As represented in Figure 1.2, trans- and post-impact periods of disasters provide natural laboratories for observing how people actually cope with stressful events. As a result, post-disaster fieldwork has been a hallmark of hazards and disaster research since the origins of the field (Tierney, 2002). Indeed, post-disaster investigations are seen as so important to advancing knowledge that special institutional arrangements have been adopted and special funding has sometimes been made available (particularly for earthquake research) to enable social scientists and other researchers to enter the field to collect perishable data or conduct more systematic research.

NSF has a long history of providing support for post-disaster investigations. For many years, for example, NSF has provided support for EERI's earthquake reconnaissance work, which involves the collection of perishable data from damaging earthquakes in the United States and abroad by multidisciplinary teams organized by the institute. Social scientists serve on EERI's Learning from Earthquakes Committee and participate (although in a limited way) in EERI post-earthquake reconnaissance teams. EERI has also recently formed a social science committee to better integrate social scientists into its activities, especially those involving the collection of perishable data following earthquakes. NSF has also supported a more modest effort at the University of Colorado's Natural Hazards Center, one that covers travel costs primarily for social science researchers to study a variety of disasters. NSF also provides funding directly to researchers for post-disaster studies through its standard grants program, its Small Grants for Exploratory Research Program, and timely supplements to existing grants. In addition to post-earthquake studies, NSF has also used such mechanisms to fund post-disaster research on other natural disasters, including floods, tornadoes, tsunamis, and hurricanes. And after September 11, 2001, NSF funded a major portfolio of post-disaster studies in New York and Washington, D.C., on the terrorist attacks (Natural Hazards Research and Applications Information Center, 2003).

NSF and its other NEHRP partners have cooperated on post-earthquake investigations, including research carried out after the 1989 Loma Prieta and 1994 Northridge earthquakes. Recently, the Plan to Coordinate NEHRP Post-earthquake Investigations (Holzer et al., 2003) was released to promote greater coordination among agencies and to specify their expected research roles. The plan emphasizes that along with other relevant disciplines, the social sciences make important contributions to post-earthquake investigations in the United States and abroad.

A key issue is how to exploit state-of-the-art technologies and methods in maximizing the value of post-impact investigations. The 2003 NEHRP plan makes it clear that improved collection, management, and dissemination of perishable data are essential. For example, the NEHRP plan speaks to the need for searchable web-based data systems, but is not precise about how these systems should be constructed, the kinds of data that should be included in them, when these data should be collected and stored, and how the demands for information from multiple audiences will be met. Post-impact studies also provide a window for documenting what did or did not take place pre-disaster with respect to hazard vulnerability assessment, hazard mitigation, and disaster preparedness actions.

The use of state-of-the-art technologies and methodologies is no less important for pre-impact investigations of hazards and the risks association with them. While the interpretation of perishable data is different for hazards as opposed to disaster research, the technical issues of building and maintaining databases are equally nontrivial as are requirements for data sharing and providing user-friendly data presentation and dissemination techniques to multiple audiences. Thus, a wealth of innovative technologies and methodologies (e.g., advanced survey research techniques, geospatial and temporal tools and methods, various types of remote-sensing technologies, data integration and fusion techniques, automated scanning of documents collected in the field, automated compilation of data from standardized field protocols, parallel computing equipment and software, computer modeling and simulation, gaming experiments) are relevant to hazards as well as disaster research. Both research and guidance are needed in determining how best to exploit these and other tools as matters of research and application. For want of a better phrase, "hazards and disasters informatics" is both a challenge and an opportunity for the future.

*Dissemination of Social Science Findings on Hazards and Disasters.* At the Disasters Roundtable workshop on NEHRP, some participants talked about an "implementation gap," arguing that significantly more is known about solving hazard and disaster problems than is being applied. One expectation those in other disciplines have had of the social sciences is that the latter would contribute major insights about how to improve the implementation process. Essentially, the expectation is that the social sciences play a key role in evaluation research that can lead to the development of best practices on the dissemination of findings to policy makers, practitioners, and individual citizens. Social scientists have conducted some evaluation research on the dissemination of findings from hazards and disaster research (Yin and Moore, 1985; Yin and Andranovich, 1987). Additionally, the Natural Hazards Center is known for its leadership in furthering the application of social science research results through its information dissemination activities and other programs that link researchers and practitioners. There are at least some social science research groups that share their knowledge with practitioners through close and sustained relationships. Another sign of progress, sometimes involving collaboration between social scientists and other stakeholders such as FEMA, is the development of college courses and degree programs on hazards and disaster management. Currently there are several dozen such programs at colleges and universities in the country. Finally, a new initiative has been undertaken by the Natural Hazards Center and the University of Colorado at Denver, in partnership with FEMA's Higher Education Program and with support from NSF, to advance such efforts by formulating a national model for emergency management college curricula. The committee therefore has ample foundations for developing evaluation research strategies in the field.

*Meeting Future Hazards and Disaster Research Workforce Needs.* The sustainability of social science research on hazards and disasters depends on its most vital resource, the next generation of researchers. The period of the 1960s and 1970s was arguably a high-water mark for the training of young scholars entering the field, first at such institutions as the University of Chicago and the Ohio State University and then at such institutions as Clark University and the University of Colorado. It was a period, for example, when the Disaster Research Center was created at the Ohio State University and the landmark First Assessment on Natural Hazards was carried out at the University of Colorado. Flush with outstanding faculty, innovative research activities, and funds, the above institutions produced their largest cohorts of Ph.D.s committed to careers in hazards and disaster research. The resulting advancement of hazards and disaster research during the latter three decades of the twentieth century has been associated with the evolving careers of this 1960s-1970s cohort, complemented by the contributions of scholars whose involvement in the field has been more episodic than sustained. Simply put, however, formerly young scholars have now aged and need to be replaced.

The traditional way of developing future generations of researchers has been to identify promising students, enroll them in graduate programs, and involve them in meaningful ways in ongoing research activities. The hopeful result is an expanding pool of newly committed scholars, in this case hazards and disaster researchers. Some senior social scientists in the field now argue that this traditional approach is no longer adequate to meet workforce needs. In response to this argument, NSF funded the highly innovative Enabling Project in 1996 (administered through Texas A&M University) as an alternative way to increase the number of younger professionals entering the field. The project was designed to attract junior faculty from doctoral degree-granting universities who showed promise and expressed an interest in hazard, disaster, and risk research. Thirteen junior faculty members were selected competitively to participate as fellows in the two-year program. The fellows were assigned senior mentors, given an overview of the field, and provided the opportunity to sharpen their proposal writing skills, among other things. The project proved a success, with some of the fellows initiating promising hazards and disaster research activities with funding from NSF after their proposals had undergone the agency's rigorous peer review process. As a result, a follow-up project was funded by NSF for a two-year period starting in 2003 (administered through the University of North Carolina at Chapel Hill). Several of the fellows who took part in the second "enabling" project have already received NSF awards.

Finally, as a result of a grant from NSF, the Natural Hazards Center and the Public Entity Research Institute collaborated on a dissertation fellowship program for young scholars from the social and behavioral sciences, engineering, and the physical sciences. This was a two-year pilot program to provide supplemental support for dissertation work on hazards and disasters as yet another way of bringing new researchers into the field. The intention was to evaluate the program at the end of the pilot period to determine if it would be continued.

Notwithstanding concerns about the size of the research workforce, the diversity of the field is also an issue. Women have made significant strides in hazards and disaster research in recent years, both in terms of their numbers and their success in assuming leadership roles. Unfortunately, there has been little progress in terms of the involvement of minorities in disaster research, including African Americans and Hispanics. This circumstance persists despite the fact that minorities have a higher representation in the social and behavioral sciences than they do in many other research disciplines.

## A VISION OF SOCIAL SCIENCE CONTRIBUTIONS TO KNOWLEDGE AND A SAFER WORLD

While NSF social science studies supported through NEHRP are summarized in some detail in the report that follows, the committee's overall vision of future hazards and disaster research underlies the summary recommendations that have been developed. The committee envisions a future in which:

• the origins, dynamics, and impacts of hazards and disasters become much more prominent in mainstream as well as specialty research interests throughout the social sciences;
• traditional social science investigations of post-disaster responses become more integrated with no less essential studies of hazard vulnerability, hazard mitigation, disaster preparedness, and post-disaster recovery;
• disciplinary studies of the five core topics of hazards and disaster research within the social sciences increasingly become complemented by interdisciplinary collaborations among social scientists themselves and between social scientists and their colleagues in the natural sciences and engineering;
• there is continuing attention throughout the hazards and disaster research community on resolving interdisciplinary issues of data standardization, data management and archiving, and data sharing;
• there is continuing attention throughout hazards and disaster research on the dissemination of research findings and assessments by social scientists of their impacts on hazards and disaster management practices at local, regional, and national levels;
• each generation of hazards and disaster researchers makes every effort to recruit and train the next generation; and
• the funding of hazards and disaster research by social scientists, natural scientists, and engineers is a cooperative effort involving the NSF, its partner agencies within NEHRP, the Department of Homeland Security, and other government stakeholders.

The committee feels that such recent disasters as Hurricanes Katrina and Rita significantly reinforces the relevance of its vision.

## STRUCTURE OF THE REPORT

The above discussions follow directly from the earlier statement of tasks mutually agreed upon by the NSF and this NRC committee. The chapters to follow are organized in terms of those tasks and are informed by the framework developed in this lead chapter.

Chapter 2 addresses environmental, technological, and willful disasters within a broader discussion of key demographic, technological, economic, social, and political changes in the United States and internationally. Chapters 3 and 4 document the social science knowledge base on the five mainstream topics of the field, as defined within Figure 1.1 and modeled in Figure 1.2. Both of these key chapters highlight social science contributions under NEHRP, thus meeting one major task associated with the committee's charge. Chapter 3 focuses primarily on hazard vulnerability, disaster event characteristics, pre-impact interventions, and how they interact in determining disaster impacts from a cross-hazards perspective. Chapter 4 focuses primarily on post-impact responses and their interactions with pre-impact interventions, as both relate to the determination of disaster impacts from a cross-hazards perspective. Fortunately, the more recent Second Assessment (led by Mileti, 1999b) includes several published volumes that provide detailed summaries of knowledge. The committee's intent is not to "reinvent the wheel" but rather to highlight major themes and findings and, in particular (as required by Figure 1.2), to document what is known and not known about their relationships. This approach allows the committee to identify major gaps in social science knowledge and opportunities to reduce them in the early decades of the twenty-first century.

Building on the foundation of the initial four chapters, subsequent chapters address the remaining tasks assigned to the committee. Chapter 5 considers both multidisciplinary and interdisciplinary studies within the social sciences and cross-disciplinary studies that link social science with natural science and engineering fields. Its aim is to document exemplars of successful collaborations and, in so doing, document various challenges that must be overcome in the future. Chapter 6 examines relationships between hazards and disasters and economic development from an international perspective, drawing on ideas of sustainability and resilience in framing development issues. Chapter 7 highlights the role of new technologies and methodologies for enhancing pre-, trans-, and post-disaster studies. Chapter 8 gives attention to practical problems of disseminating research findings and then develops a conceptual framework as the basis for framing future research questions on dissemination. Finally, Chapter 9 provides a summary of research workforce challenges and offers specific steps to solve them.

Committee recommendations and their rationales are offered in Chapters 3 through 9. A majority of the recommendations relate to the need for comparative studies of societal responses to natural, technological and willful hazards and disasters. No explicit priorities among the recommendations have been set forth by the committee, primarily because traditional topics within, respectively, hazards and disaster research necessarily are interrelated. The committee also wishes to ensure that stakeholders have the flexibility to consider the broad range of research and application issues specified in its statement of task.

# 2

# Societal Changes Influencing the Context of Research

The explosion in Halifax harbor on December 6, 1917 precipitated social science interest in disasters, hazards, and their associated risks. The SS Mont Blanc was laden with munitions that completely leveled approximately two square kilometers of northern Halifax when they exploded. More than 2,000 people were killed in the blast or lost to the subsequent tsunami, which inundated a Native American encampment in an upstream cove. Thousands more were injured. This singular event inspired a sociology doctoral student, Samuel Prince, to write his dissertation on the collective behavior of the community in response to the disaster (Prince, 1920).

Elsewhere, geographer Harlan Barrows suggested that his discipline was particularly well-suited to examine the relationship between natural environmental processes (e.g., hazards) and societal responses to them. Paralleling developments in the biological sciences toward integrative approaches to understanding organisms and their environment (the nascent field of ecology), Barrows took the occasion of his presidential address to the Association of American Geographers to argue for a new view of geography as human ecology—understanding the interaction between natural events and human agency and response (Barrows, 1923). His ideas resonated with one of his students, Gilbert F. White, and the social scientific study of natural hazards began in earnest.

World War II and, in particular, the United States Strategic Bombing Surveys (Fritz, 1961) had a strong influence on sociology and to a lesser extent on psychology with respect to the types of events studied in the ensuing decades (e.g., rapid-onset, big-bang types of natural and human-induced disasters that roughly parallel the effects of explosions). World War II and Cold War public policy concerns about nuclear weapons had a profound effect on the directions of these two disciplines. Thus, parallel development of the hazards field in geography before, during, and after the war was absolutely critical for achieving a balanced perspective, as was Gilbert White's leadership in natural hazards, generally, and flood hazards, more specifically, in terms of public policy.

It is not coincident then that hazards and disaster research coevolved at roughly the same time. Studies of disasters, hazards, and their associated risks have always been grounded in the everyday and guided by the prevailing social, economic, and political conditions in specific historical periods. The context within which disasters, hazards, and risks are studied and the ways in which society responds to them are often a function of demographic, economic, and political changes not only in the United States, but throughout the world (see Chapter 6). The nature of the subject matter addressed by social scientists—whether events that arise from the interaction of natural systems and human systems, willful or human-induced threats, or technological failures—means that it is impossible to understand the human response without understanding the larger context within which that response takes place. Thus, to understand the types of events studied and the substantive topics addressed by hazards and disaster researchers, some of the macro- and meso-level societal changes that have influenced social science research on hazards, disasters, and risk must be reviewed.

Accordingly, this chapter provides an overview of societal changes that influence how and what hazards and disaster researchers study. The chapter begins with discussions of basic demographic shifts and economic developments in the post-World War II era. A general discussion follows on geopolitics at home and abroad and its implications for hazards and disaster management policies and practices. The reactive nature of these policies and practices in the United States is then characterized as are subtleties related to the enactment of specific mitigation, preparedness, and response initiatives. Settlement patterns are given specific attention in this regard because of their direct and highly complex relationships to hazard vulnerability as well as land-use planning and other forms of hazard mitigation. A discussion of the influences of societal changes would not be complete without a consideration of quality-of-life and social equity patterns and issues as they relate to social vulnerability. To complete its context-setting function for the report, this chapter closes with discussions of technological change and global environmental patterns. The questions that are raised in the conclusion illustrate the uncertainties and continuing importance of societal change for hazards and disaster research.

## DEMOGRAPHIC SHIFTS

The demographic character of the United States and the world has changed significantly during the past 50 years. The basic composition of American society, as viewed by its age structure, increasing ethnic and linguistic diversity, and disparities in socioeconomic status creates regional patterns of demands for housing, employment, and quality of life. Not surprisingly, large-scale population shifts experienced during the past 50 years, such as the out-migration from the industrial Northeast to the Sun Belt cities in the South and West, and the movement of people from rural to suburban and urban places and to coastal areas, has exacerbated the vulnerability of many of the nation's citizens to environmental hazards (see Chapter 6). Changes in the age structure of the American population, its racial and ethnic diversity, and patterns of socioeconomic status also provide an important context for social science research in the field.

Life expectancy has increased dramatically over the past 50 years. In 1950, for example, a person born in the United States had a life expectancy of 68 years. By 2000, that life expectancy had increased to 77 years, leading to an increasingly large portion of the population who are over the age of 65—many of them women whose life expectancy is 5.4 years longer than that of men. By the year 2020, it is expected that 20 percent of the U.S. population will be over 65. This demographic transition is common among industrialized nations, especially those that experienced a baby boom immediately after World War II, but a generation later, fewer births occur. Unlike most countries in Western Europe, the United States has maintained birthrates near the replacement level of 2.1 children per woman of childbearing age. Despite this, the U.S. population continues to grow, largely due to immigration.

As the population ages, more demands are placed on health care services, affordable housing, and the special needs of the elderly population during disasters. The impacts of Hurricane Charley in August 2004 (see Box 2.1) illustrate how the changing age structure of Americans affects what hazards and disaster researchers study.

There is greater diversity in terms of race, ethnicity, and culture (including language) in the United States at present than at any other time in its history. In 1950, for example, the U.S. population was approximately 150 million, with 89 percent racially classified as white and 11 percent nonwhite. The faces of America continue to diversify, as the 2000 Census confirms: With a population of 291 million people, 80 percent were classified as white; 13 percent African American; 4 percent Asian; 1 percent American Indian and Alaska Native or Native Hawaiian and Other Pacific Islander; and 1 percent claiming to be of one or more races. Among the white population, 17 percent claim Hispanic or Latino origin (U.S. Census,

---

**BOX 2.1**
**Hurricane Charley in Punta Gorda, FL**

Charlotte County, located on Florida's southwest coast between Fort Myers to the south and Sarasota to the north, is an ideal location for retirees. The calmer waters of the Gulf of Mexico, less development, and a good quality of life appealed to many snowbirds as they sought retirement communities. In fact, Charlotte County has the highest median age of any county in the mainland United States—54.6 years.

For many of the county's new residents, affordable housing meant manufactured housing. After selling homes in the north, retirees moved to the Sunshine State and put their nest eggs in mobile homes. Because the homes were purchased with equity from the previous home rather than through a mortgage, some of the elderly chose not to carry hazard insurance on their homes (a mandated requirement if the home was financed through a bank). The home became the nest egg for the elderly, but on August 9-14, 2004 much of that changed as Hurricane Charley, a category 4 storm, slammed into the Punta Gorda area, catching many residents off-guard because the storm was predicted to make landfall 100 miles to the north. The mobile homes (especially those purchased prior to 1992) did not weather the hurricane force winds and were totally destroyed.

Not only have the elderly lost their life savings, but the longer-term impact on their physical and mental health is uncertain as they try to recover from the devastating effects of Hurricane Charley.

---

2004:Table 21). As the nation has become more racially and ethnically diverse, the race and ethnic classifications employed by the decennial census have changed as well—posing significant challenges for the research community, especially those interested in longitudinal studies. In the 2000 Census, for example, six racial categories were used: white; black or African American; American Indian and Alaska Native; Asian; Native Hawaiian and Other Pacific Islander; and One or more races. Also, two ethnicity categories were used: Hispanic or Latino and Not Hispanic or Latino (Brewer and Suchan, 2001).

These demographic changes present important challenges for disaster mitigation, preparedness, response and recovery, in part because they often result in differential impacts on various social groups as Box 2.2 illustrates. The geographic distribution of this racially and ethnically diverse population has also influenced the kinds of research that hazards and disaster researchers pursue. For example, there is increasing research interest in racial and ethnic disparities in disaster impacts as well as differences in coping responses and longer term recovery capabilities based on race and ethnicity (Bolin and Bolton, 1986; Bolin and Klenow, 1988; Peacock et al.,

---

**BOX 2.2**
**Response to Crisis:**
**Linguistic Diversity and the Northridge Earthquake**

Southern California is one of the most ethnically diverse metropolitan areas in the nation and one of the most racially differentiated. The vulnerability of Los Angelinos has been shaped by the post-war patterns of immigration, urbanization, and environmental transformations that have reshaped the natural landscape. Disasters are no longer unusual events, but are embedded in the region's psyche (Davis, 1998; Ulin, 2004). Despite this, there remain some interesting challenges in warning residents of dangers and in assisting them following a natural, techno-logical, or willful disaster event. For example, there are more than 224 identified spoken languages and dialects in the Los Angeles region, and 180 different language publications. Within the Los Angeles Unified School District, there are 92 recognized languages (Los Angeles Almanac, 2004). This linguistic diversity poses severe problems in communicating warning information and ways to protect themselves to the residents. It may impede rescue, relief, and recovery efforts in the aftermath of a disaster as was seen in the 1994 Northridge earthquake (Bolin and Stanford, 1998). Finally, interesting questions arise from the differential use of foreign language media by emergency managers and the receptivity of different language media to disseminate warning messages.

---

1997; Fothergill et al. 1999). Changes in the racial and ethnic identities of Americans as well as modifications in the way we measure them, have affected hazards and disasters research. For example, prior to 1970 there was no Census variable for Hispanic populations, so tracking regional or local changes in this ethnic population can only occur for the past 30 years.

The gap between the rich and privileged and the poor and disadvantaged has widened in the past 50 years. The key measures of socioeconomic status (education, occupation, and income) all have changed dramatically. In 1950, only 6 percent of the population over the age of 25 had completed 4 or more years of college; by 1970 this had risen to 11 percent; and by 2000, nearly 25 percent of the population over 25 had a college degree. In 1950, almost half (47 percent) of the population had completed eight or fewer years of formal education, but by 2000, most Americans graduate from high school (80 percent). However, seven percent of Americans still only have eight or fewer years of formal educational training. There is significant variability in educational achievement by race, ethnicity, and gender. High school completion rates are highest among white females and lowest among Hispanic females. Regionally, Texas, Louisiana, Alabama, and West Virginia have had the lowest percentage of high school graduates (less than 80 percent), while the Great Plains states (especially Wyoming,

Minnesota, Nebraska, and Montana) along with Alaska and New Hampshire have the highest (around 90 percent) (U.S. Census, 2004:Tables 213 and 216).

The poverty rate has also improved since the 1950s, when approximately 30 percent of the population lived below the poverty level. By 1970, only 12.6 percent of the American population lived below the poverty level. However, despite the economic growth over the past 30 years, the percentage of Americans living below the poverty level (12.4 percent) in 2000 was the same as in 1970. Given the increase in overall population, this means that there were 16 million more people living in poverty in 2000 than there were in 1970. Again, there is variability in poverty levels based on age (20 percent of all children under 18 live in poverty as do 10 percent of the elderly persons over 65) and race (where more than 50 percent of black and Hispanic populations live in poverty). Geographically, the highest levels of poverty are found in the District of Columbia and New Mexico, while the lowest levels are found in Wisconsin and Colorado.

Although the socioeconomic status picture has improved generally, these improvements are not consistent across all portions of the population or by geographic region. Such differences are important to hazards and disaster researchers, because they can lead to an understanding of how communities and their diverse residents prepare for, respond to, and recover from disasters (see Chapters 3 and 4). For example, there is a disproportionate relationship between death rates and economic costs when comparing developing (see Chapter 6) to developed societies. Deaths following disasters are higher in the former and economic losses are greater in the latter. In the United States, death rates related to disasters have declined over time, while economic losses have increased (Cutter, 2001).

## U.S. ECONOMIC CONDITIONS AND PROSPERITY IN THE POST-WAR ERA

Unparalleled economic growth and prosperity have characterized the past half-century in the United States. The effects of changing economic conditions not only influence our understanding of the economics and social dimensions of disasters, but also fundamentally alter the social science research agenda, offering challenges, opportunities, and constraints regarding what hazards and disaster researchers study.

The post-World War II era was characterized by increasing economic growth fueled by technological innovations, world dominance as an economic power, and increased demand by American consumers for goods and services. For example, per capita gross national product (GNP) in the United States in 1960 was $2,929 (current dollars), but it has nearly doubled in every decade since then. At present, GNP per capita is $33,898 (U.S. Census,

2004:Table 648). Personal incomes have risen as well, but as noted earlier, this trend is not evident in all regions or among all social groups.

The shift from primary sector employment (extractive industries such as agriculture, mining, fisheries) to secondary sector employment (manufacturing) helped fuel the economic engine of the United States. However, in the past several decades, more and more of the economy became service sector based. For example, in 1970, one-third of all employees in the United States were producing goods, but by the end of the century, this had fallen to around 20 percent. This means that the United States relies more on consumer spending and the provision of services as the basis of economic growth than on manufacturing or extractive industries. This shift to a service economy has resulted in the closure of manufacturing plants throughout the industrialized Northeast and Great Lakes Rust Belt and helped fuel the explosive growth in the Sun Belt states—growth predicated on service, not manufacturing jobs. Not only are there regional variations in these patterns, but they also affect workers differently. Generally speaking, the majority of service sector jobs that are low wage often fall to racial and ethnic minorities and women. At the same time, manufacturing jobs (particularly those insured by strong labor unions) are found in the traditional manufacturing belt in the Northeast and Midwest, but not in "right-to-work" states in the South where unions have less traction. Thus, the changing economic structure in which employment and output in services has expanded faster than manufacturing or agriculture influences hazard vulnerability, although some of the effect is likely due to the change in location of economic activity that has accompanied these sectoral shifts, rather than to fluctuations in the size of the sectors themselves (Berry et al., 1996; Clark et al., 2000).

The rise of multinational corporations and their diversification through mergers and acquisitions in the 1970s and 1980s paved the way for exploitative practices (domestically and globally) and a situation in which markets for goods and services are controlled by world supply and demand rather than at the national level (Cutter and Renwick, 2004). The general trend of rising interregional trade in intermediate products has made producers in one region more dependent on inputs from other regions. This could have two very different implications for the effects of disasters. On the one hand, it could mean that disasters in one region have greater effects on output in other regions because of growing global interdependence. Alternatively, greater interregional trade could mean that producers in one region can be supplied from multiple regions. Disasters might interrupt supply from one region, but substitute suppliers would be available so that the overall effects on production in undamaged regions would be mitigated.

During the 1980s and 1990s, deregulation of business, especially the deregulation of transportation and power production, introduced competition, lowered prices, and raised efficiency. However, part of the increase in efficiency has been achieved by eliminating "redundant" capacity (Cutter and Renwick, 2004). Should average load factors rise, then there is less of a margin between "normal" production and maximum capacity production. Accordingly, the impacts of disasters that damage a portion of the capital stock are more likely to reduce available capacity below normal operating levels and hence to force cutbacks in production. Put another way, what is considered redundant during normal operations may be essential when disaster strikes.

Domestically, consumer confidence still plays a big role in the economic growth of the United States, especially given the shift from a manufacturing to a service sector economy. At the same time, reductions in consumer confidence such as those fostered by the savings and loan crisis of the 1980s and by corporate malfeasance (such as Enron) slow economic growth and often result in periods of economic decline. Well-paying jobs or affordable housing—once the American dream—are beyond the attainment of many. Recessionary periods often hit those at the lower ends of the economic ladder the hardest, so that when disasters occur there is no economic cushion or savings for victims to draw upon during the hard times. Recovery from disasters often takes much longer, especially among those who were barely meeting their basic needs prior to the event.

In addition to the broad trends noted above, there have been changes in the economy that are having and will continue to have significant implications for future disaster problems. In general, these changes have been given little formal analysis. First, the rising spatial concentration of population and economic activity in large urbanized areas places more buildings and infrastructure at risk with a greater potential for catastrophic losses should a natural, technological, or terrorist event occur in a major metropolitan area.

Second, the rising rate of homeownership and ownership of second or vacation homes produces a context in which decisions about location, mitigation, insurance, and other types of disaster preparedness measures are being made by individuals with little expertise in real estate and management. Moreover, the rise of the second (vacation) home has put more real estate in harm's way because such development tends to be concentrated along shorelines where flood and wind damage are more likely or in woodlands where fire hazard is likely. There is evidence that owner-occupants fail to renew National Flood Insurance Program (NFIP) insurance even when there is a significant subsidy in the pricing of that insurance.

Third, there has been a general trend for inventory-to-output ratios to fall over time, particularly in the past decade. This trend extends over a broad range of industries. The change in the inventory-to-output rate reflects changing manufacturing practices, such as just-in-time materials management, which necessitates considerable coordination between suppliers and end users or sellers. The benefit of this coordination is lower costs of warehousing and managing inventory. However, it is not clear what effect this lower ratio of inventory to output has on disaster losses and business disruption. On the one hand, there may be fewer goods to suffer damage, but the potential vulnerability of economic activity to disasters in other locations may arise. If inventories of inputs are low, supply interruptions will have more dramatic effects on output.

Finally, the economic repercussions from willful events such as terrorist acts have impacts not only at the local level, but nationally and internationally. The 2001 attacks on the World Trade Center had enormous economic impacts locally, but more importantly, the ripple effects throughout the United States and global economies are being felt years later (Bram et al., 2002; Hughes and Nelson, 2002; Hewings and Okuyama, 2003). In highly industrialized nations such as the United States, there is more capacity in the economic system to absorb short-term direct impacts from hazards and disasters than there is in the developing world, although this is a researchable question. We know little about many aspects of the economics of natural hazards, especially the role of indirect impacts and information effects from disasters on local, state, regional, and national economies (Kunreuther and Rose, 2004). Moreover, the existing research is often at the aggregate level (state or nation) so less is known about disruptions in the supply chain (or spatial nodes) that could interrupt the flow of materials and goods (Park et al., 2005). The economic consequences and disruptions caused by terrorist activity and other unexpected extreme events constitute an important avenue of research for the hazards and disaster research community.

## GEOPOLITICS AT HOME AND ABROAD

Like all public policy issues, hazards and disaster policies influence and are influenced by national and international trends and events. In the United States, there have been substantial shifts in national priorities and the "national mood" (Kingdon, 1995) since the 1950s, and these shifts have influenced the nature of social science research on hazards and disasters. The national priorities are a function of changes in the administration and political leadership. Moreover, these macro trends have influenced the evolution of the emergency management system in the United States.

The emergency management system in the United States evolved from preparations taken during World War II and postwar concerns about nuclear weapons (Kreps, 1990). The Federal Civil Defense Act of 1950 was enacted when foreign policy, national security, and civil defense policies were made under the "Cold War consensus" that the Soviet Union was the most important threat to the nation. The first Soviet nuclear (fission) weapon test in 1949 led to the realization that nuclear war with the USSR was a possibility, and fears of nuclear attack increased with the development of a Soviet hydrogen (fusion) weapon in 1955 and the launch of *Sputnik* in 1957. The Cuban missile crisis in 1962 further raised fears of nuclear war. The organization of federal efforts to address and alleviate the harms done by hazards and disasters reflects broader civil security concerns at the time, as indicated in Figure 2.1. This timeline reflects the fact that for much of the last 50 years, federal policy dealing with natural hazards has been part of broader civil security or, today, "homeland security" functions. The organization of federal disaster policy often subordinated natural and technological hazards preparedness and mitigation to broader national security goals. This is how the system was developed in the early 1950s, although in the Kennedy administration two hurricanes—Donna in Florida in 1960 and Carla in Texas in 1961—led to the establishment of the civilian Office of Emergency Planning (later Office of Emergency Preparedness) in the White House.

### Great Society Programs and Hazard and Disaster Policy

The 1960s were a period of substantial social change. The Great Society programs of the Johnson administration sought to revitalize cities and to relieve poverty. Greater efforts to provide federal disaster relief were consistent with the intent of these programs. In the mid-1960s, greater attention was beginning to be paid to natural disasters, including the 1964 Alaska earthquake and Hurricane Camille in 1969 (Waugh, 2000). Camille led to the Disaster Relief Act of 1969 while Hurricane Agnes (1972), which resulted in substantial inland flooding in Pennsylvania and New York, led to the Disaster Relief Act of 1974, which provided for relief assistance to local governments and to individuals (May, 1985).

By 1976, after the souring of some Great Society programs and the Watergate scandal, the national mood had turned against what some called "big government." The election of President Jimmy Carter began a period of deregulation and government contraction that continued under the Reagan administration. This contraction was generally on domestic spending; after the Soviet invasion of Afghanistan in 1979, the United States began to spend more on defense after the substantial cuts in defense spending following the Vietnam War. By the mid-1970s, it was clear that multiple agencies shared and had overlapping disaster management responsibilities. There were more than 100 federal agencies with responsibility for some aspect of hazards and risks and at least five federal agencies with direct responsibility for emergency management response functions (Haddow and Bullock,

| Function | 1950 1951 1952 1953 ==== ===> 1957 1958 1959 1960 1961 ==== ==== ===> 1972 1973 ==== ===> 1978 1979 ==== ==== ===> 2002 2003 2004 2005 | | | | | | |
|---|---|---|---|---|---|---|---|
| Disaster Relief | Housing and Home Finance Administration (independent) | | | | Federal Disaster Assistance Administration (FDAA) in HUD | | DHS (FEMA becomes part) |
| Civil Defense | Federal Civil Defense Administration (Independent) | Federal Civil Defense Administration | Office of Civil Defense Mobilization (EOP) | Office of Emergency Planning (1968: Renamed Office of Emergency Preparedness) | Office of Preparedness, later Federal Civil Preparedness Agency (GSA) | Federal Emergency Management Administration (FEMA) (Independent) | DHS |
| Defense Mobilization | Office of Defense Mobilization (Executive Office of the President [EOP]) | | | DoD (Defense Civil Preparedness Agency) | DoD (Defense Civil Preparedness Agency) | | DOD |

FIGURE 2.1 Organization of federal disaster, civil defense, and defense mobilization functions, 1950-present. NOTE: DHS = Department of Homeland Security; DoD = Department of Defense; GSA = General Service Administration; HUD = Department of Housing and Urban Development.

2003). The creation of the Federal Emergency Management Agency (FEMA) was intended to address some of this overlap (Figure 2.1).

The Reagan administration rejected the policy of détente with the Soviet Union in favor of a more confrontational approach; defense spending increased; and administration officials began to speak of nuclear war survivability (Leaning, 1984). FEMA was not focused very intensively on natural disasters during this time period because there were relatively few of them. Instead, FEMA's leadership reflected Reagan era commitments to civil defense and preparedness for limited as well as full-scale nuclear war. By the late 1980s, relations with the Soviet Union had improved somewhat, and emergency management moved from a civil defense mentality to again focus attention on natural hazards. Morale problems and charges of political misbehavior at FEMA led to "an agency in trouble" from 1989 to 1992, as the agency was unable to effectively respond to the Loma Prieta earthquake, Hurricane Hugo, and Hurricane Andrew (GAO, 1992, 1993a, b; NAPA, 1993; Haddow and Bullock, 2003).

The collapse of the Soviet Union in 1991 removed, to a considerable extent, the threat of nuclear war. FEMA needed a change of leadership and direction, which came in the form of what Haddow and Bullock (2003) call the "Witt Revolution," named for the leadership of James Lee Witt, FEMA director from 1993 to 2001. Witt, an emergency management professional in Arkansas, was familiar with state concerns at a time when FEMA had lost its credibility with its state partners. He also enjoyed the confidence of President Clinton. The agency became transformed from a haven of political patronage to a modern, professional government organization.

The late 1990s and early 2000s continued to be marked by considerable political polarization and the influence of this polarization on hazards and disaster policy is still not entirely clear. While the distribution of relief in the name of either compassion or of constituent service was generally not ideologically based, changes in the emergency management system itself have become part of partisan politics. Singular events such as the September 11, 2001 attacks can profoundly alter the organization of emergency management in the United States (Box 2.3). The absorption of FEMA within the Department of Homeland Security (DHS) is a notable case in point.

## THE REACTIVE NATURE OF HAZARDS AND DISASTER POLICY

Federal policy in hazards and disaster management is reactive in nature and responsive to singular disaster events. This was particularly true through the 1970s when individual disaster events prompted post-event legislative responses. The establishment of the National Earthquake Hazards Reduction Program (NEHRP) in the wake of the 1964 Alaska and 1971 San Fernando Valley earthquakes is one example, the passage of the National Oil and Hazardous Substances Pollution Contingency Plan in the aftermath of the 1967 *Torrey Canyon* tanker spill is another. Cutter (1993), Platt (1994), Godschalk et al. (1999), Rubin (1999), and Rubin et al. (2003) provide examples of hazardous conditions and disasters and policy responses to them. However, policy responses are often more nuanced than this simple reaction suggests.

For example, some legislation has often been event specific and, as typifies distributive policy (Ripley and Franklin, 1984) characterized by "logrolling" (i.e., pledges to support each other's preferred legislation) and accommodation of particular areas' needs. May (1985) notes that not only was such logrolling predicated on potential future disasters, but it was also based on past disasters. Legislation may have been languishing without the requisite political support to make its way through Congress, and the particular hazard event or disaster provided the impetus to "push the legislation" through. Moreover, federal governmental efforts to alleviate suffering in the wake of disasters traditionally concentrated on disaster relief. Aid provisions retroactive to prior disasters were often written into new relief measures to ensure broader support. The Disaster Relief Act of 1950 (P.L. 81-875), coincident with severe flooding on the lower Missouri River, re-

**BOX 2.3**
**Reinventing Government Redux**

Events that led to the creation of federal agencies provide the focal point for their activities. The Federal Emergency Management Agency (FEMA) was created in 1978 as the result of years of federal experience with disaster preparedness and response that suggested reorganization would more clearly focus federal efforts in one place. FEMA combined about five functions into one agency; functions previously performed in the Departments of Defense, Commerce, and Housing and Urban Development, as well as programs located in the Executive Office of the President. The Department of Homeland Security (DHS) was created by the Homeland Security Act of 2002, which was passed in response to the terrorist attacks of September 11, 2001. DHS encompasses no fewer than 22 functions, including such disparate functions as the United States Coast Guard, FEMA, the Transportation Security Administration (TSA), border patrol functions, and former U.S. Department of Agriculture (USDA) responsibilities for managing plant and animal diseases. The fact that DHS was created in response to the event was as much a response to political demands as it was a careful consideration of the organization of government to meet homeland security challenges. This is reflected in the fact that important homeland security functions, such as intelligence gathering, remain in the Department of Defense, the Federal Bureau of Investigations (FBI), and the Central Intelligence Agency (CIA), even as experts on homeland security, and the September 11 Commission, concluded that the integration and dissemination of information is key to homeland security (National Commission on Terrorist Attacks upon the United States, 2004).

The key question for social science is whether the Department of Homeland Security will be able to address the new homeland security challenges while still attending to the "traditional" role in disasters that FEMA assumed, with some success, in the 1990s. Regardless of FEMA's location in the federal bureaucracy, its response to disasters will be under careful scrutiny from victims, their elected officials, the news media, and researchers. Social science research is needed to address a number of issues about the new organizational structure. For example, have FEMA's programs on hazard mitigation been compromised by its new administrative structure? Has the organizational culture changed the focus of the agency away from older, known threats to the identification of newer, unknown threats, and how does this affect preparedness programs? Would the nation be better off if FEMA had not been absorbed within DHS, but had maintained its independent agency status? These are testable questions that the committee believes should be addressed by social scientists.

placed ad hoc, event-specific aid packages with a general disaster relief law.

FEMA's shift away from preparedness for nuclear war and toward a disaster relief and hazard mitigation orientation was foreshadowed by the enactment, in 1988, of the Robert T. Stafford Disaster Relief and Emergency Assistance Act (hereinafter the Stafford Act). Indeed, the passage of the Stafford Act was an important milestone in American disaster policy for several reasons. First, the Stafford Act essentially served as FEMA's enabling statute. Second, the Stafford Act created a routine system of disaster declaration and relief, which, while still not perfect, is more predictable than the ad hoc policies that had preceded it. Third, the Stafford act was extremely important because it provided much more attention to mitigation.

Mitigation has traditionally received less attention because of the routine pressures on government officials and citizens to deal with many other problems that are much more salient until there is a catastrophic disaster (Rossi et al., 1982; May, 1985; Kreps and Drabek, 1996; Waugh, 2000). It is simply easier to declare a Presidential Disaster and provide relief. Hazard mitigation, according to scientific and technical consensus, should be a pre-disaster program to reduce the ultimate costs of relief and recovery.

The original Stafford Act provided a new program for hazard mitigation, which allowed the federal government to allocate 10 percent of federal moneys granted to states after disasters on "repair and restoration of facilities" (Section 406). The mitigation funds, under a program called the Hazard Mitigation Grant Program (HMGP), could be granted to states only if they had prepared a mitigation plan. The results of these mitigation programs were not as promising as their proponents had hoped. There were some positive developments; in particular, FEMA created a Mitigation Directorate to manage the HMGP and promote the idea of mitigation among state and local governments. Yet mitigation has not become an important part of broader natural hazards policy (Godschalk et al., 1999) and remains a post-event program. Little changed in the Hazard Mitigation and Relocation Assistance Act of 1993, legislation passed in direct response to the 1993 Midwest floods. The 1993 act did contain policy improvements by providing the means by which property owners in flood-prone areas could sell their property to state governments, which would mitigate flood hazards. However the act—even with an increase in HMGP moneys from 10 to 15 percent of federal disaster relief per disaster—remained a post-disaster program, not the sort of proactive, pre-disaster program for which experts had argued.

The continued shortcomings of the Stafford Act led to the enactment of the first explicit pre-disaster all-hazards mitigation program. The Disaster Mitigation Act of 2000 (DMA 2000) created the National Pre-disaster Mitigation Fund; states and localities would be eligible to apply for funds through a proposal process. According to the legislation, funds were to "(1) support effective public-private partnerships; (2) improve the assessment of a community's natural hazards vulnerabilities; or (3) establish a community's mitigation priorities." Where mitigation planning and implementation are taken seriously, they yield mitigation benefits and involve states and localities (Burby, 1994, 1998). DMA 2000 also required local governments to develop local mitigation plans to complement the state mitigation plans (Srinivasan, 2003). This is particularly important if localities wish to receive pre-disaster mitigation funds made available through DMA 2000. In 2002, FEMA extended the deadline for preparation of these plans to December 2004. Such plans are required if a community wishes to be eligible to receive post-disaster HMGP funds provided under the Stafford Act. However, considerable challenges confront policy makers who seek to change individual and community behaviors to mitigate disasters. Some political constituencies deny the need for more disaster mitigation efforts (Rossi et al., 1982; Alesch and Petak, 1986; Briechle, 1999) or believe that traditional mitigation policies, such as levees or other engineered solutions, are as effective as land-use planning mitigation in protecting lives and property.

Many of the activities called for in DMA 2000 were consistent with FEMA's now defunct Project Impact, which was created in 1997 to build public-private partnerships and broad levels of local commitment to hazard mitigation. However, there have been very few disasters that have tested the effectiveness of Project Impact. The most often cited example was the 2001 Nisqually earthquake that struck near Olympia, Washington, and was widely felt in western Washington, British Columbia, and Oregon. The relatively low level of damage done in Seattle (a Project Impact community) was attributed by Project Impact advocates as an example of the success of the program (Akaka, 2001; Chang and Falit-Baiamonte, 2002; Chang, 2003). Yet others cited the characteristics of the event and seismic building codes as reasons for the low level of damage. Despite making some headway in encouraging local action to mitigate disasters, in 2001 the Project Impact program was terminated, because the new administration had other priorities. However, despite discontinuation of the Project Impact initiative at the federal level, many local communities have continued with projects originally undertaken with federal Project Impact support. Tulsa Partners in Tulsa, Oklahoma, is an example of a relatively large-scale effort that is continuing with community mitigation and preparedness activities that were begun as part of Project Impact. Research on both Project Impact and its spin-off programs is needed to assess their effectiveness.

Federal policies can also unintentionally undermine local support for mitigation (Platt, 1999, 2004), even those done unintentionally. Prudent planning for and regulation of urban development often take a secondary role when the federal government pays for protection of private property from loss by building hazard control structures and offering disaster relief expenses that cover losses when they occur (Burby et al., 1999). Local governments, as the regulators of land use and building construction, are politically susceptible to blame for restricting land development and requiring flood control or earthquake resistance measures that increase local development costs. States have attempted to support local governments while meeting federal requirements in many different ways, including traditional land use requirements, but also by mandating or encouraging local governments to use capital investment policies and land-use planning for hazard mitigation purposes (Burby et al., 1997; Berke, 1998).

As the costs of disasters have risen, the private sector has become increasingly interested in hazard mitigation and preparedness. Some insurers have pulled out of particular hazard-prone areas, and the industry as a whole has begun to promote mitigation for households and businesses, as well as disaster planning for business functioning after disasters. The Institute for Business and Home Safety (IBHS), an insurance industry coordinating organization, has been a leader in this effort through its Showcase Community Program and Public Private Partnerships 2000. Despite these efforts, hazard mitigation faces important legal challenges in the United States (Box 2.4). Social scientists have a major role to play in providing information on the tradeoffs and costs and benefits of various mitigation options, including takings, available to decision makers.

Finally, the September 11, 2001 terrorist attacks on the United States led to a wide range of policy changes that may affect all phases of emergency management. In addition to the newly created DHS (see Box 2.3), the

---

**BOX 2.4**
**Takings: Good or Bad for Hazard Mitigation?**

Public interest versus private property rights has long been a controversial topic among planners, environmental managers, and local residents. Several "regulatory takings" cases have been heard in the U.S. Supreme Court, the first of which was *Lucas v. South Carolina Coastal Council* (112 S.Ct., at 2886, 1992). These cases sought to clarify the conditions under which localities can regulate the use of private property in order to accomplish a public purpose and when governments must provide compensation for "taking" the value of property. The net effect of these cases has been to limit, but not eliminate, the ability of local governments to regulate land use for hazard mitigation. Governments must not remove all value of a property ("total taking") without compensation, regardless of the purpose of the law. *Dolan v. City of Tigard, Oregon* (114 S.Ct., at 2309, 1994) established a "rough proportionality" between the burden on the property owner and the benefit to the public. More recently, the U.S. Supreme Court's decision in *Tahoe-Sierra Preservation Council, Inc. v. Tahoe Regional Planning Agency* (122 S. Ct. at 1465, 2002) reversed a two-decade-long trend that favored private property rights over the public interest. Consequently, there now is some uncertainty about the way in which public benefits can be balanced against private property rights.

U.S. Patriot Act (broad law enforcement powers to monitor terrorist activity), the Aviation and Transportation Security Act (which created the Transportation Security Administration to assist in aviation security), and the issuance of a series of Homeland Security Presidential Directives (HSPDs), which created the Homeland Security Advisory System and the National Incident Management System are of considerable concern to social scientists (see Chapter 8). In the aftermath of the inadequate response to Hurricane Katrina, Congress is considering new organizational changes to improve the nation's ability to cope with future threats (Congressional Research Service, 2006). Such proposed changes will also be of interest to social scientists.

## SETTLEMENT PATTERNS AND LAND USE

The United States as well as most of the world has become increasingly urban over the last five decades. By 1950 roughly two-thirds of the population of the United States lived in an urban area and by 2000 that proportion was close to 79 percent. In addition to increasing urbanization, there is a marked tendency toward settlement in coastal counties throughout the United States where 53 percent of the nation's population currently resides (U.S. Census, 2004:Table 23). Human settlements are subject to continuous change in response to trends in land use, advances in technology, and appearance of new urban design innovations. Trends at the beginning of the twenty-first century continue and extend those of the recent past. Conventional low-density development patterns (or sprawl) have dominated the landscape, while the concepts of Smart Growth and New Urbanism have emerged to counter the impacts of sprawl. The trends and new visions have important implications in coping with and responding to future threats of hazards and disasters.

### The Dominant Pattern of Twentieth Century Development

Metropolitan areas throughout the country are increasing their vulnerability to disasters because development continues unabated in many hazard-prone areas. Most of the vulnerability is associated with sprawling low-density development patterns caused by the outward expansion of suburban development on the urban fringe and commercial strip development along highways leading into and out of cities and suburbs. For example, between 1982 and 1997, the percentage increase in urban land dramatically outpaced the increase in population growth in all regions of the country. These land consumption rates place intense pressure on environmentally sensitive lands, including floodplains, earthquake fault zones, and unstable slopes.

This twentieth century model of the sprawling American metropolis has fostered a massive buildup of development in hazard-prone areas. Data on the buildup and subsequent disaster losses are abundant. Natural hazards cause average annual economic losses of about $25 billion to $30 billion in the United States, and losses have been rising rather than falling relative to increases in population and gross national product (Mileti, 1999b; Cutter, 2001; Cutter and Emrich, 2005). This model of sprawl has fostered the exposure of development to hazards in several ways. First, urban planning approaches to hazard mitigation are viewed by economic interests and local governments pursuing economic growth as a "good" to be fostered rather than a "bad" to be avoided. Hazard areas tend to be viewed as sufficiently safe, profitable places for development, especially by many players in the real estate market (appraisers, developers, and real estate investors) who are increasingly syndicated nationally and internationally. They have little stake in the local consequences of their actions. Community values aimed at creating safe, affordable, and livable places often have a lower priority for investors than protecting property values and profit gains.

Second, federal policies that facilitate the consumer-based model of city space are designed to stimulate investment in hazardous areas. Federal mitigation policies generally ignore risk avoidance (public land acquisition in hazardous areas or relocation from hazard areas) and, instead, have focused on risk reduction (building codes, seawalls) and risk sharing (disaster relief, tax write-offs, and flood insurance) (Burby et al., 1999). This approach makes sense if the goal is to foster development in hazardous areas. In the process of pursuing this goal, the federal government has severely limited the range of land-use options for local governments. In particular, it has crippled their ability to pursue risk avoidance policy goals. The ease of securing federal intervention to aid in the development of areas exposed to hazards establishes disincentives for local governments to plan for the most appropriate uses of these areas and to develop risk elimination programs to reduce losses of existing development. This situation cries out for more social science research to provide policy guidance to decision makers.

Unfortunately, the economic organization of the nation and the globalization of the economy constitute major impediments to the construction of safe places to live and work. Locations of urban land uses are arranged for maximizing property values, not as habitations that meet civic values such as avoidance of risk from hazards as noted earlier. While the trend is to create communities that are safe economic spaces, this does not always translate into creating safe living spaces (Box 2.5).

Two concepts prevalent in contemporary planning—New Urbanism and Smart Growth—are increasingly receiving attention as ways of coun-

---

**BOX 2.5**
**Living Too Close to the Edge**

In many parts of the West, suburban sprawl and the desire to live closer to nature have led to the development of residential areas in wildfire-prone regions. Fire is part of the natural ecosystem functioning and helps to regenerate the forest and rangelands. On the other hand, when people encroach into forested lands, these wildfires can cause tremendous damage to residential property and result in lives lost. The increasing movement of subdivisions into these fire-prone mountainous areas will increase the losses in these urban-wildland interfaces. Experiences in the last decade show that this pattern is increasing not decreasing: the Oakland Hills fire in 1991 that destroyed 2,900 structures and killed 25 people; the Flagler, Florida, fire in 1998 where thousands were evacuated; the Cerro Grande fire, which destroyed portions of the Los Alamos National Lab in 2000 (www.nifc.gov/stats/); and the multiple fires in Southern California in 2003, where seven people died, more than 5,000 buildings were destroyed, and 3,700 vehicles were destroyed or damaged. Insured losses exceed $2 billion (Guy Carpenter, 2004). Continued expansion of the urban fringe into forested areas will exacerbate the wildfire hazard in the United States not only in the West, but in the Southeast as well.

---

tering the societal ills associated with sprawl. Both have important implications for the way society copes with future threats posed by environmental hazards and the types of needed research from the social science community.

### New Urbanism

The urban design concept of New Urbanism is intended to counter the adverse effects of sprawl. This pattern of development is designed to create compact, mixed-use urban forms to foster social communities by enhancing civic engagement and interactions between public and private spaces, as well as to increase pedestrian (not auto) movement through use of a grid layout to shorten trip lengths, in contrast to the looped cul de sac pattern of conventional suburban developments. Linkages are created among commercial, office, residential, and transit facilities (as opposed to the spatial segregation of land uses under sprawl), and each development pattern is designed at the half-mile-wide "village scale." Individual New Urban developments are conceived as fundamental building blocks of New Urbanism at the regional scale (Calthorpe and Fulton, 2001; Duany and Talen, 2002). They form an interconnected network of mixed-use, high-density nodes of development linked by transit corridors. Within this network, regional open spaces create landscape-scale commons that serve as parks, act as barriers to limit outward expansion of urban development, and protect farmlands and environmentally sensitive areas. The New Urban version of metropolis builds on a long tradition of planning promulgated most prominently by the British planners Patrick Geddes and Ebenezer Howard in the late-nineteenth century, and the Regional Planning Association of America in the 1920s.

New Urban developments have the potential to further compound the growing risk to hazards by adding more higher-density development than in the past. High-density developments associated with New Urban forms can place more people, residential and commercial buildings, and infrastructure at risk than conventional development on an equivalent land unit exposed to hazards. This pattern of development also potentially exacerbates evacuation and emergency shelter needs for populations in hazard-prone areas. Future losses from New Urban developments due to natural disasters can be reduced if hazards are recognized in advance of exposure and appropriate disaster preparedness, structural, site design, and land-use planning practices are taken. Emergency preparedness and hazard mitigation practices are costly, however, and they are not likely to be applied to individual development projects without ample evidence of the threat from New Urban developments (Box 2.6). This evidence from individual cases, of course, is circumstantial absent the ability to control for other factors that can contribute to risks of hazards. On the other hand, New Urbanism can cluster development on safer lands, keeping those parcels most at risk in parks or in open space. The human-scale neighborhoods could actually reduce vulnerability, especially as communities rebuild in the aftermath of disasters such as Hurricane Katrina. Instead of the rush to rebuild in a hodge-podge fashion, New Urbanism principles of social interaction and environmental sustainability are now being considered in the rebuilding of the Mississippi coast (www.mississippirenewal.com).

Research is needed to examine the effect of New Urban design as a compact urban form on the disaster resiliency of urban development. For example, how well do the New Urban developments integrate hazard mitigation practices compared to the dominant mode of urbanization in the United States—the conventional low-density sprawl developments? Because New Urban communities are typically designed to be large, high-density developments, project reviews generate much higher levels of citizen reaction and opposition compared to project reviews of conventional developments. Does this high level of participation generate increased opportunity for public awareness of hazards and hazards mitigation practices? These are but a few of the questions that social science perspectives can contribute.

---

### BOX 2.6
### The New Urbanism: Risk Amplification or Risk Reduction?

Since its inception in the mid-1980s, the New Urbanism movement has been expanding rapidly. Data from the Congress of New Urbanism indicate that local governments in 41 states are currently experimenting with specific plans, policies, codes, and development standards that promote New Urban projects (Congress of New Urbanism, 2004). The data further indicate that between 1986 and 2002, about 474 New Urban projects that include 571,262 dwelling units housing more than 1.47 million residents have been completed or are under construction (estimates of residents are based on the national average household size of 2.59 people taken from the 2000 U.S. Census). Anecdotal evidence about various New Urban development projects supports the potential severity of the risk impacts. Consider the following examples.

***Envision Utah.*** This regional planning effort covers the 100 mile long Wasatch region that contains a widespread presence of earthquake faults, liquefaction, and landslides (Berke and Beatley, 1992). The region currently holds 1.7 million people (including Salt Lake City) and has experienced rapid expansion of conventional low-density development patterns to accommodate explosive population growth. The Envision Utah initiative channels future growth into a series of New Urban developments along the entire region, which are denser than conventional developments (Calthorpe and Fulton, 2001). However, given the higher densities, these New Urban developments may be at higher risk. Only 12 of the 24 major local governments in this region currently use U.S. Geological Survey maps that delineate fault, liquefaction, and landslide hazards in their land-use regulations, with the remainder not accounting for the threat in their land regulatory framework (interview with Gary Christensen, Geologic Manager, Utah Geologic Survey, September 26, 2003).

***Birkdale Village, North Carolina.*** This New Urban project is a case of locating a major stormwater pollution treatment and sediment control facility in the floodplain. It is moderately small by New Urban standards, consisting of 320 dwelling units, but the commercial core is designed to be a regional center with a large amount of commercial and office space (about 500,000 square feet). Since the stormwater treatment pond system is built in the main channel of the McDowell Creek floodplain, it is subject to floods that could flush out pollutants and sediment, which places a nearby downstream drinking water supply reservoir at risk.

---

### Smart Growth

Compared to New Urbanism, Smart Growth is based on land-use and development guidance policy frameworks but is less architecturally prescriptive and detailed in specifying the physical layout of a community. Since the early-1990s, 10 states have adopted "smart growth" legislation that requires or encourages local governments to adopt community planning programs to alter development practices dominated by conventional low-density patterns of urbanization and create more compact urban forms (Godschalk, 2000).

Smart Growth programs seek to identify a common ground where communities can explore ways to accommodate growth based on consensus on development decisions through inclusive and participatory processes. Smart Growth promotes compact, mixed-use development that encourage choices among different travel modes (walking, cycling, transit, and autos) by coordinating transportation and land use, requires less open space, and gives priority to maintaining and revitalizing existing neighborhoods and business centers. State and local Smart Growth initiatives include incentives and requirements to direct public and private investment away from the creation of new infrastructure and development that spreads out from existing areas. While Smart Growth's central concern has been to reform state growth management legislation, its concepts have also influenced local plans and been endorsed in the policy statements of professional and business interest groups, such as the American Planning Association, the International City County Management Association, the National Association of Homebuilders, and the Urban Land Institute.

Similar to New Urbanism, Smart Growth projects can lead to greater risks than low-density sprawl. The higher densities promoted by Smart Growth, state and local plans, and legislation can place more people and property at risk unless advanced planning is put in place. State Smart Growth legislation has to date offered limited guidance on how to integrate emergency management and hazard mitigation practices into local land-use plans and development ordinances that promote Smart Growth.

## WELL-BEING AND QUALITY OF LIFE

The health of populations and the provision of health care have both changed significantly in the past 50 years. By the 1950s the United States and Western Europe had both benefited from the public health advances in sanitation and nutrition that had begun in the late nineteenth century. Most developed nations had already undergone the epidemiologic shift (similar to and contributing to the demographic shift described previously) and now found that the leading causes of death and morbidity were "life-style" diseases (i.e., stroke, heart disease, cancer) instead of infectious diseases. This shift was furthered by the introduction of antibiotics in the 1940s to treat bacterial diseases and by the widespread use of vaccines to prevent viral diseases such as polio and measles beginning in the 1950s. Today in virtually all developed nations and many developing countries, the leading causes of childhood mortality and morbidity are unintentional injuries and life-style-related diseases such as obesity.

The provision of health care in the United States has also changed dramatically in the past 50 years. Previously, the practice of medicine was primarily under the direction of general practitioners. In 1950 there were 142 physicians per 100,000 people. Changes in training and health care began shortly thereafter, so that by 1965 there were equal numbers of generalists and specialists. By 1995, there were almost twice the number of specialists as generalists and 274 physicians per 100,000 people. While there are now more physicians per population, most of them are specialists, and they are geographically concentrated. Furthermore, most physicians practice in large metropolitan areas leaving the smaller cities and rural areas drastically underserved by health care workers. Likewise, inner-city areas also suffer from physician shortages.

One of the new "specialties" that began around 1970 was emergency medicine. This specialty area of medicine has contributed to the practice of disaster medicine as well. The World Association of Disaster and Emergency Medicine began in the 1970s as a gathering of physicians (primarily anesthesiologists) who were interested in bringing the lifesaving and resuscitation techniques of the surgical suite to the field in post-disaster situations (Frey, 1978).

The growth of specialty areas has benefited from the tremendous scientific and technological advances that have occurred in the past 50 years. Many of the standards of medicine that we take for granted today, such as MRI's (magnetic resonance imaging) and CATscans (computerized axial tomography) are relatively new advances. Likewise, the ability to quickly characterize infectious disease agents such as SARS (sudden acute respiratory syndrome) is the result of scientific advances made in the last decade (Marra et al., 2003).

While many of the changes in health care have been either positive or mixed in their effects, one significant change that has universally had a detrimental effect on both the population and the system has been the cost of health care. In the last 20 years, the costs of health care have skyrocketed. It is estimated that in 1950, per capita spending on health care was $497; by 2002, that amount (in constant dollars) was $5,241 (U.S. Census, 2004:Table 117). While some of the increase in costs can be attributed to the aging of the population, most of these costs are attributed to innovations in health care. Increasing costs of new pharmaceuticals contributes the lion's share of these increasing costs of innovation. Between 1992 and 2002, the share of health care dollars spent on prescription drugs rose from 5.8 to 10.5 percent. Most of this increase is due to new pharmaceuticals, but more importantly, most of these costs are borne by a small portion of the population. Five percent of the population accounts for more than half of the health care spending.

There are also changes in the trends of who pays for these increasing costs. The share of health care costs covered by government sources has increased during the last decades. Out-of-pocket expenses for health care have diminished, but the costs of private insurance coverage have increased greatly, with much of the increase taking place within the last 10 years. The rising costs of health care and health insurance to both individuals and employers have led to an increasingly large portion of the population being uninsured. Over the past 20 years, the percentage of uninsured grew from 11.8 percent to 17.3 percent. Some states (especially in the Southwest) have rates of uninsured that exceed 20 percent of the population. This puts an extraordinary burden on health care providers, especially those in hospital emergency departments, to provide essential medical care that is uncompensated (Henry J. Kaiser Family Foundation, 2002).

There are significant implications of these health care trends for the hazards and disaster research community. First, a moribund and aging Public Health Service will be unable to meet the emergency preparedness needs in the future, especially those involving willful acts such as bioterrorism, despite efforts such as the national network of Centers for Public Health Preparedness (funded by the Centers for Disease Control and Prevention); DHS-initiated projects such as the urban surveillance and monitoring of atmospheric pathogens and biothreats (Project BioWatch); centralized information depositories and rapid decision making (Project BioSense); and the development of the next generation of medical countermeasures (Project BioShield). Second, health care capacity (e.g., hospitals, extended care facilities) is growing at a slower rate than the population and, during times of crises, may be severely overextended. Finally, the emergence of new infectious diseases and the reappearance of older strains necessitate additional understanding of the origin and diffusion of diseases especially among high-risk populations.

The implications of these changes in health care and its cost directly influence the availability of services to highly diverse population groups, as noted earlier in the chapter. The differential in access to emergency services between urban and rural places, among different racial or ethnic groups, or based on socioeconomic status portends significant emergency preparedness and disaster response problems for the future.

## SOCIAL JUSTICE AND EQUITY

The historical evolution of the Civil Rights Movement, the Great Society programs, the War on Poverty, women's liberation, the environmental movement, and U.S. involvement in Vietnam gave rise to societal concern and actions for social and environmental justice during the 1960s and 1970s. These broader based social movements occurred at a time when the vast majority of hazards and disaster researchers were beginning their research careers and thus provided the context for the ways in which research problems were defined and studied. Not only did the subject matter change (expansion of hazards from natural hazards to technological events), but so too did the subjects of analyses. For example, hazards and disaster researchers have determined that women and members of racial and ethnic minorities sometimes suffer disproportionately from disasters (Beady and Bolin, 1986; Schroeder, 1987; Cannon, 2002; Cutter, 1995; Fothergill, 2003), particularly given the relationships between race, gender, poverty, and community vulnerability.

This is understandable given that hazards and their associated risks are embedded in our political, economic, and social institutions. Disasters are not only "acts of God," but also "acts of people." Two key issues that arise from social activism govern contemporary hazards and disaster research. First, hazards and their associated risks are social constructs. As such, they are the products of failures in technological, political, social, and economic systems that govern the use of technology, on the one hand, and influence response to disasters, on the other. This social construction leads to different perceptions of the "nature of the problem" and thus a politicized response, especially in the area of human-induced or technological hazards and risks. The driving forces behind the environmental stressors (e.g., materialism, poverty) that place people at risk are rarely considered in the policy world where there is often a preference for resolving the immediate impact, not the longer-term causes (Cutter, 1993).

Second, hazards and risks of disasters place uneven burdens and risks on people and the places in which they live. Concern about the distributional impacts of risks has a long tradition both in academe and within the federal government (NRC, 1999c). For example, the pioneering empirical work on distributional impacts focused on pollution in cities (Kruvant, 1974; Berry, 1977). This work was followed by claims focusing on environmental injustices based on the disproportionate burden of toxic waste on minority communities that were offered by the landmark General Accounting Office (GAO, 1983) and United Church of Christ (UCC, 1987) reports and social science research (Bullard, 1990; Lester et al., 2001). Most of the recent literature (1993–present) on inequity and environmental justice focused on activism and advocacy, on the legal and civil rights aspects of the environmental justice movement, or on more theoretically based discussions on the meaning of equity (Szasz and Meuser, 1997; Bowen, 2001, 2002; Liu, 2001; Rhodes, 2003; English, 2004).

What empirical research exists is fragmented, inconclusive, and inconsistent in its results. While there has been a marked increase in the number of methodologically sophisticated articles, especially those employing spatial analytical techniques (Stockwell et al., 1993; Chakraborty and Armstrong, 1997; McMaster et al., 1997; Cutter et al., 2001; Mennis, 2002; Pine et al., 2002), or historical demographic methods for measuring the evolution of inequities (Oakes et al., 1996; Yandle and Burton, 1996; Been and Gupta, 1997; Mitchell et al., 1999), the science of measurement and modeling is still in its infancy. There are also fundamental questions regarding the appropriate geographic scale for proving the existence of inequity (Greenberg, 1993; Zimmerman, 1994; Cutter et al., 1996; Sexton et al., 2002), as well as the role of environmental justice in the larger context of public policy decision making (Sexton and Adgate, 1999; Bowen, 2001; Margai, 2001; Bowen and Wells, 2002; Miranda et al., 2002).

As a partial federal response to the disproportionate impact of hazardous waste on poor and minority communities, Executive Order 12898 (signed February 11, 1994) was implemented. The language of Executive Order 12898 states:

> To the greatest extent practicable and permitted by law, and consistent with the principles set forth in the report on the National Performance Review, each Federal agency shall make achieving environmental justice part of its mission by identifying and addressing, as appropriate, disproportionately high and adverse human health or environmental effects of its programs, policies, and activities on minority populations and low-income populations in the United States and its territories and possessions, the District of Columbia, the Commonwealth of Puerto Rico, and the Commonwealth of the Mariana Islands.

This order forces the federal government to examine all of its policies and their implementation to ensure that they do not affect one social or economic group more than others. While largely focused on toxic releases and hazardous waste, all federal agencies are required to examine their programs in this light. Whether disaster assistance has been equally distributed to all affected communities, or whether such assistance is also a reflection of environmental injustice, is an important and understudied area in the hazards and disaster community. Moreover, what is the relationship between the county-level pattern of direct losses and the demography of counties and has this changed over time or across space? Are poor minority communities disproportionately affected (e.g. incur a greater relative loss) than wealthier nonminority communities (whose capacity to absorb losses is greater)? These are a few questions that will challenge social science research on hazards and disasters in the future.

## TECHNOLOGICAL CHANGE

During the past 50 years there have been tremendous technological advancements that have profoundly influenced our daily lives. We live in an age with complex and tightly coupled systems that govern the water we drink, the food we eat, the energy we use, and how we commute to and from work (Perrow, 1984). While the technological advances illustrated below certainly support hazards and disaster research (see Chapter 7), the technologies themselves may prove to be hazards (Perrow, 1984; Cutter, 1993). The 2003 electrical grid failure in the eastern United States is a recent example of how a failure in technology can lead to potentially disastrous situations. Thus, technologies can make societies both less and more vulnerable to environmental threats and willful acts. For example, in terms of the latter, as technology advances, societies may be particularly vulnerable to terrorism for a number of reasons (NRC, 2002b). One reason is that technological systems are so closely connected that disruptions in one system can spread to others, causing catastrophic failures. Furthermore, the means of mass destruction are potentially more available due to technological advancements. Thus, nation-states or small terrorists groups—either locally or internationally based—may gain access to materials used to produce nuclear, chemical, and biological weapons. The openness of countries, like the United States, also makes them more vulnerable to attacks because terrorist groups have easier access to potential targets, and they have relatively free use of communication technologies that can be used in planning and carrying out attacks.

The measurement or acquisition of information about an object or phenomena that is not in direct contact with that object is called remote sensing. The earliest use of aerial photography, a form of remote sensing, began with a French balloonist in 1859 and then progressed to fixed winged aircraft in 1909 (NASA, 2005). Aerially photography was used extensively in both World Wars. With the invention of radar, and thermal infrared remote sensing, remote-sensing technologies greatly advanced in the 1950s and 1960s (Jensen, 2000). Coupled with the postwar space program (and its associated satellites) remote sensing moved from exclusive military applications to civilian ones in the early 1960s first with the launch of experimental weather satellites and then with the Earth Resources Technology Satellite (later renamed Landsat). Today, remote sensing is widely used in surveillance and monitoring of hazards and disasters (e.g., hurricane tracking and tornado formation on Next Generation Radar [NEXRAD] Doppler; wildfire monitoring using satellites that carry Advanced Very High Resolution Radiometers [AVHRR], hazard zone delineations such as floodplains; assessment of post-event damages. The newest generation of nonmilitary satellites has the ability to "see" 1 meter by 1 meter from space, greatly enabling the precise monitoring of hazards and disasters and their impacts.

In addition to satellite remote sensing, the increased use of sensors and robotics has facilitated hazard and disaster threat detection and monitoring. The in situ sensors are the most useful and have been used to monitor ground motion and tsunami waves in the open ocean and, more recently, to monitor and model offshore coastal conditions in the advance of tropical storms (Caro-Coops, 2005). In the area of willful disasters, sensor systems are now widely deployed to monitor bioterrorist agents.

Americans enjoy more modes of telecommunication today than at any other point in the nation's history, and can watch events as they are unfolding on live television. The influence of television and round-the-clock cable news has not only affected the perceptions of risk by individuals, but also their responses to warnings. For example, the often-watched Weather Channel is now one of the primary sources of hurricane risk and warning information. Similarly, the use of advanced warning technologies (such as Doppler Weather Radar) by local weather forecasters has proven effective during tornado season.

Yet as Hurricane Katrina demonstrated, the use of advanced warning technologies alone does not guarantee an effective organizational and public response to an impending disaster, particularly when major regional impacts are possible. During times of impending disaster, along with using the technological resources that are available to them, decision makers at all levels need to consider the social, economic, and political dynamics that come into play in these situations. Social science expertise is vital at such times.

The management of disaster response has been aided by improvements in computing and computer systems. Easy to use software, laptop computers, and wireless communications are now the norm in post-event responses. Coupled with enhanced performance of pre-impact planning, computers have fundamentally altered the ways in which we study hazards and disasters and also how practitioners respond to them.

Wireless communications also have produced changes in warning and response to disasters, as well in surveys as a tool for disaster research. Two recent trends in survey research—falling response rates and emergence of new data technologies—will have longer-term consequences for disaster research. Public reactions to telemarketing (such as the Do Not Call list), aging of the U.S. population, and the rise in non-English speaking immigrants all contribute to declining survey response rates. Further, the switch from land-line phones to cellular phones as the primary contact number has significantly altered response rates among certain segments of the population. Survey research has also undergone technological changes related to the increasing use of information technologies, including Web-based data collection tools that reduce or eliminate the need for an interviewer. At the same time, not everyone has access to the Web, so certain segments of the population may not be adequately sampled or may be impossible to reach through these new technologies.

## ENVIRONMENTAL CHANGE

Local and global environmental changes and our understanding of them in chronological and social time have influenced what hazards and disaster researchers study. The earlier focus on extreme natural disasters in the early years (floods, earthquakes, severe weather, hurricanes) has been replaced by research on more common natural events such as coastal erosion, heat, and urban snow hazards. At the same time, slow-onset disasters (persistent drought cycles, deforestation) offer new perspectives on preparedness, warning, and response. Large-scale global processes such as those embodied in global climate change as well as more cyclic phenomena such as El Niño-La Niña illustrate the need for understanding the interactions of the biophysical system with human systems and how these effects manifest themselves over chronological and social time and across different regions.

The impacts of climate change are no longer hypothetical and will include temperature increases, changes in temperature regimes, changes in storm tracks and intensities, and sea level rise. The effects of global changes on local places, generally, and the uneven distributions of these impacts, especially as they relate to vulnerable populations provide an additional research context for hazards and disaster research (AAG GCLP, 2003). They also provide an opportunity to link social science hazards and disaster research to the human dimensions of the global change community in developing more robust understandings of the interactions between human systems and natural systems through advancements in sustainability science (Kates et al., 2001; Turner et al., 2003a) and vulnerability science (Cutter, 2003a).

Complex emergencies, such as the Rwandan refugee crisis, or the genocide and starvation in Darfur—which result in humanitarian crises and international relief efforts—are also important domains for pre-, trans-, and post-disaster investigations (Alexander, 2000). The precursors of these crisis occasions, such as environmentally induced changes in land use by poor and ethnically diverse populations, coupled with dysfunctional social and political systems, require more detailed analyses by hazards and disaster researchers than has hitherto been the case.

Finally, social scientists continue to study toxic substances and their production and influence on human and environmental health. Signal crisis events such as the *Torrey Canyon* tanker (1967) and later the *Exxon Valdez* (1989) spills, Three Mile Island (1979), Love Canal (mid-1970s), Bhopal (1984), and Chernobyl (1986) have resulted in both hazards and disaster policy initiatives and considerable research within the hazards and disaster research community (Kates et al., 1985; Kleindorfer and Kunreuther, 1987; Kasperson et al., 1988; Cutter, 1993; Freudenburg and Gramling, 1994).

## CONCLUSIONS

It is clear that the evolution of hazards and disaster research has taken a parallel path that parallels changes in American society and world events. The very nature of the problems that are studied and the approaches that social scientists take are set within this broader context of change. Researchers are able to respond to opportunities to extract lessons from particular disaster experiences as well as to draw theoretical, conceptual, and methodological understanding of human adjustments to hazard vulnerability.

The economic, political, and social changes during the past five decades cited above provide a rich array of researchable questions, many of which, as reflected in the following chapters, have been pursued by social scientists.

What are the vulnerabilities associated with settlement and occupant patterns, and how have these changed over time and across space? Do uneven distributions of impacts (which raise questions of equity in a much more diverse society) affect policy responses at local, state, and federal levels? What is the significance or importance of scale as we move from the local to the global, and how can we understand the cascading impacts of hazards and disasters as we move from one scale to another? How can we assist elected and appointed officials to make decisions under uncertain conditions and with incomplete information? How will the changes in American society (e.g., access to health care, greater ethnic diversity) influence disaster response in the future?

The salience of the terrorism threat following the September 11, 2001 attacks also raises a number of fundamental questions for researchers to consider. For example, in what ways are terrorist threats similar to and different from risks posed by natural and technological hazards? How has the increased salience of willful disasters shaped the emergency management system in the United States? Also how prepared are local communities and the nation as a whole for possible future attacks.

These are but a few of the questions derived from the context within which this research takes place. Many questions remain unanswered, providing opportunities for further research by current and future generations of hazards and disaster researchers in the social sciences. In some cases, this will require collaboration with colleagues from other disciplines such as earth sciences and engineering as discussed in Chapter 5 and with international colleagues as discussed in Chapter 6.

# 3

# Social Science Research on Hazard Mitigation, Emergency Preparedness, and Recovery Preparedness

The committee's goal in Chapters 3 and 4 is to document social science contributions under the National Earthquake Hazards Reduction Program (NEHRP) to the development of knowledge about the five core topics of hazards and disaster research and their interactions (see Figure 1.1.). As an organizing tool, the conceptual model of societal response to disaster, also introduced in Chapter 1 (see Figure 1.2), is employed. Within that conceptual model the catalytic impacts of disaster events are determined by conditions of systemic vulnerability, disaster event characteristics, and the actions of what the committee has termed the hazards and disaster management system. This chapter reviews research related to hazard vulnerability, disaster event characteristics and pre-impact emergency management interventions as determinants of disaster impacts. Chapter 4 then reviews research related to planned and improvised post-impact responses as determinants of disaster impacts. Each chapter concludes with recommendations for future research within the framework provided by the conceptual model.

## FURTHER COMMENTS ON THE CONCEPTUAL MODEL OF SOCIETAL RESPONSE TO DISASTER

Understanding the causal processes by which disasters affect social systems (i.e., communities, regions, societies) is important for at least four reasons. First, research on these processes is needed to identify the pre-impact conditions that render social systems vulnerable (hazard exposure, physical vulnerability) to disaster impacts (physical and social) in both chronological and social time. Second, research on these processes can be used to identify specific segments of threatened social systems that could suffer disaster impacts disproportionately, such as low-income households, ethnic minorities, or specific types of businesses (social vulnerability). Third, research on these processes can be used to identify disaster event-specific conditions (length of forewarning, predictability, controllability, and magnitude, scope, and duration of impact) that influence the level of disaster impacts. Fourth, findings on the interrelationships among characteristics of hazard vulnerability and disaster event characteristics allow documentation of the roles and interaction of pre-impact interventions (mitigation, emergency preparedness, and recovery preparedness practices) and post-impact responses (emergency and recovery activities) in influencing the level of disaster impacts. The causal processes by which disasters produce systemic effects in chronological and social time is informed generally within theorizing by Kreps (1985, 1989b) and Quarantelli (1989), and more specifically by causal models proposed by Cutter (1996), Lindell and Prater (2003), and Prater et al. (2004).

## HAZARD VULNERABILITY

The preexisting conditions most directly relevant to disaster impacts are hazard exposure, physical vulnerability, and social vulnerability.

### Hazard Exposure

Hazard exposure is defined by the probability of occurrence (or, equivalently, the recurrence interval) of events of a given physical magnitude and scope occurring in different locations. Hazard exposure arises from people's occupancy of geographical areas where they could be affected by extreme events that threaten their lives or property. Social scientists have made contributions to understanding hazard exposure principally by examining the distribution of hazardous conditions and the human occupancy of hazardous zones (Burton et al., 1993; Monmonier, 1997).

### Physical Vulnerability

A major component of physical vulnerability is structural vulnerability, which arises when buildings are constructed using designs and materials that are incapable of resisting extreme energy levels (e.g., high wind, hydrodynamic pressures of water, seismic shaking) or that allow the infiltration of hazardous materials. Thus, structural vulnerability can be defined by the likelihood that an event of a given magnitude will cause various damage states, ranging from slight damage through immediate total failure, to buildings and infrastructure. The construction of most buildings is governed by building codes intended to protect the life safety of building occupants from the dead load of the building material themselves and the live load of the occupants and furnishings, but they do not necessarily provide protection from extreme wind, seismic, or hydrostatic loads. Nor do they provide an impermeable barrier to the infiltration of toxic air pollutants. Adopting hazard-related building codes for the purpose of providing protection in the event of earthquakes, hurricanes, and other types of disaster is not just a technological matter. It is a complex process involving a number of significant social, economic and political issues. Social scientists in the hazards and disaster field that study such issues are in a position to provide guidance to policy makers and practitioners who make decisions about how to protect life and property in at-risk communities.

## Social Vulnerability

Social vulnerability can be defined by the probability of identifiable persons or groups lacking the "capacity to anticipate, cope with, resist and recover from the impacts of a . . . hazard" (Blakie et al., 1994). Vulnerable population segments might (1) have greater rates of hazard zone occupancy; (2) live and work in less hazard-resistant structures within those zones; (3) have lower rates of pre-impact interventions (hazard mitigation, emergency preparedness, and recovery preparedness); or (4) have lower rates of post-impact emergency and disaster recovery responses. Thus, these population segments are more likely to experience casualties, property damage, psychological impacts, demographic impacts, economic impacts, or political impacts—as direct, indirect, or informational effects.

## Hazard Vulnerability Analysis

It is important to recognize the difference between social vulnerability as a construct and demographic indicators of social vulnerability. The latter are characteristics of individuals and households that are *associated with* social vulnerability. These characteristics, which include gender, age, education, profession, income, ethnicity, and number of dependents, are associated with the above four components of hazard vulnerability. The broad factors (or driving forces) that contribute to social vulnerability include a lack of access to resources, limited access to political power and representation (Mustafa, 2002), certain beliefs and customs, demographic characteristics, the nature of the built environment, infrastructure (lifelines), and urbanization (Watts and Bohle, 1993; Heinz Center, 2002; Bankoff, 2004). Social science research contributions, including those made by NEHRP–

supported investigators, have demonstrated that gender (Fothergill, 1996; Enarson and Morrow, 1998; Fordham, 1999), race and class (Perry and Lindell, 1991; Peacock et al., 2000; Cutter et al., 2001), and age (Ngo, 2001) are among the most important indicators of vulnerable individuals and social groups.

The integration of hazard exposure, structural vulnerability, and social vulnerability indicators into systematic procedures for hazard vulnerability analysis (HVA) has progressed significantly from the regional ecology of hazards first proposed by Hewitt and Burton (1971), and this progress has been made possible by improvements in data and mapping technologies such as geographic information systems (GIS) and remote sensing (Lougeay et al., 1994; Monmonier, 1997; King, 2001; Greene 2002; Tobin and Montz, 2004). GIS-based approaches to vulnerability assessments were initially developed under NEHRP by social scientists (Mitchell et al., 1997; Morrow, 1999; Cutter et al., 2000) and are now a standard procedure for many state and local governments conducting hazard vulnerability analyses under the Disaster Mitigation Act of 2000. These advances in GIS-based modeling have been instrumental in advancing our understanding of exposure to a wide range of hazards (Carrara and Guzzetti, 1995; Mejia-Navarro et al., 1994; Hepner and Finco, 1995; Chakraborty, 2001; Rashed and Weeks, 2003). Once data have been collected on hazard exposure, physical vulnerability, and social vulnerability, GIS analyses can either overlay or mathematically combine the data to assess the overall vulnerability of a jurisdiction (e.g., a county) or to identify social vulnerability "hot spots" within that jurisdiction. Emergency managers and land-use planners can use the results of these HVAs to adapt their hazard mitigation policies, emergency response plans, and disaster recovery plans to meet the special needs of vulnerable community segments.

Maps of hazard exposure, structural vulnerability, and social vulnerability produced by HVA are expensive, require significant expertise to produce, and can become outdated over time as a community grows (Burby, 1998). These potential impediments to the development of hazard management policy make it important to identify the sources of data on hazard exposure, physical vulnerability, and social vulnerability that emergency managers and land-use planners use to formulate local policies for mitigation, response preparedness, and recovery preparedness. In addition, it is important to determine the staff capabilities of local governments to conduct HVAs and whether their capabilities are adequate to provide a sufficient fact basis to support the formulation of policies that will be effective in reducing hazard vulnerability and withstanding legal scrutiny (Deyle et al., 1998).

## DISASTER EVENT CHARACTERISTICS

There are many ways to classify threats based on the causal nature of the event but the most popular dichotomy has been natural versus technological hazards. One assumed implication of this distinction is that the societal response to disasters is fundamentally different for each of these categories. For example, some social scientists supported under NEHRP have argued that technological hazards are fundamentally different from natural hazards in their impacts on the human, natural, and built environments (Kroll-Smith and Couch, 1991), whereas others have suggested that natural disasters elicit a therapeutic community response and technological hazards elicit a nontherapeutic response. Since the events of September 11, 2001 some have suggested that another category of events be defined as intentional or willful acts—implicitly assuming that the response to such events will be different from the response to natural or technological events. Column A in Table 3.1 provides a list that is consistent with the way in which many government agencies define their missions, many physical scientists define the physical phenomena they research, and information is provided to the public about how to prepare and respond to environmental hazards.

The classification of disasters simply as natural, technological, and willful does recognize the distinctions among them in terms of human agency, but this should not be overdrawn. There is little dispute that terrorism differs from natural and technological hazards in some ways. For example, the social dynamics that generate terrorist hazard agents are clearly different from the physical dynamics that generate natural hazard agents. However, technological hazard agents are determined by both physical and social dynamics (Perrow, 1984), so the differences are smaller than some might believe. Even if the unreasoning laws of nature and the faulty reasoning of human error are different from deliberate intent to harm, these different causal processes can produce equivalent results. Thus, it is important to recognize underlying dimensions of similarity among hazard agents. As Table 3.1 indicates, these are the threats (column B), and agent and impact characteristics (column C), with the latter addressed by such scholars as Dynes (1970); Cvetkovich and Earle (1985); Kreps (1985, 1989a); Sorensen and Mileti (1987); Burton et al. (1993); Lindell (1994); and Noji (1997).

To date, however, there has been no systematic scientific characterization of the ways in which different hazard agents (column A) vary in their threats (column B) and characteristics (column C) and, thus, requiring different pre-impact interventions and post-impact responses by households, businesses, and community hazard management organizations. In the absence of systematic scientific hazard characterization, it is difficult to determine whether—at one extreme—natural, technological, and willful hazard agents impose essentially identical disaster demands on stricken communities or—

**TABLE 3.1** Hazard Typologies

| A<br>Hazard Agents | B<br>Threats | C<br>Characteristics |
|---|---|---|
| **Natural Hazards** | **Materials** | **Agent Characteristics** |
| Extraterrestrial | Chemical | Frequency or likelihood |
| Meteorological | Biological | Predictability |
| Geophysical | Radiological | Controllability |
| Biological | Nuclear | |
| Hydrological | | **Impact Characteristics** |
| | **Energy** | Speed of onset or forewarning |
| **Technological Hazards** | Explosive | Impact magnitude |
| Structural failure | Flammable | Impact scope |
| Environmental pollution | | Impact duration |
| Resource depletion | **Information** | |
| | Corruption | |
| **Willful Hazards** | Theft | |
| Sabotage | Deception | |
| Terrorism | | |

at the other extreme—each hazard is unique. Thorough examination of the similarities and differences among hazard agents would have significant implications for guiding the societal management of these hazards.

## DISASTER IMPACTS

### Physical Impacts

Damage to the built environment can be classified broadly as affecting residential, commercial, industrial, infrastructure, or community services sectors. Moreover, damage within each of these sectors can be divided into damage to structures and damage to contents. It usually is the case that damage to contents results from collapsing structures (e.g., hurricane winds that cause the building envelope to fail and allow rain to destroy the contents). Because collapsing buildings are a major cause of casualties as well, this suggests that strengthening the structure will protect the contents and occupants. However, some hazard agents can damage building contents without affecting the structure itself (e.g., earthquakes striking seismically resistant buildings whose contents are not securely fastened). Thus, risk area residents may have to adopt additional hazard adjustments to protect contents and occupants even if they already have structural protection.

As a result of a solid body of research, much of it sponsored by NEHRP, one of the best understood structural impacts of disasters is the destruction of dwellings. According to Quarantelli (1982), people typically pass through four stages of housing recovery—emergency shelter, temporary shelter, temporary housing, and permanent housing. Nonetheless, households vary in the progression and duration of each type of housing, and the transition from one stage to another can be delayed unpredictably (Bolin, 1993). Particularly significant are the problems faced by low-income households, which tend to be headed disproportionately by females and racial or ethnic minorities. Consistent with the social vulnerability perspective, such households are more likely to experience damage or destruction of their homes because of their location in areas of high hazard exposure. This is especially true in developing countries such as Guatemala (Bates and Peacock, 1987; Peacock et al., 1987), but also has been reported in the United States (Peacock and Girard, 1997). Low-income households also are more likely to be affected because they tend to occupy structures that were built according to older, less stringent building codes; used lower-quality construction materials and methods; and were less well maintained (Bolin and Bolton, 1986). Because low-income households have fewer resources on which to draw for recovery, they also take longer to transition through the stages of housing, sometimes remaining for extended periods of time in severely damaged homes (Peacock and Girard, 1997). In other cases, they are forced to accept as permanent what originally was intended as temporary housing (Peacock et al., 1987). Consequently, there may still be low-income households in temporary sheltering and temporary housing even after high-income households all have relocated to permanent housing (Berke et al., 1993; Rubin et al., 1985).

There has been little systematic research thus far under NEHRP on the rates of post-disaster reconstruction in the commercial, industrial, infrastructure, and community service sectors; and the reason for this are unclear. Research on housing recovery has identified a number of problems and, although the broad outlines of housing recovery are reasonably well understood, there is little research on the rate at which households (of different demographic categories) progress through the stages of housing. Such information would be very useful in forecasting the demand for temporary shelter and temporary housing after disasters. Some initial efforts in this regard have been incorporated into HAZUS (FEMA, 2004; NIBS-FEMA, 1999) and further efforts have been undertaken by Prater et al. (2004), but more needs to be done.

## Social Impacts

Social impacts—which can be psychological, demographic, economic, or political—can result directly from physical impact and be seen immediately or can arise indirectly and develop over shorter to longer periods of chronological and social time. For many years, research on the social impacts of disasters consisted of an accumulation of case studies, but two research teams conducted comprehensive statistical analyses of extensive databases to assess the long-term effects of disasters on stricken communities (Friesma et al., 1979; Wright et al., 1979). These studies both concluded no long-term social effects of disasters could be detected at the community level. In discussing their findings, the authors acknowledged that their results were dominated by the most frequent disasters—tornadoes, floods, and hurricanes. Moreover, most of the disasters they studied had a relatively small scope of impact and thus caused only minimal disruption to communities even in the short term. Finally, their findings did not preclude the possibility of significant long-term impacts upon lower levels of aggregation such as the neighborhood, business, or household, or over periods of time shorter than the 10-year interval between censuses.

One significant limitation of previous studies before and after the creation of NEHRP is that they have defined the research question as whether there are long-term social effects at the community level, but a more fruitful objective would be to determine the distribution of the chronological and social time periods during which disruption is experienced at different scales of analysis (e.g., household or business, neighborhood, community, region) in disasters of different magnitudes. Such research could reveal how long it takes for the horizontal and vertical linkages in American society to produce disaster recovery resources for those in need.

## Psychological Impacts

One type of social impact not measured by census data consists of measurements of psychosocial impacts and, indeed, research reviews conducted over a period of 25 years have concluded that disasters can cause a wide range of negative psychosocial responses (Perry and Lindell, 1978; Bolin, 1985; Gerrity and Flynn, 1997; Houts et al., 1988). In most cases, the effects that are observed are mild and transitory—the result of "normal people, responding normally, to a very abnormal situation" (Gerrity and Flynn, 1997:108). Few disaster victims require psychiatric diagnosis and most benefit more from a "crisis counseling" orientation than from a "mental health treatment" orientation, especially if their normal social support networks of friends, relatives, neighbors, and coworkers remain largely intact. However, there are population segments that require special attention and active outreach. These include children, frail elderly people with preexisting mental illness, racial and ethnic minorities, and families of those who have died in the disaster. Emergency workers also need special attention because they often work long hours without rest, have witnessed horrific sights, and are members of organizations in which discussion of emotional issues may be regarded as a sign of weakness (Rubin, 1991).

The negative psychosocial impacts described above, which Lazarus and Folkman (1984) call "emotion-focused coping" responses, generally disrupt the social functioning of only a very small portion of the victim population. Instead, the majority of disaster victims engage in adaptive "problem-focused coping" activities to save their own lives and those of their closest associates. Further, there is an increased incidence in pro-social behaviors such as donating material aid and a decreased incidence of antisocial behaviors such as crime (Mileti et al., 1975; Drabek, 1986; Siegel et al., 1999). In some cases, people even engage in altruistic behaviors that risk their own lives to save others (Tierney et al., 2001).

In addition, there are psychological impacts, which are called informational effects in Chapter 1. These impacts can have long-term adaptive consequences, such as changes in risk perception (beliefs in the likelihood of the occurrence of a disaster and its personal consequences for the individual) and increased hazard intrusiveness (frequency of thought and discussion about a hazard). In turn, these adaptive informational effects can increase risk area residents' adoption of household hazard adjustments that reduce their vulnerability to future disasters. However, such positive informational effects of disaster experience do not appear to be large in the aggregate—resulting in modest effects on household hazard adjustment (see Lindell and Perry, 2000, for a review of the literature on seismic hazard adjustment, and Lindell and Prater, 2000, and Lindell and Whitney, 2000, for more recent empirical research).

The findings from the research on psychological impacts of disasters indicate that there is no need for communities to revise their recovery plans to include widespread assessments of direct and indirect psychological impacts following disasters, nor does there appear to be a major need for research on interventions for the general population. However, there is a need for research on appropriate interventions for children, and perhaps other vulnerable populations, *before* disasters strike. These could help them develop emotion-focused coping strategies or, as discussed later in the section on risk communication, acquire personally relevant information about hazards and hazard adjustments.

## Demographic Impacts

The demographic impact of a disaster can be assessed by adapting the *demographic balancing equation*

$$<P_a - P_b = B - D + IM - OM>$$

where $P_a$ is the population size after the disaster, $P_b$ is the population size before the disaster, $B$ is the number of births, $D$ is the number of deaths, $IM$

is the number of immigrants, and $OM$ is the number of emigrants (Smith et al., 2001). In practice, population data are available for census divisions (census blocks, block groups, or tracts) rather than disaster impact areas, so GISs must be used to estimate the population change. Moreover, population data are most readily available from decennial censuses, so the overall population change and its individual demographic components—births, deaths, immigration, and emigration—are likely to be estimated from that source (e.g., Wright et al., 1979). On rare occasions, special surveys have been conducted in the aftermath of disaster (e.g., Peacock et al., 2000). The limited research available on demographic impacts (Friesma et al., 1979; Wright et al., 1979) suggests that disasters have negligible demographic impacts on American *communities* but there are documented exceptions such as Lecomte and Gahagen's (1998) report of 50,000 out-migrants from south Dade County in the aftermath of Hurricane Andrew. It is widely anticipated that the aftermath of Hurricane Katrina in the case of New Orleans will also be an exception. As noted earlier, the highly aggregated level of analysis in the Friesma and Wright studies does not preclude the possibility of significant impacts at lower levels of analysis such as the census tract, block group, or block levels. The major demographic impacts of disasters are likely to be the (temporary) immigration of construction workers after major disasters and the emigration of population segments that have lost housing. In many cases, this emigration is also temporary, but there are documented cases in which housing reconstruction has been delayed indefinitely—leading to "ghost towns" (Comerio, 1998). Other potential causes of emigration are psychological effects (belief that the likelihood of disaster recurrence is unacceptably high), economic effects (loss of jobs or community services), or political effects (increased neighborhood or community conflict)—all of which could produce significant demographic impacts at the neighborhood level.

Most of the research under NEHRP that has addressed household behavior in the aftermath of disaster has examined the recovery of households that decided to return and rebuild. A few studies have examined highly aggregated data that could only discern *net* migration, not in-migration and out-migration separately. Thus, research is needed to assess the extent to which households decide to leave after disaster and the ways in which these migrating households differ from those who remain as well as from the in-migrants who replace them.

## Economic Impacts

Economic impacts can be divided into direct and indirect losses. The property damage produced by disasters results in direct losses that can be thought of as losses in asset value (NRC, 1999c), measured by the cost of repair or replacement. Disaster losses in the United States are borne initially by the affected households, businesses, and local government agencies whose property is damaged or destroyed, but some of these losses are redistributed during the disaster recovery process through insurance, grants, or subsidized loans. There have been many attempts to estimate the magnitude of direct losses from individual disasters and the annual average losses from particular types of hazards (e.g., Mileti, 1999a). Unfortunately, these losses are difficult to determine precisely because there is no organization that tracks all of the relevant data and some data are not recorded at all (Charvériat, 2000; NRC, 1999c). For insured property, the insurers record the amount of the deductible and the reimbursed loss, but uninsured losses are not recorded so they must be estimated—often with questionable accuracy.

The ultimate economic impacts of direct losses depend upon the disposition of the damaged assets. Some of these assets are not replaced, so their loss causes a reduction in consumption (and, thus, a decrease in the quality of life) or a reduction in investment (and, thus, a decrease in economic productivity). Other assets are replaced—through either in-kind donations (e.g., food, clothing) or commercial purchases. In the latter case, the cost of replacement must come from some source of recovery funding, which generally can be characterized as either intertemporal transfers (to the present time from past savings or future loan payments) or interpersonal transfers (from one group to another at a given time). Disaster relief is an interpersonal transfer, whereas hazard insurance involves both interpersonal and intertemporal transfers.

In addition to direct economic losses, there are indirect losses that arise from the interdependence of community subunits. Research, including that supported by NEHRP, on the socioeconomic impacts of disasters (Dacy and Kunreuther, 1969; Durkin, 1984; Kroll et al., 1991; Alesch et al., 1993; Gordon et al., 1995; Dalhamer and D'Sousa, 1997) suggests that the relationships among the social units within a community can be described as a state of dynamic equilibrium involving a steady flow of resources, especially money (Lindell and Prater, 2003). Specifically, a household's linkages with the rest of the community are defined by the money that it must pay for products, services, and infrastructure support. This money is obtained from the wages that employers pay for the household's labor. Similarly, the linkages that a business has with the community are defined by the money it provides to its employees, suppliers, and infrastructure in exchange for inputs such as labor, materials and services, electric power, fuel, water or wastewater, telecommunications, and transportation. Conversely, it provides products or services to customers in exchange for the money it uses to pay its inputs.

Businesses' operational vulnerability arises from their proximity to the point of maximum impact and the structural vulnerability of the buildings in which they are located (Lindell and Perry, 1998; Tierney, 1997a,b). Other sources of operational vulnerability arise from dependency upon inputs as well as those who purchase its outputs—distributors and customers. Evidence of businesses' operational vulnerability to input disruptions can be seen in data provided by Nigg (1995), who reported that business managers' median estimate of the amount of time they could continue to operate without infrastructure was 0 hours for electric power, 4 hours for telephones, 48 hours for water or sewer, and 120 hours for fuel. If this infrastructure support is unavailable for periods longer than these, the businesses must suspend operations even if they have suffered no damage to their structures or contents. These findings at the level of individual firms are consistent with data from regional economic models showing that disruption of transportation and utility infrastructure services causes particularly widespread and substantial economic loss (e.g., Gordon et al., 1995) and major disasters can also cause long-term loss of sales and competitiveness (Chang, 2000, 2001).

Since certain sectors and business types are more dependent on infrastructure, they are more vulnerable to economic loss. Small businesses, those that are in the retail sector (and to a lesser extent the services sector), and those that rent rather than own their space tend to be most vulnerable (Kroll et al., 1991; Tierney, 1997a,b; Alesch and Holly, 1998; Webb et al., 2000; Chang and Falit-Baiamonte, 2002; Meszaros and Fiegener, 2002; Zhang et al., 2004b). Tourism is also often slow to recover from disaster. Consistent with earlier conclusions about communities (Wright et al., 1979), economic sectors in decline before disaster are especially vulnerable to structural change that accelerates pre-disaster trends.

It also is important to recognize the financial impacts of recovery (in addition to the financial impacts of emergency response) on local government. Costs must be incurred for damage assessment, emergency demolition, debris removal, infrastructure restoration, and replanning stricken areas. These additional costs must be incurred at a time when there are decreased revenues due to loss or deferral of sales taxes, business taxes, property taxes, personal income taxes, and user fees. The federal government will reduce the financial burden if the disaster is severe enough to warrant a Presidential Disaster Declaration (PDD), but communities that do not receive a PDD must bear the burden of the recovery themselves.

There have been significant advances under NEHRP in modeling the regional economic impacts of disasters. Thirty years ago, the literature consisted of a single conceptual discussion of the applicability of input-output models to disasters (Cochrane, 1974). Twenty years later, several studies had suggested or applied several methods of regional economic modeling to the disaster problem (NEHRP, 1992; Jones and Chang, 1995). Researchers now recognize that disasters pose fundamental challenges for economic modeling including the dynamics of economic systems in disequilibrium, the linkages between physical damage and economic disruption, the representation of physical infrastructure networks in largely aspatial models, and the incorporation of resilience and behavioral adjustments into economic models (Okuyama and Chang, 2004). With recent advances in modeling, analysts are now able to quantitatively describe the anticipated economic impacts of future disasters—identifying sectors that would be hard hit and those that will benefit. They are also able to assess, but to a much more limited degree, the potential economic benefits of specific pre-disaster mitigations and post-disaster responses.

Although there is an emerging technology for projecting the economic impacts of a disaster in the immediate aftermath of physical impact—or even for a disaster hypothesized in advance—local emergency managers and community economic development planners need to be able to identify the specific types of businesses in different sectors of the disaster impact area (or even in unaffected areas nearby; see Zhang et al., 2004b) that are at risk of failure. Moreover, it is unclear if business owners can assess their future vulnerability to indirect impacts of disasters with enough accuracy to forecast their need for the disaster recovery resources made available by government agencies.

## Political Impacts

As documented through NEHRP supported research, disasters can lead to community conflict resulting in social activism and political disruption during recovery periods in the United States (Bolin, 1982, 1993a) and abroad (Bates and Peacock, 1987). Victims often experience a decrease in the quality of life associated with their housing, with the following complaints being most frequent. First, availability of housing is a problem because there are inadequate numbers of housing units and delays in movement from temporary shelter to temporary housing and on to permanent housing. Second, site characteristics are a problem because temporary shelter and temporary housing are often far from work, school, shopping, and preferred neighbors. Third, victims usually attempt to re-create pre-impact housing patterns, but this can be problematic for their neighbors if victims attempt to site mobile homes on their own lots while awaiting the reconstruction of permanent housing. Conflicts arise because such housing usually is considered a blight on the neighborhood and neighbors are afraid that the "temporary" housing will become permanent. Fourth, building characteristics are a problem because of lack of affordability, inadequate size, poor quality, and designs that are incompatible with personal or cultural preferences. Fifth, neighbors also are pitted against each other when developers attempt to buy up damaged or destroyed properties and build multifamily units on lots previously zoned for single family dwellings. Such rezoning attempts are a major threat to the market value of owner-occupied homes but tend to have less impact on renters because they have less incentive to remain in the neighborhood. There are exceptions to this generalization because some ethnic groups have very close ties to their neighborhoods, even if they rent rather than own.

Sixth, conditions of allocation are a problem because recovery agencies impose financial conditions, reporting requirements, and onsite inspections. All of these complaints can cause political impacts by mobilizing victim groups, especially if victims with grievances have a shared identity (e.g., age, ethnicity) or a history of past activism (Tierney et al., 2001). The situation is especially problematic when the beliefs, values, artifacts, and behavior shared by members of a subgroup differ from those of other groups, especially the majority. Seventh, such cultural conflicts are compounded when people differ in their beliefs about the *goals* of recovery— their ultimate values regarding the kind of community in which they want to live. Many members of a community seek to reestablish conditions just as they were before the disaster, while others envision the disaster as "instant urban renewal" that provides an opportunity to achieve a radically different community (Rubin, 1991; Dash et al., 1997). Eighth, there is a contrast between a personalistic culture in many victim communities, which is based on bonds of affection, and the universalistic culture of the alien relief bureaucracy, which values rationality and efficiency over personal loyalty even when engaged in humanitarian activity (Bolin, 1982; Tierney et al., 2001). This conflict typically manifests itself in differences in emphasis regarding a task (material/economic) versus social-emotional (interpersonal relationships/emotional well-being) orientation toward recovery activities. In many cases, recovery is facilitated when outside organizations hire local "boundary spanners" to provide a link between these two disparate cultures (Berke et al., 1993).

Attempts to change prevailing patterns of civil governance can arise when individuals sharing a grievance about the handling of the recovery process seek to redress that grievance through collective action. Consistent with Dynes's (1970) typology of organizations, existing community groups with an explicit political agenda may *expand* their membership to increase their strength, whereas community groups without an explicit political agenda may *extend* their domains to include disaster-related grievances. Alternatively, new groups can *emerge* to influence local, state, or federal government agencies and legislators to take actions that they support and to terminate actions of which they disapprove. Indeed, such was the case for Latinos in Watsonville following the Loma Prieta earthquake (Tierney et al., 2001). Usually, community action groups pressure government to provide additional resources for recovering from disaster impact, but might oppose candidates' reelections or even seek to recall some politicians from office (Olson and Drury, 1997; Shefner, 1999; Prater and Lindell, 2000). In short, disasters do not produce political behavior that is qualitatively different from that encountered in normal life. Rather, disaster impacts might only produce a different set of victims and grievances and, therefore, a shift in the prevailing political agenda (Morrow and Peacock, 1997) that is enacted mostly in the recovery period after emergency conditions have stabilized.

There is a limited amount of research on the political information effects of disasters, and it is not entirely clear how existing research findings would apply to future events because there has been a clear pattern over time of disaster victims' decreasing tolerance for extended disruptions to their daily lives. Whether or not victims believe natural disasters are "acts of God," there seems to be an increasing tendency for them to hold government responsible for effective emergency response and rapid disaster recovery. Such attributions of government responsibility might also extend to terrorist attacks. Thus, further research is needed to assess the extent to which victims' future expectations of government performance are increasing, which could create a need for higher standards in pre-impact emergency management actions.

As the above discussion indicates, there is a small, but important, body of work on the politics of disaster, including research funded through NEHRP. Consistent with the committee's conception of disaster, Olson (2000) observes that disasters are political in nature, and expresses concern that this political dimension is too often neglected or given insufficient attention by researchers. He attributes this neglect to the small number of political scientists currently engaged in research on hazards and disasters and the view held by some that disasters should elicit a nonpartisan response. Nevertheless, politics is an essential feature of disasters and should be taken seriously by scholars.

Hurricane Katrina is providing a new opportunity to advance knowledge on the politics of disaster, including its nonconsensual aspects. For example, as the Gulf Coast region moves into the recovery period, many political dimensions that often have been observed following previous events appear to be emerging, including instances of intraorganizational and interorganizational conflict. Olson (2000) notes that such conflicts can often be expected to occur over the evaluation of the performance of organizations during the emergency and recovery periods, over who will set the political agenda for recovery, and over whom to blame for perceived lapses in the provision of pre-disaster protection and post-disaster assistance. Such conflicts have, indeed, emerged at the intergovernmental level as local and state agencies in the impacted Gulf Coast region and federal agencies have offered competing strategies for advancing the region's recovery and protection.

These conflicts have been related to such matters as debris clearance, assistance for rebuilding homes and businesses, and the design of flood protection works, including levees. It is crucial that social science investigators, especially political scientists, systematically study the political context of catastrophic disasters such as Hurricane Katrina.

Finally, another line of needed research is the comparison of the politics of natural and technological disasters and the politics of terrorism. Given the attention that the threat of terrorism has received since the September 11, 2001 attacks, a number of intriguing questions relating to the comparative politics of disasters could be investigated. To mention only two of many interesting questions: Has the threat of terrorism led to more partisan politics than other types of threats because acts of terrorism involve both criminal acts and can be seen as more of a national threat than natural or technological disasters? How does the allocation of government and other resources for countering terrorism compare with resource allocations for other types of disasters, and what accounts for any differences observed?

## PRE-IMPACT EMERGENCY MANAGEMENT INTERVENTIONS

The left-hand side of Figure 1.2 points to four types of pre-impact interventions that can, in effect, reduce the impacts of disasters. As noted above, HVA examines the preexisting conditions within a community to assess hazard exposure, physical vulnerability, and social vulnerability. Accordingly, it provides the foundation for hazard mitigation, emergency preparedness, and disaster recovery preparedness. Hazard mitigation consists of practices that are implemented before impact and provide *passive* protection at the time impact occurs. By contrast, emergency preparedness practices involve the development of plans and procedures, the recruitment and training of staff, and the acquisition of facilities, equipment, and materials needed to provide *active* protection during emergency response. Disaster recovery preparedness practices involve the development of plans and procedures, the recruitment and training of staff, and acquisition of facilities, equipment, and materials needed to provide rapid and equitable disaster recovery after an incident no longer poses an imminent threat to health and safety.

### Community-Level Hazard/Vulnerability Analysis

According to federal guidance (e.g., FEMA, 1996), community emergency operations plans (EOPs) should be based on an explicit statement of *Situation and Assumptions* derived from hazard/vulnerability analyses and should also have hazard-specific appendixes that address any distinctive disaster demands imposed by specific hazard agents. There are a number of sources for this information including the Federal Emergency Management Agency's (FEMA) (1997) *Multi-Hazard Identification and Risk Analysis* and HAZUS-MH (National Institute of Building Sciences, 1998; FEMA, 2004). However, there appears to be no research that has examined whether EOPs do contain appendixes for the appropriate hazards and whether the distinctive demands of these hazards are correctly identified. With NEHRP support, Hwang et al. (2001) found that there generally was a poor correspondence between a state's exposure to a hazard and the information addressing that hazard on the state emergency management agency's Web site. This finding suggests there will be a poor correspondence between local hazard exposure and the degree to which hazard-specific demands are addressed in local EOP appendixes, but research will be needed to determine if this is the case.

## Community-Level Hazard Mitigation Practices

There has been important social science research on hazard mitigation practices, including a significant amount sponsored by NEHRP. Hazard mitigation practices include hazard source control, community protection works, land-use practices, and building construction practices (Lindell and Perry, 2000). Hazard source control involves intervention at the point of hazard generation. For example, flood source control can be achieved by using reforestation and wetland preservation. Community protection works, which limit the impact of a hazard agent on the entire community, include dams and levees that protect against floodwater and seawalls that protect against storm surge. Land-use practices reduce hazard vulnerability by limiting development in areas that are susceptible to hazard impact. Such restrictions range from excluding especially vulnerable population segments (e.g., schools, hospitals, nursing homes, jails) to excluding all development. Finally, hazard mitigation can be achieved through building construction practices that make individual structures less vulnerable to natural hazards. These include elevating structures out of floodplains, designing them to respond more effectively to lateral and upward stresses from wind and seismic forces, and providing window shutters to protect against wind pressure and debris impacts.

Sometimes the distinction is made between structural and nonstructural mitigation, with structural mitigation being defined by the use of engineered works such as dams, levees, and other permanently constructed barriers to disaster impact. Unfortunately, the term "nonstructural mitigation" has limited utility because it includes an extremely diverse set of mitigation measures such as land-use planning and development controls in urban areas, on the one hand, and securing room contents to walls in earthquake zones, on the other. The ambiguity of this term is especially pronounced in connection with some technological hazards because non-structural also describes engineering measures such as changing production processes in hazardous materials facilities (e.g., substituting less toxic or volatile chemicals, reducing temperatures and pressures).

One important finding under NEHRP about community protection works such as dams and levees is that they are commonly misperceived as providing complete protection, so they actually increase development—and thus vulnerability—in hazard-prone areas (Burby, 1998). However, because the design basis for these structures will eventually be exceeded (i.e., a flood will eventually overtop the levee), the long-term effect of this particular mitigation strategy is to eliminate small frequent losses and increase the magnitude of rare catastrophic losses. In addition, some protection works such as stream channeling and levees do not even eliminate the small losses so much as displace them onto downstream jurisdictions—thus creating a social dilemma in which a community benefits if it is the only one to adopt this form of flood protection but all lose if they all build such structures.

Land-use practices and building construction practices are especially important methods of hazard mitigation because these are the ones most commonly used by local jurisdictions. It is important to recognize that the term *land-use practices* is broader than *land-use regulation,* and *building construction practices* is broader than *building codes* because *regulations* and *codes* involve setting standards and establishing sanctions (punishments) for failure to comply with those standards. Planning scholars have identified a number of planning tools that can be used to manage growth and development of land within a community (Nelson and Duncan, 1995; Olshansky and Kartez, 1998). These include land acquisition, development regulations, critical facilities policies, capital investment programs (providing roads, power lines, and water and sewer lines only in less hazardous areas), and incentives (providing subsidies for mitigation actions). Other policies include taxation or fiscal incentives and risk communication (informing people about the risks and benefits of development in locations throughout the community as well as the costs and benefits of mitigation measures).

Berke and Beatley (1992) examined a range of seismic hazard mitigation measures and ranked them according to effectiveness, political feasibility, cost (both public and private), administrative cost, and ease of enforcement. The most effective measures are land acquisition, density reduction, clustering of development, building codes for new construction, and mandatory retrofit of existing structures, but some of these are more politically and financially feasible than others. Land acquisition programs are very effective, but their high cost makes them unattractive to local governments. Mandatory retrofit programs are expensive for property owners, who often make it their business to thwart or delay such programs (Olson and Olson,

1993, 1994). Godschalk et al. (1998) noted some of the negative effects of such programs, but a definitive assessment is needed.

Social science research has yet to assess the extent to which each of the above tools is actually used by local planners for hazard mitigation, the community conditions that are necessary for successful use, and local planners' perceptions of the suitability of each tool for hazard mitigation. It is especially important to assess these factors over a wide range of hazards.

Another key research gap involves the lack of systematic knowledge on the costs and benefits of different mitigation strategies, such as land-use and building construction practices. A major study has just been released by the Multihazard Mitigation Council for FEMA assessing future savings from hazard mitigation. This is an important start to addressing this research need. The cost of mitigation efforts is usually straightforward, but the benefits of mitigation are more difficult to determine. Recent work on benefits of improvements in the U.S. electricity transmission network indicates that the benefits accruing because of decreased vulnerability to hazards— including lower required reserve capacity to deal with service interruptions and savings to customers from lower rates of service interruptions—may be much larger than any other source of benefits. Yet the current regulatory process determining investment in transmission capacity tends to ignore these benefits and utilities may similarly discount this source of benefits because it does not result in a revenue stream to them. Similarly, a study that compared three seismic mitigation options for an urban water system found that reduced business interruption to water consumers in future earth-quakes was much greater than any other category of benefits. If included in the economic analysis, a moderate-cost upgrade option would be optimal; if excluded, the optimal choice would be "no mitigation" (Chang, 2003). Other types of costs (e.g., potential increases in risk taking by the public) and benefits (e.g., reduced psychological stress in future disasters) are also commonly excluded from economic analysis of mitigation efforts. Further research is needed to develop methods for more comprehensively assessing the full costs and benefits of different mitigation actions, to build a knowledge base of the relative cost-effectiveness of different types of pre- and post-disaster interventions, and to develop approaches for incorporating such methods and knowledge into a decision-making process that reflects the needs of all stakeholders.

## The Process of Local Hazard Mitigation

Scholars, including many supported by NEHRP, have long noted the potential for disaster mitigation to be highly politicized, especially when multiple layers of government and multiple jurisdictions at a given level (e.g., states, counties, or cities) are involved in implementing a particular mitigation policy—for example, in the management of a large watershed such as the Mississippi River. A significant amount of the NEHRP research on the process of adopting hazard mitigation measures has focused on the hierarchical relationships among federal, state, and local government (see Figure 3.1, adapted from Lindell et al., 1997).

The core of the figure provides solid arrows to indicate the (downward) direction in which much of the power is exerted in these relationships among government levels. In addition, as May and Williams (1986) have documented, local government can thwart the efforts of state government, and states in turn can do the same to the federal government. It is important to note that conflicts among governmental levels for influence over the land-use practices of households and businesses is compounded by the multiple stakeholders within each community. In addition to the influence government has over households and businesses, these stakeholders are also affected by social influentials (e.g., knowledgeable peers), who are in turn influenced by social associations (e.g., environmental organizations) and economic influentials who are in turn influenced by industry associations (e.g., bankers, developers).

Finally, local governments and businesses are influenced by hazards practitioners who, in turn, can be influenced by their professional associations. All of these stakeholders interact with the government system to promote their preferred definitions of, and solutions to, problems of environmental hazard management (Stallings, 1995). Thus, this figure indicates that hazard mitigation is a much more complex process than government mandates "trickling down" from the federal government. Rather, environmental hazard management involves a complex web of interlinked bidirectional power relationships among stakeholders with widely differing characteristics.

Figure 3.1 is useful as a structural model that describes the relationships among stakeholders, but like all structural models, it cannot describe the process by which hazard mitigation is enacted. This process can be described by Anderson's (1994) *policy process model* that includes five stages—agenda setting, policy formulation, policy adoption, policy implementation, and policy evaluation of outcomes. In stage 1, *agenda setting*, different stakeholders (and coalitions of stakeholders) attempt to bring the matters that concern them most to the attention of public officials. Agendas are unstable over time and disasters can affect them by serving as focusing events (Birkland, 1997), concentrating public and official attention for a certain time, resulting in a window of opportunity (Kingdon, 1984). Because of the short amount of time available to effect policy change, individual actors known as policy entrepreneurs must work actively to get or keep issues on the agenda because the window of opportunity will not stay open forever. At present, it is unknown how long such a window will stay open or precisely what factors will make it close under a given set of condi-

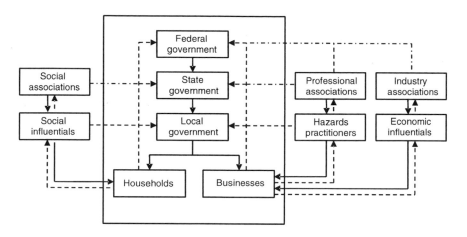

**FIGURE 3.1** Power relationships among emergency management stakeholders.

tions, although Kingdon offers a number of possible reasons. These include the taking of action on a problem or, alternatively, the failure to take any action. Windows also can close when another event occurs (shifting the systemic agenda to other matters), when key people leave their positions in a policy-making body, or when no possible course of action presents itself for consideration.

Starting with agenda setting and progressing through all stages of the policy (or planning) process, the media have an important role in the policy development process, particularly in the matter of issue framing—the words used to describe an issue. Scholars have noted that political issues are not necessarily defined immediately as political problems. Rather, they can exist as conditions for some time before the emergence of feasible coping strategies moves them into the realm of public discussion as problems that are amenable to solutions (Rochefort and Cobb, 1994). Thus, the first stakeholder to frame an issue can seize a significant political advantage, especially if he or she is successful in linking a proposed policy to widely shared public values. Emergency managers and land-use planners can place environmental hazards on the agenda by documenting community hazard exposure, physical vulnerability, and social vulnerability in a way that generates the *fact basis* for policy formulation (Cutter et al., 2001). Although anecdotal evidence attests to the effectiveness of GIS in accomplishing this objective, there is little systematic research available that documents the degree to which hazard/vulnerability analyses affect political agendas or the social and psychological mechanisms by which these effects are achieved.

During stage 2, *policy formulation*, hazards policy entrepreneurs develop proposed courses of action for dealing with community hazard vulnerability. These include the mitigation alternatives listed earlier—source control, community protection works, land-use practices, and building construction practices—as well as emergency response preparedness and disaster recovery preparedness. It is well understood that a proposed mitigation policy should make a significant contribution to solving the problem of hazard vulnerability yet must avoid generating significant opposition by other stakeholders. In fact, this is a major dilemma because hazard mitigation policies typically benefit a diffuse constituency (taxpayers at large) over the long term but impose costs on a definite group of stakeholders (especially developers) in the short term. Unfortunately, it is not known if attributes other than efficacy and cost are important in the development and framing of hazard mitigation policies and specifically how policy entrepreneurs must account for the local political context. Specifically, how important are environmental protection and economic development in shaping local hazard mitigation policy?

During stage 3, *policy adoption*, hazard policy entrepreneurs mobilize support for a specific proposal so it can be authorized by elected officials. If the policy entrepreneurs have been successful in setting the agenda, framing the issues, and formulating the policy to maximize the strength of the proponents and minimize the strength of the opponents, policy adoption will be relatively simple. However, the process of policy adoption will be slow and possibly even unsuccessful if they have performed inadequately at earlier steps. Unfortunately, existing research provides little *specific* guidance for emergency managers and land-use planners on how to mobilize support for mitigation policies.

The fourth stage, *policy implementation*, is defined by the events and activities that occur after a policy is adopted and include the administration of the policy and its actual effects (Mazmanian and Sabatier, 1989). During policy implementation, bureaucrats use the government's administrative machinery to apply the policy. In a federal structure, this means that the federal government can impose unfunded mandates on state and local governments, which in turn can either facilitate or thwart the implementation of federal policy depending on its compatibility with their capacity and commitment.

During stage 5, *policy evaluation of outcomes*, agency personnel determine whether the policy was effective and what adjustments are needed to achieve desired outcomes. Despite the many reasons for conducting them, it appears that hazards policy evaluations are infrequent and the reasons for this neglect are largely anecdotal and speculative. Some contend that practitioners are so convinced of program efficacy that they are unwilling to spare any expense for evaluation, but there appears to be no research to confirm this speculation.

One important aspect of hazard management policy concerns the effect of state mandates. Previous research has examined the effect of mandate design on policy implementation (Van Meter and Van Horn, 1975; Mazmanian and Sabatier, 1989; Goggin et al., 1990). Accordingly, May (1994) compared data from five states (California, North Carolina, Florida, Texas, and Washington) to discover the links between the design of hazards relevant aspects of land-use mandates and the implementation of hazards mitigation policy. May's analyses found partial support for Mazmanian and Sabatier's model, thus indicating successful implementation of hazard mitigation policy is facilitated when the state agency charged with implementation has a high level of commitment to the policy, a high level of technical expertise, a low level of personnel turnover, and when there are adequate facilitating features and controls built into the mandate. Nonetheless, state mandates and guidance to local government increase the adoption and implementation of effective land-use practices (May and Deyle, 1998), these mandates have a measurable impact on the reduction of disaster losses. Moreover, a cross-sectional analysis of disaster recovery of communities after the Northridge earthquake found that the quality of mitigation elements in local comprehensive plans has a positive influence on implementation of mitigation practices and on the reduction of property loss (Burby et al., 1998).

Another important aspect of hazard management policy concerns the mobilization of local support because this raises questions about how governments can use hazard awareness campaigns to make households and businesses aware of the risks they face and of suitable hazard adjustments for reducing their vulnerability. Information campaigns relying on voluntary compliance tend to be politically acceptable but have not been based on contemporary scientific theories of social influence and, to date, have had limited success (Lindell et al., 1997). Alternatively, governments can motivate the adoption of hazard-resistant land-use and construction practices by providing economic incentives such as low-interest loans or tax credits. Of course, the money for such incentives must come from somewhere and cash-strapped local jurisdictions may not be able to provide it. Finally, governments can require hazard-resistant land-use and construction practices as a condition for construction permits. The verification of compliance requires onsite inspections, and the problems with such inspections have been noted elsewhere (Lindell et al., 1997).

Considerations other than the cost of mitigation should be studied as well. Agencies such as public works departments might be accustomed to dealing with hazards but feel threatened when the decision-making process is expanded to include meetings with neighborhood groups. As anonymous bureaucrats, they may not be accustomed to being held personally accountable for technical decisions and may equate citizen participation with needlessly looking for trouble. Conversely, some neighborhoods that are especially vulnerable to hazard impact may have a large proportion of lower-income or ethnic minority residents who lack knowledge about, or mistrust, the political system. All of these concerns need to be balanced because any perceived unfairness in the policy itself or its adoption is likely to cause problems in the implementation phase. Even after a policy has been developed, there are many veto points at which interests can block the implementation of policies they consider undesirable.

There has been a significant amount of research under NEHRP on the adoption of hazard mitigation measures, but there are also significant limitations to that research. Figure 3.1 explicitly addresses the linkage between local and state government, but neglects the role of regional authorities, such as councils of government and metropolitan planning organizations, in promoting hazard mitigation through shared hazard/vulnerability analyses and development of coordinated hazard mitigation policies. Such organizations could provide an important role in establishing the horizontal and vertical linkages that local jurisdictions need to acquire critical but infrequently used skills at a reasonable cost.

The policy process model provides an important complement to lists of factors affecting the adoption of hazard mitigation tools (e.g., Godschalk et al., 1998:171-191). Such lists identify broad principles, but more specific guidance is needed on how to become an effective policy entrepreneur, how to frame issues, and other specific activities in which local land-use planners and emergency managers must become involved. Conversely, planning research has identified critical limitations of stand-alone mitigation plans prepared by emergency managers who are disconnected from comprehensive land-use planning. Thus, research is needed on planning processes that involve emergency managers with land-use planners in integrating hazard mitigation objectives into community comprehensive plans. Moreover, the policy process model outlines a process that differs in some significant respects from planners' recommendations. For example, Burby (1998) recommended establishing a hazard mitigation committee, conducting an HVA, analyzing mitigation options, preparing a plan, and implementing that plan. Research is needed to determine if there are any important ways in which the hazard planning model differs from the policy process model.

There is also a need to more systematically examine the effects of nongovernmental (e.g., social and economic) stakeholders in the mitigation process. For example, an International City/County Management Association nationwide survey of local governments reported support for hazard mitigation was higher among utilities, news media, insurance companies, and building owner/property managers than among neighborhood/civic groups and professional associations and was surprisingly strong among financial/mortgage companies and realtors (Briechle, 1999). Moreover, professional associations have been found to be potentially, but not actually, useful to line professionals in government (Bingham et al., 1981), so research is needed to identify methods of enhancing the effectiveness of professional organizations.

Research also is needed on building construction practices because 25 percent of the destruction in Hurricane Andrew was attributed to poor design, materials, or construction techniques (Lecomte and Gahagan, 1998). There are significant obstacles to getting engineering knowledge incorporated into model building codes, getting the provisions of these codes adopted at the local level, and getting local codes enforced.

A notable feature of social science research on hazard mitigation is the lack of integration among planners, sociologists, and political scientists studying overlapping aspects of the policy adoption process. Much of the research by political scientists has examined the conflicts among government layers, whereas sociologists have focused on conflicts among community groups, and planners have tended to address the substantive content of the mitigation measures. Research that links all of the elements of Figure 3.1 is needed.

## COMMUNITY-LEVEL EMERGENCY RESPONSE PREPAREDNESS PRACTICES

According to the systems perspective proposed in federal guidance (FEMA, 1996), the first step in emergency response preparedness is to identify the demands that different types of disasters will place upon the community and, thus, the need to perform four basic emergency response functions—emergency assessment, expedient hazard mitigation, population protection, and incident management (Lindell and Perry, 1992, 1996). Emergency assessment consists of those actions that define the potential scope of the disaster effects (e.g., projecting hurricane wind speed), expedient hazard mitigation consists of last-minute actions to protect property (e.g., sandbagging around structures), population protection consists of actions to protect people from death or injury (e.g., warning and evacuation), and incident management consists of actions to initiate and coordinate the emergency response (e.g., communication among responding agencies). The next step in community emergency preparedness is to determine which community organizations will be responsible for accomplishing each of the functions (FEMA, 1996). Households and businesses have substantial capabilities for self-protection, especially in performing expedient hazard mitigation and population protection, but government agencies must usually address the emergency assessment and incident management functions. In addition, some households and businesses have such limited response capabilities (e.g., limited mobility, lack of personal vehicles) that they need external assistance. Sometimes this assistance is provided by peers (friends, relatives, neighbors, or coworkers), but government agencies or nongovernmental organizations must also be prepared to meet these needs. Thus, functional responsibilities must be assigned to each agency, which then must develop procedures for accomplishing the assigned functions. Moreover, these agencies must acquire response resources (personnel, facilities, and equipment) to implement their plans. Finally, they need to establish, test, and maintain preparedness for emergency response through continued planning, training, drills, and exercises (Daines, 1991).

Major failures occurred in the provision of evacuation assistance by both governmental and nongovernmental organizations to citizens with limited capacity to evacuate on their own prior to Hurricane Katrina's destructive blow to New Orleans. These failures occurred despite the fact that this problem had been anticipated for quite some time. Social scientists are now investigating these and related preparedness and response problems exposed by Hurricane Katrina for lessons that might be learned.

## Emergency Response Preparedness Functions

**Planning Processes.** NEHRP-sponsored studies of preparedness planning processes have addressed a range of topics, including the extent of local support for disaster preparedness (Rossi et al., 1982) and management strategies for improving the effectiveness of community preparedness efforts (Drabek, 1987, 1990). Other work has focused on the structure of community emergency preparedness networks (Gillespie and Streeter, 1987; Gillespie, 1991; Gillespie and Colignon, 1993; Gillespie et al., 1993) and formalized organizational networks, such as those developed to prepare for chemical hazards (Lindell and Meier, 1994; Lindell et al., 1996; Lindell and Brandt, 2000; Whitney and Lindell, 2000). Lindell and Perry (2001) recently summarized this literature as indicating network effectiveness, and especially the effectiveness of formalized emergency management committees, can be defined in term of individual (job satisfaction, organizational commitment, effort/attendance) and organizational (product quality, product timeliness, product cost) outcomes. In turn, these outcomes are affected by extra-community resources (e.g., professional associations, government agencies), the planning process (e.g., planning activities, team climate, situational analysis, and strategic choices), and the local emergency response organization's staffing and organization (e.g., staffing levels, organizational structure, technology). More distal influences include the community's hazard exposure and vulnerability (e.g., emergency/disaster experience and hazard/

vulnerability analyses), community support (e.g., from senior elected and appointed officials, the news media, and the public), and community resources (e.g., staff, budget)

**Training/Equipment Needs Assessment.** There is a long history of research in psychology on training and training needs assessment, but this work has not been addressed by explicit research on emergency preparedness for environmental hazards. Training scholars have recommended a systematic assessment of organizational needs as the basis for training programs and for the evaluation of training programs in terms of trainees learning, performance on the job, and the outcomes of the training (Goldstein and Ford, 2001), but it does not appear that these issues have been examined in research on emergency management organizations. Moreover, a recent review of research on training has called attention to the unique challenges of training for emergency response—including retention of infrequently practiced skills over long periods of time (Ford and Schmidt, 2000). In addition, there is a burgeoning research literature on team training that has examined the effects of taskwork (knowledge and skill related to the work itself) and teamwork (knowledge and skill related to other team members) in resolving issues involving the coordination of individual efforts, distribution of workload, and selection of task performance strategies (Guzzo and Dickson, 1996; Salas and Cannon-Bowers, 2001; Campbell and Kuncel, 2002; Kraiger, 2003; Hollenbeck et al., 1998; Arthur et al., in press). This research has clear relevance for some of the classic issues addressed by disaster sociologists (Dynes, 1974; Kreps, 1978; Quarantelli, 1978; Stallings, 1978; Wenger, 1978). An integration of these different perspectives is needed.

**Drills, Exercises, and Incident Critiques.** There is agency guidance on drills, exercises, and incident critiques (e.g., FEMA, 2003b; National Response Team, 1990) that appears to be derived from practitioner experience. However, there appears to be no social science research on emergency response organizations' performance of these tasks. This is unfortunate because there is research on individual and team training that is relevant to this problem. For example, Hackman and Wageman (2005) proposed a model of team coaching that contains relevant concepts. An assessment of the applicability of their model to emergency response organizations is needed.

**Emergency Response Functions.** As discussed in Chapter 4, there is a long history of social science research on some aspects of disaster response, especially population protection and incident management. However, none of this research has addressed the extent to which practitioners use the findings from disaster research in developing community emergency preparedness.

**Emergency Assessment and Expedient Hazard Mitigation.** There appears to be no research that has explicitly addressed emergency response or preparedness for either of these emergency response functions. Nor does there appear to be any research on the extent to which practitioners use the findings of social science research in developing community emergency preparedness. Nonetheless, emergency assessment involves important social science issues regarding threat detection and classification and population monitoring and assessment. Expedient hazard mitigation involves important issues regarding the evaluation of alternative methods of hazard source control and impact mitigation.

**Population Protection.** Many, if not most, major emergencies require local officials to initiate protective actions for the population at risk. This requires protective action selection (usually between evacuation and sheltering in-place), warning, protective action implementation, impact zone access control and security, reception and care of victims, search and rescue, emergency medical care and morgues, and hazard exposure control (Lindell and Perry, 1992). The population protection function is distinctive in that it has generated the greatest amount of social science research on disaster response, including that supported by NEHRP—undoubtedly due to the fact that this function involves the risk area population's degree of compliance with emergency responders' protective action recommendations. However, there is virtually no research on *preparedness* for population protection.

First, the emergency response organization must be prepared to select an appropriate protective action recommendation. Sheltering in-place is preferable to evacuation in cases when exposure to the hazard conditions while in an evacuating vehicle would be more dangerous than remaining in a substantial structure (however, for many hazards, remaining in a mobile home is more dangerous than leaving). Sheltering in-place is the most common protective action recommendation for some hazards (e.g., tornadoes), but the criteria for choosing between evacuation and sheltering in-place can be complex (Lindell and Perry, 1992). Regrettably, there appears to be no research assessing emergency managers' planning concepts and decision criteria for choosing between evacuation and sheltering in-place. Nor is there adequate research on risk area populations' likely compliance and timeliness in implementing protective action recommendations. Much of NEHRP-sponsored research on warning response has sought to identify the factors associated with compliance, but little research has sought to develop guidelines that could inform emergency managers about likely levels of compliance when a protective action recommendation is issued, early evacuation before one is issued, and spontaneous evacuation in locations near the risk area for which a protection action recommendation was issued.

Second, emergency managers must be prepared to warn those in the risk area about the hazard—which can be easy in some situations (e.g., a small area can be warned by emergency responders going door-to-door) and difficult in others (large areas when people are asleep at night). Warnings can use any or all of seven primary warning mechanisms—face-to-face warnings, mobile loudspeakers, sirens, commercial radio and television, tone alert radio, newspapers, and telephones (a given warning mechanism can have multiple warning channels, as when there are multiple radio stations within a community, Lindell and Perry, 1992). These warning mechanisms differ with respect to their precision of dissemination, penetration of normal activities, specificity of the message, susceptibility to message distortion, rate of dissemination over time, receiver requirements, sender requirements, and feedback (verification of receipt). In principle, communities can select the most appropriate warning mechanisms based on the characteristics of the hazards to which they are exposed (especially speed of onset and scope of impact) and the characteristics of the jurisdiction (e.g., population density, wealth). However, no research to date has examined the process by which communities develop warning systems (but see Gruntfest and Huber, 1989).

Emergency response organizations also must be prepared to transmit warning messages that describe the threat, an appropriate protective action, and sources of additional information. Here also, there appears to be little or no research on the extent to which practitioners use the findings of social science research in developing community emergency preparedness.

There is a small but important research literature on protective action implementation, some of it resulting from NEHRP sponsorship. The large scope of evacuations from hurricanes and from accidents at nuclear power plants (and some chemical plants) has made clear the need for advance estimates of the time required to implement an evacuation because it can take many hours to clear a risk area when the population density is high in relation to the capacity of the evacuation route system. Indeed, hurricane evacuation time estimates for some major urban areas along the Atlantic and Gulf coasts exceed 30 hours. There have been some significant advances in empirical estimation of warning times in four floods and the eruption of Mt. St. Helens (Lindell and Perry, 1987) and two hazardous materials accidents (Sorensen and Rogers, 1989). Further, Rogers and Sorensen (1988) proposed methods of mathematically modeling the warning process in terms of two components, an official ("broadcast") component and an informal ("contagion") component, and Sorensen (1991) identified some predictors of household warning reception times in the Nanticoke chemical incident. Warning time distributions for floods and volcanic eruptions in Japan have been reported by Asada et al. (2001) and Katada and Kodama (2001), respectively. Because of the limited availability of data on warning time distributions, further studies are needed over a variety of rapid-onset incidents so that generalizable warning time distributions can be obtained. In particular, these studies should assess the warning time distributions associated with different warning mechanisms such as electronic news media (television and radio), sirens or fixed loudspeakers, and route alert vehicles (Lindell and Perry, 1987, 1992). As is the case for local emergency management agency notification, research is needed to estimate warning time distributions that would be found under a variety of conditions for hurricanes and tsunamis.

There also is only a limited amount of research on preparation time distributions. Recently, Lindell et al. (2002) estimated hurricane evacuation preparation times by summing coastal residents' expectations about the time they would need to perform six evacuation preparation tasks: (1) prepare to leave from work; (2) travel from work to home; (3) gather all persons who would evacuate with the household; (4) pack items needed while gone; (5) protect property from storm damage; (6) shut off utilities, secure the home, and leave. Later, Lu et al. (in press) reported data on preparation time distributions derived from data collected in Hurricane Lili. Moreover, Kang et al. (2004) found that the individual reports from the Hurricane Lili evacuation were significantly correlated with respondents' expectations of these time components collected during the earlier coastal survey by Lindell et al. (2002). The prediction of actual preparation times from expected preparation times was statistically significant and practically useful, but it was far from perfect. Moreover, the time components used in this research were limited to what can be called logistical preparation and did not specifically address what can be called psychological preparation. NEHRP supported research on warning response (Drabek, 1986; Lindell and Perry, 2004) clearly indicates that people engage in milling during which time they seek confirmation that a danger exists, obtain further information about the threat and alternative protective actions, and relay warnings to peers. Thus, although research conducted to date has distinguished between logistical preparation and psychological preparation, no estimates are available concerning the amount of time spent in each of these two types of activities. Thus, quantitative data are needed over a variety of incidents to assess the extent to which the preparation time components identified for hurricane evacuations generalize to other hazards and the extent to which evacuees' expectations of rapid-onset hazards would reduce each of the preparation time components. In addition, research is needed to assess the extent to which these preparation time distributions are predictable from households' demographic characteristics.

Third, quantitative models have been proposed for computing evacuation time estimates (ETE) (Tweedie et al., 1986; Urbanik, Moeller, and Barnes, 1988; Abkowitz and Meyer, 1996; Cova and Church, 1997; Safwat and Youssef, 1997; Hobeika and Kim, 1998; Barrett et al., 2000; Cova and

Johnson, 2002; Lindell et al., 2004). At one extreme are macroanalytic models such as EMBLEM, which has been used to compute the hurricane ETEs for all 22 Texas coastal counties for the Texas Governor's Division of Emergency Management (Lindell et al., 2002). At the other extreme are microanalytic models such as OREMS (Oak Ridge National Laboratories, 2003), which are designed to generate precise ETEs for small areas threatened by toxic chemical releases but require very detailed data on local evacuation route systems. Although published response time models vary significantly in their mathematical sophistication and the apparent precision of their estimates, there has been little effort to validate either the models or their input data to determine if analysts' assumptions about evacuees' behavior are accurate. One major uncertainty concerns the rate of traffic flow when the demand on evacuation routes in a risk area exceeds its capacity—especially when queues take many hours to clear (see Urbanik, 1994, 2000; Homberger et al., 1996; Transportation Research Board, 1998). It is important to know the duration of the queues but it also is important to know where they are located because queues inside risk areas are potentially life threatening whereas those outside them are merely inconvenient. Research on response time models is needed to assess these models' abilities to produce realistic ETEs (see Box 3.1) in a variety of situations ranging from those in which evacuees must travel such long distances that they need to use motor vehicles (for most hurricane evacuations) to those in which a significant portion of the population at risk could walk to a higher elevation or safe haven (for tsunami evacuations in some Pacific coast communities).

There appears to have been no research on preparedness for protective action selection, impact zone access control and security, search and rescue, emergency medical care and morgues, or hazard exposure control. Moreover, there appears to be no research on the ways in which these topics are addressed by local emergency managers in their EOPs, procedures, and training. However, there is some anecdotal information about the utilization of research on the reception and care of victims; Mileti et al.'s (1992) review of the research on this topic was used as the basis for planning hurricane emergency response, primarily because hazards researchers drafted the planning documents for the emergency management agency.

**Incident Management.** There has been a significant amount of social science research supported under NEHRP on incident management during disasters, but here also, there is little or no research on *preparedness* for incident management or on the *utilization* of disaster research findings in the development of community EOPs, procedures, or training. Research is needed to examine the degree to which the adoption of the Incident Command System successfully addresses the patterns of intra- and inter-

---

**BOX 3.1**
**Hurricane Evacuation Time Estimates for the Texas Coast**

Over the past 25 years, analysts have attempted to estimate hurricane evacuation times for coastal counties but have typically made inaccurate assumptions about evacuee behavior out of ignorance of the findings from NEHRP-funded research. Lindell et al. performed a project for the Texas Division of Emergency Management (DEM) that developed EMBLEM, an algorithm that updated Safwat and Youssef's (1997) evacuation time estimate (ETE) model to correct deficiencies in the traffic flow model noted by Urbanik (1994, 2000). EMBLEM also used data from NEHRP-funded research to improve the evacuee behavior model. For example, Lindell et al. (2002) used data from the eruption of Mt. St. Helens to estimate the time distribution for warning diffusion in a hurricane with a late-changing track. They combined the warning distribution with data on coastal residents' expectations about their evacuation behavior to produce trip generation time distributions for residents and transients. A later NEHRP-funded study of the Hurricane Lili evacuation (Kang et al., 2004; Lu et al., in press) has incorporated these behavioral data into EMDSS2 (Lindell et al., 2004), which further refines the models of evacuee behavior and traffic flow. The work for Texas DEM also included empirical data on shelter utilization reported in a meta-analysis conducted by Mileti et al. (1992). These data replaced grossly inaccurate estimates that had been used previously.

---

organizational coordination identified by disaster researchers (e.g., Dynes, 1977; Drabek et al., 1981; Kreps, 1989a,b, 1991a,b).

## COMMUNITY DISASTER RECOVERY PREPAREDNESS PRACTICES

After a disaster, many tasks need to be accomplished very quickly, and virtually simultaneously, so pre-impact planning for disaster recovery is as critical as planning for disaster response (Schwab et al., 1998). Emergency response and disaster recovery frequently overlap because some sectors of the community are in emergency response mode, while others are moving into disaster recovery, and some organizations might be carrying on both types of activity at the same time. Moreover, senior elected and appointed officials are likely to be inundated with policy decisions to implement the emergency response at the same time that they have to plan for the disaster recovery. Consequently, there is increasing recognition that pre-impact emergency response planning should be linked to pre-impact disaster recovery planning.

In principle, resources can be allocated more effectively and efficiently—

increasing the probability of a full and rapid recovery—if recovery planning is begun before disaster impact. That is, coordinated planning for emergency response and disaster recovery can avoid delays while decisions are made about procedures and resource utilization. Coordinated pre-impact planning can also decrease the probability of conflicts arising due to competition over scarce resources during the recovery period. The necessary coordination between pre-impact emergency response planning and pre-impact disaster recovery planning can be achieved by establishing organizational contacts, and perhaps overlapping membership, between the entities responsible for these two activities. However, such coordinated planning involves some significant challenges because the agencies that are most often involved with the development of the EOP (e.g., police, fire, emergency medical services) and those that need to be involved in the development of the disaster recovery plan (e.g., land-use planning, economic development, public works) have significantly different organizations and organizational cultures. Thus, it will take a determined effort in most jurisdictions to achieve the needed coordination. To date, only a limited amount of research has examined the effectiveness of pre-impact disaster recovery planning (see Box 3.2). Both studies employed weak research designs, so further research is needed to verify its effectiveness and to identify its most important elements and processes.

---

### BOX 3.2
### Pre-Impact Recovery Planning

The City of Los Angeles, under the leadership of the former planning director, prepared and adopted a "Recovery and Reconstruction" element of the City's Emergency Operations Plan. It dealt with recovery management, redevelopment, intergovernmental relations, and financing. Following the 1994 Northridge earthquake, the National Science Foundation funded an evaluation of how the innovative element was used, the effectiveness of its use as a post-disaster decision support tool, and what lessons were learned that could be applied to similar planning efforts. Eight years later, the data from that study were used in conjunction with other archival data on the aftermath of the Northridge earthquake and also with data from the 1999 Chi-Chi earthquake in Taiwan in a comparative case study that concluded pre-impact recovery planning accelerated the rate of housing recovery and also increased the extent to which hazard mitigation was incorporated into the recovery process. For further details, see Spangle Associates Urban Planning and Research (1997), and Wu and Lindell (2004).

## ADOPTION OF HAZARD ADJUSTMENTS WITHIN COMMUNITIES

NEHRP-supported research has also produced an extensive body of research on the adoption of what have been termed *hazard adjustments* by households, businesses, and government agencies. The term *hazard adjustments* is adopted from Burton et al. (1993) to refer to all actions that reduce hazard vulnerability—hazard mitigation, emergency response preparedness, and disaster recovery preparedness. The reason for addressing all three types of hazard adjustments as a single category is that the adoption process appears to be relatively similar for all of them. Research on the adoption and implementation of hazard adjustments has consistently found support for the notion that hazard *awareness* might well be high among affected populations and within organizations and government agencies, but *action* to reduce hazard vulnerability does not necessarily follow. Regardless of the social unit involved, studies suggest that the relationship between risk perception and hazard adjustment is a complex one (see Lindell and Perry, 2004).

### Household Hazard Adjustments

There is an extensive set of studies on household seismic adjustment, with Lindell and Perry (2000) finding 23 studies that attempted to correlate household adjustment to earthquake hazard with at least one or more explanatory variables. Data from these studies confirmed theoretical predictions that households' adoption of earthquake hazard adjustments is correlated with their perceptions of the hazard and alternative adjustments and, to a lesser extent, with demographic characteristics and social influences. Specifically, hazard adjustment tended to be more highly correlated with beliefs about the probability of personal consequences (death, injury, property damage, and disruption to job and daily activities) than with beliefs about the probability of the event itself. That is, for action to take place, general knowledge about hazards must translate into beliefs about personal vulnerability (Turner et al., 1986; Showalter, 1993). Moreover, hazard intrusiveness—the frequency with which people think and talk about hazards and hazard adjustments—appears to be as important in predicting hazard adjustment adoption as people's perceptions of personal risk (Lindell and Prater, 2000).

Similarly there is evidence, subsequently confirmed by Lindell and Whitney (2000) and Lindell and Prater (2002), that adoption intentions and actual adoption are higher for hazard adjustments that are higher in hazard-related attributes (efficacy in protecting persons, efficacy in protecting property, and suitability for other purposes) and—to a lesser degree—lower in resource-related attributes (cost, time and effort, knowledge and skill, and required cooperation with others). The studies reviewed by Lindell and Perry (2000) also indicate that hazard adjustment adoption is correlated with perceived personal responsibility, a finding confirmed by Lindell and Whitney (2000) and Arlikatti et al. (in press). Preparedness is also correlated with feelings of self-efficacy with respect to hazard adjustments (Mulilis and Duval, 1995).

Lindell and Perry (2000) reported that the correlations of hazard adjustment with demographic variables are consistently small, although this research consistently pointed to the importance of *resources* of various kinds in the preparedness process. The concept of resources is used broadly here to encompass access to money and information, as well as ties to community institutions. For households, higher levels of hazard adjustment are generally associated with higher levels of income, education, and home ownership (Turner et al., 1986; Edwards, 1993; Russell et al., 1995; Lindell and Prater, 2000). Of course, the effect of home ownership might reflect higher levels of personal vulnerability (a greater level of personal assets at risk), as well as greater access to the resources needed to adopt and implement hazard adjustments.

Lindell and Perry's review also found that the effect of previous earthquake experience was inconsistent across studies, probably because this variable was measured in so many different ways (see Baker, 1991, for a similar finding with respect to the correlation of hurricane evacuation with previous storm experience). Finally, there were significant effects of social influences on hazard adjustment adoption, through both information receipt (see the discussion of risk communication below) and observation of others' behavior (social modeling).

Despite the significant contributions of NEHRP-funded research to the understanding of the hazard adjustment process, further research is needed. Most of the research on household hazard adjustment has addressed the adoption of hazard adjustments by households in California—a high-hazard zone. Much less is known about household adoption of hazard adjustments in lower-frequency hazard zones such as the Cascadia, Wasatch, and New Madrid seismic zones. Moreover, much of the existing research has neglected the problems of *erroneous beliefs* (Turner et al., 1986; Whitney et al., 2004) and *pseudo-attitudes* (Converse, 1964; Schuman and Kalton, 1985; Lindell and Perry, 1990). The neglect of erroneous beliefs is serious because these variables are usually not measured by researchers and, to the degree that they are relevant to people's adjustment adoption decisions, depress the prediction of hazard adjustment. Pseudo-attitudes can arise when researchers attempt to assess respondents' beliefs using standardized rating scales, but respondents' answers can be unstable when attitude objects are rated on dimensions that have no meaning to the respondents. The neglect of pseudo-attitudes is significant because this can produce a spuriously high level of predictive accuracy in models of hazard adjustment decisions, but there is some evidence that careful analysis of survey data can identify pseudo-attitudes (Lindell and Perry, 1990).

**Households.** Lindell and Perry (2000) identified needs for further research in six major areas: hazard adjustments, perceived hazard characteristics, perceived adjustment characteristics, household characteristics, past experience, and social influences. The first of these is a pressing need to adopt a consistent typology of pre-impact hazard adjustments, to develop standardized scales for measuring these adjustments, and to assess the psychometric adequacy of these scales (e.g., Mulilis and Lippa, 1990). Future studies also should systematically develop and test scales measuring the information-seeking activities that have been reported to be highly correlated with adjustment (Mileti and Fitzpatrick, 1993; Mileti and Darlington, 1997). These information-seeking scales should distinguish between information about a hazard and information about hazard adjustments. Another important task for future research is to assess the perceived interdependencies among hazard adjustments. If the information and other resources acquired in the process of adopting one adjustment make it easier to adopt others, then an adjustment perceived as having more efficacy *and* lower resource requirements might serve as a "gateway" to the adoption of adjustments that are perceived to be lower in efficacy and more resource demanding.

Previous research generally has reported statistically significant relations between perceived hazard characteristics and hazard adjustment, but the size of the correlation coefficients is modest. One potential explanation for the small correlations is that researchers have failed to accurately capture risk area residents' cognitive representations of the hazard. Most research on hazard adjustment has measured perceived characteristics of hazards in terms of respondents' judgments of the probability and severity of personal consequences, but other beliefs also are relevant. Mileti and Fitzpatrick (1993) assessed respondents' perceptions of the probability of a major earthquake, property damage, injury, and death and, moreover, assessed perceptions of these consequences over two different time periods.

Researchers should also assess the linkages among people's beliefs about hazards to identify the preconditions for risk personalization. Palm and Hodgson's (1992) work suggests assessing the locational, structural, and demographic components of perceived vulnerability. With respect to perceived locational vulnerability, studies should examine people's actual and perceived proximity to hazard sources (Palm and Hodgson, 1992; Zhang et al., 2004a; Arlikatti et al., in press). Perceived structural vulnerability should be assessed by asking respondents to compare the vulnerability of the structures in which they live and work to the vulnerability of the average home, whereas perceived demographic vulnerability could be assessed by obtaining respondents' comparisons of their household members' vulnerability to that of the average household.

Some researchers have measured risk area residents' hazard concern in terms of a single global item (e.g., Dooley et al., 1992), whereas others have measured threat personalization by multi-item scales addressing the perceived likelihood of specific impacts (Mileti and Fitzpatrick, 1993). A global item would be a more accurate characterization of risk area residents' beliefs if they have only very diffuse conceptions of the threat, whereas specific impact dimensions would be appropriate if they have differentiated beliefs. The problem is that asking about very specific impact dimensions could create pseudo-attitudes if people have only very diffuse conceptions. Thus, further research is needed to determine what proportions of risk area populations have specific beliefs, global beliefs, and no beliefs at all about the environmental hazards to which they are exposed.

The relationships of hazard adjustments to the hazard and to household resources imply that attributes of hazard adjustments can be categorized as hazard-related or resource-related (Lindell and Perry, 2000). Hazard-related characteristics include efficacy for protecting persons and property and utility for other purposes. By contrast, resource-related characteristics are defined by demands on household resources such as money, knowledge, skill, time, effort, and interpersonal cooperation. Such characteristics are closely linked to household members' self-efficacy, which refers to a belief in the adequacy of one's knowledge and skills as well as access to any materials, equipment, and money that also are needed. There has been limited research to date on perceptions of adjustment-related attributes, and more is needed to understand the trade-offs people make among these dimensions in selecting hazard adjustments.

Some studies have suggested that perceived protection responsibility is an important variable in determining household hazard adjustment, but the research base is quite limited. Early research on seismic hazard adjustment indicated that many risk area residents held government responsible for reducing their seismic vulnerability (Jackson, 1981; Turner et al., 1986; Mulilis and Duval, 1995), but more recent research has shown a greater acceptance of personal responsibility (Arlikatti et al., in press). However, this research has addressed only one hazard, and most of it has been conducted on the Pacific Coast. Research on a variety of hazards should examine risk area residents' perceptions of the protection responsibility of different levels of government in relation to informal sources such as the news media, employers, friends, and family to determine if this variable is important for predicting household adoption of adjustments for other hazards as well.

Future research should examine the role of community bonded-ness, whose significant correlations with seismic hazard adjustment was originally reported by Turner and his colleagues (1986) and replicated by some (Dooley et al., 1992) but not other researchers (e.g., Palm et al., 1990). These inconsistencies cannot be explained by sampling fluctuations because of the studies' large sample sizes. It is likely that the magnitude of the correlations between household characteristics and hazard adjustments depends on which household characteristics and hazard adjustments are being correlated. Correlations of demographic variables with adjustment adoption might be valuable in allowing hazard managers to target population segments that are most disposed to adopt seismic adjustments. For example, the presence of school-aged children in the home might signal a need to focus on schools as a channel for disseminating hazard information, while correlations of income with overall adjustment might suggest an emphasis on the least expensive adjustments, at least until risk area residents become more committed to the seismic adjustment process.

Future research should also examine the ways in which past hazard experience affects future expectations of vulnerability and hazard adjustments. One possible explanation for the lack of consistency in previous findings is that this variable has been measured in many different ways. The variations in the measurement of earthquake experience, which are similar to those found in research on hurricane adjustments (Baker, 1991), suggest that hazard experience has to be conceptualized carefully and measured consistently. One important contribution of future studies would be to assess hazard experience in terms of multiple indicators of experience. An important task for future researchers will be to identify what it is about direct experience that increases seismic adjustment and develop methods of providing these critical elements vicariously rather than directly.

Researchers long have recognized that hazard adjustment takes place in a social context. Accordingly, social influence has been examined in many studies of hazard adjustment, but most of these have focused on persuasive influences. Consistent with the classical communication model, these studies have addressed source, message, channel, receiver, and effect variables. Future research should complement investigation of influence sources with an examination of the *basis* of influence. Raven (1993; French and Raven, 1959) has concluded that sources use six bases of influence—legitimate, referent, expert, information, reward, and coercive. A slightly different typology arises from the literature on persuasive communications, which indicates that sources are perceived in terms of their credibility (e.g., expertise, trustworthiness), attractiveness, and power (Eagly and Chaiken, 1993). Further examination of the characteristics of information sources and their bases of influence could substantially advance our understanding of this aspect of the hazard adjustment process.

Message characteristics—information quality (specificity, consistency, and source certainty) and information reinforcement (number of warnings)—have a significant impact on adoption of seismic adjustments. However, only a few studies have examined this component of the seismic adjustment process (Mileti and Fitzpatrick, 1993; Mileti and O'Brien, 1992). Future research should examine whether there are other message characteristics that also affect adjustment. In particular, there is a need to develop objective measures of these characteristics.

The differential impact of communication channels has been examined, with Turner et al. (1986) finding that television had a greater impact than other media. However, other research reported stronger effects for print media (Mileti and Darlington, 1997; Mileti and Fitzpatrick, 1993). Still other research found that residents of a rural area vulnerable to volcano hazard had complex patterns of communications channel use (Perry and Lindell, 1990) and that channel use varied by community and ethnicity (Lindell and Perry, 1992). Moreover, risk area residents use channels for different purposes: radio and television are useful for immediate updates, meetings are useful for clarifying questions, and newspapers and brochures are useful for retaining information that might be needed later. The ways in which residents of risk areas exposed to other hazards use the mass media need similar scrutiny.

### Business Hazard Adjustments

Private sector disaster preparedness had not previously been studied extensively, but NEHRP-sponsored studies have made considerable progress in understanding the extent to which businesses prepare for disasters and the factors that influence this process (Drabek, 1991a, 1995; Dahlhamer and D'Souza, 1997; Tierney, 1997a,b; Lindell and Perry, 1998; Webb et al., 2000). Research on business preparedness for earthquakes in Los Angeles found that while awareness of the threat was high among business owners, preparedness levels tended to be quite low—even after the 1994 Northridge earthquake (Dahlhamer and Reshaur, 1996). Disaster experience appears to have made the threat more salient to these businesses because those that had sustained damage in the Northridge earthquake, were forced to close, or experienced lifeline service interruptions, subsequently increased their levels of preparedness (Dahlhamer and Reshaur, 1996). Companies that handled hazardous materials also increased their preparedness efforts after that earthquake (Lindell and Perry, 1998).

As is the case for households, the access of businesses to resources is generally associated with higher levels of hazard adjustment. Larger businesses are significantly more likely to engage in preparedness activities than smaller ones—a pattern that is thought to be related to the fact that larger firms are more likely to have additional resources to devote to loss reduction activities and more likely to have specialized positions that are specifically devoted to risk and disaster management (Webb et al., 2000; Whitney et al., 2001; Mileti et al., 2002).

The limited amount of research on business adoption of hazard adjustments has focused on the environmental and organizational conditions influencing the level of hazard adjustment adoption, but there has been little research on the process by which managers make decisions about investments in hazard adjustments other than studies by Alesch et al. (1993) and Drabek (1991a, 1995). As is the case with households, there is a need to understand what the alternative hazard adjustments considered by different businesses are and, especially, managers' perceptions of the hazard-related and resource-related attributes of the available adjustments. It would be particularly useful to examine the ways in which manager's decisions are being affected by the burgeoning business continuity industry, which has expanded into corporate emergency planning from the areas of data management, facility security, and crisis communications. To date, there has been no research to assess the effectiveness of the interventions offered by these practitioners (Tierney, in press).

## Government Agency Hazard Adjustments

There has been little research on the hazard adjustments by government agencies that do not have emergency management responsibilities but, nonetheless, will be expected to provide their normal services after a disaster strikes. Perry and Lindell (1997), who collected data on seismic preparedness by city and county agencies in a southwestern state, found the overall level of hazard adjustment was low. Hazard adjustment was correlated with agency size, perceived risk, and information seeking—findings that are similar to those for businesses—but more research is needed on other hazards and in other areas to support these initial results.

Research needs at the state level of analysis include studies on the impact of organizational and institutional arrangements on the quality and effectiveness of state preparedness, as well as studies on the extent to which federal preparedness requirements have an impact at the state level (Waugh and Sylves, 1996). Since the Disaster Mitigation Act of 2000 has expanded states' responsibilities in the area of disaster preparedness (including their role in encouraging local-level preparedness), state-level activities constitute an important area for future research. In the aftermath of 9/11, state-level preparedness initiatives also warrant study.

## Neighborhood Organization Hazard Adjustments

Local citizen-based initiatives are also becoming more common, and activities that were originally designed to decrease vulnerability to natural disasters are now being employed to prepare the public for human-induced threats. For example, Community Emergency Response Training (CERT), a program originally developed in Los Angeles and other California cities to help prepare neighborhood residents to respond to earthquakes, has been transferred to many other hazard-prone communities and adopted as a national model by FEMA (Simpson, 2001). The CERT concept is now being implemented in preparedness for terrorism-related events.

**Hazard Insurance Purchase.** The purchase of hazard insurance is a pre-impact recovery preparedness action that is addressed separately here because much of the research on this topic has been conducted almost completely independently of other work on other hazard adjustments (although see Palm and Hodgson, 1992; Palm 1998). NEHRP-sponsored research has revealed many difficulties in developing and maintaining an actuarially sound hazard insurance program. The National Flood Insurance Program has made significant strides over the past 30 years, but it continues to require operational subsidies. One of the basic problems is that those who are most likely to purchase flood insurance are, in fact, those who are most likely to file claims (Kunreuther, 1998). This problem makes it impossible to sustain a market in private flood insurance. The federal government has tried to solve this problem by requiring flood insurance for structures purchased with federally-backed mortgages that are located in the 100-year floodplain. Unfortunately, policies are frequently allowed to lapse in the years after the purchase and the program has no effect on those who purchase their homes without a mortgage. Consequently, some homes are rebuilt soon after a disaster because their owners have high-quality insurance coverage, whereas other homes take much longer because they are only partially insured or even lack *any* insurance because their occupants cannot afford quality insurance or are denied access to it because of "redlining" (Peacock and Girard, 1997).

In addition to these institutional problems, there are many cognitive obstacles to the development of a comprehensive hazard insurance program. Building on earlier hazards research (see Burton et al., 1993, for a summary) and psychological research on judgment and decision making (see Slovic et al., 1974, for an early statement, and Kahneman et al., 1982), for more a recent summary), Kunreuther and his colleagues (1998) have identified numerous logical deficiencies in the ways people process information in laboratory studies of risk. However, there remains only limited research on the extent to which heuristics and biases actually influence how households and businesses make decisions about hazard management.

There are some fascinating parallels between theories about insurance purchase and those about the adoption of other hazard adjustments. For example, what economists call moral hazard is equivalent to what psychologists refer to as a decrease in protection motivation, usually due to a felt lack of personal responsibility for protection. The concept of moral hazard or felt responsibility for personal protection has important policy implications because the Interagency Floodplain Management Review Committee (1994) report concluded that federal disaster relief policy creates this condition by relieving households of the responsibility for providing their own disaster recovery resources. This might be a significant reason why only 20 percent of structures affected by the 1993 Mississippi floods were insured. However, there appear to be no data indicating that households explicitly consider the availability of disaster relief in making decisions about whether to purchase hazard insurance and adopt other hazard adjustments.

Moreover, Kunreuther's (1998) flow chart describing a homeowner's decision to purchase hazard insurance is similar in some respects to the protective action decision model described by Lindell and Perry (1992, 2004), but there are notable differences. Future research should examine the theoretical comparability and empirical support for these two models—particularly in regard to the differences among decision makers with different levels of sophistication such as households, small businesses, and large businesses.

## Communication About Risk and Hazard Adjustments

Risk communication is an important method by which hazard managers can increase the adoption and implementation of hazard adjustments by households, businesses, neighborhood organizations, and government agencies. As used here, the term *risk communication* refers to intentional efforts on the part of one or more sources (e.g., scientific agencies, local government) to provide information about hazards and hazard adjustments through a variety of channels to different audience segments (e.g., the general public, specific at-risk groups). Research on disasters has long recognized different sources as being peers (friends, relatives, neighbors, and coworkers), news media, and authorities (Drabek, 1986). More recently, attention has been given to the ways in which these sources differ systematically in terms of such characteristics as perceived expertise, trustworthiness, and protection responsibility (Lindell and Perry, 1992; Lindell and Whitney, 2000; Arlikatti et al., in press). There are many different information channels (e.g., broadcast, print, telephone, face-to-face, Internet), but there has been no systematic investigation of the ways in which these differ in characteristics such as precision of message dissemination, penetration of normal activities, message specificity, susceptibility to message distortion, rate of dissemination over time, receiver requirements, sender requirements, and feedback (Lindell and Perry, 1992). Messages also vary in many ways, including threat specificity, guidance specificity, repetition, consistency, certainty, clarity, accuracy, and sufficiency (Mileti and Sorensen, 1987; Mileti and Peek, 2000; Lindell and Perry, 2004). More is known about the effects of these message characteristics on warning recipients, but not about the degree to which hazards professionals address them in their risk communication messages. Receiver characteristics include previous hazard experience, preexisting beliefs about the hazard and protective actions, and personality traits. In addition, there are demographic characteristics—such as gender, age, education, income, ethnicity, marital status, and family size—but these have only modest (and inconsistent) correlations with hazard adjustment.

Finally, Lindell and Perry (2004) summarized the available research as indicating message effects include pre-decisional processes (reception, attention, and comprehension), and the eight decision stages listed in Table 3.2. Each decision stage is defined by the critical question posed by the situation, the response activity, and the outcome of that activity.

There is substantial variation in the amount of time and effort people spend in each of these eight stages (indeed, people can bypass some of the stages altogether) and the order in which the stages are processed. More-

**TABLE 3.2** Warning Stages and Actions

| Stage | Question | Activity | Outcome |
|-------|----------|----------|---------|
| 1 | Is there a real threat that I need to pay attention to? | Risk identification | Threat belief |
| 2 | Do I need to take protective action? | Risk assessment | Protection motivation |
| 3 | What can be done to achieve protection? | Protective action search | Decision set (alternative actions) |
| 4 | What is the best method of protection? | Protective action assessment and selection | Adaptive plan |
| 5 | Does protective action need to be taken now? | Protective action implementation | Threat response |
| 6 | What information do I need to answer my question? | Information needs assessment | Identified information need |
| 7 | Where and how can I obtain this information? | Communication action assessment and selection | Information search plan |
| 8 | Do I need the information now? | Communication action implementation | Decision information |

Source: Lindell and Perry (2004).

over, people sometimes cycle through a decision stage repeatedly as new information is sought and received. Such extended "milling" most commonly occurs when there is conflicting or confusing information (e.g., when there are complex and uncertain scientific data about a hazard and alternative protective actions).

Two empirical studies on public risk communication campaigns are illustrative of NEHRP-sponsored research in this area. Mileti and Darlington (1995) studied responses by the public and by government and private sector organizations to new scientific information on the magnitude of the earthquake threat in the San Francisco Bay area—information that was provided to residents in a color insert they received with their Sunday newspapers. In a similar effort, Mileti and Fitzpatrick's *The Great Earthquake Experiment: Risk Communication and Public Action* (1993) analyzed the impact of government efforts to provide public information and encourage household seismic preparedness in connection with the Parkfield, California, earthquake prediction experiment. Here also, the study focused on what people in communities affected by the prediction knew about the earthquake hazard and how they responded. These two studies showed that residents did become better informed as a consequence of government risk communication, and some took steps to prepare for a coming earthquake. One key finding from both studies was that printed materials, such as the brochures residents received, were more effective in communicating risk than more ephemeral forms of communication such as television and radio. Another was that printed material—or any risk communication vehicle—is not sufficient to raise awareness and motivate action. Rather, risk-related information must be delivered through multiple channels, in different (but consistent) form, and must be repeated.

In addition to these quasi-experimental designs, some studies, including some supported by NEHRP, have also used experimental designs involving random assignment to conditions. In a well-controlled field experiment, Mulilis and Lippa (1990) provided respondents with specially prepared earthquake awareness brochures that systematically varied information about an earthquake's probability of occurrence, its severity, the efficacy of a recommended seismic adjustment, and the receiver's self-efficacy (i.e., capability) to implement the adjustment. Researchers found that brochures induced immediate changes in the receivers' perceptions of probability, severity, outcome efficacy, and self-efficacy, but these impacts were not sustained over the five to nine weeks between the administration of an immediate post-test, and a delayed post-test and there were only suggestive rather than conclusive improvements in the level of seismic adjustment.

More recently, Whitney et al. (2004) investigated the prevalence of both accurate and erroneous earthquake-related beliefs and the relationship between respondents' endorsement of earthquake beliefs and their adoption of seismic hazard adjustments. In addition, the study examined the effects of an experimental earthquake education program and the impact of a psychological trait—*need for cognition*—on this program. Data revealed a significant degree of agreement with earthquake myths, a generally low level of correlation between earthquake beliefs and level of hazard adjustments, and a significant effect of hazard information on the endorsement of accurate earthquake beliefs and increases in hazard adjustment. Compared to an *earthquake facts* format, an *earthquake myths versus facts* format was slightly more useful for dispelling erroneous beliefs.

In addition to their erroneous beliefs about hazards, some risk area residents have erroneous beliefs about such basic information as their location in a risk area. For example, Zhang et al. (2004a) found that one-third of the residents in counties threatened by Hurricane Bret were unable to correctly identify the risk area in which their home was located, even when provided with a risk area map along with the questionnaire. Moreover, Arlikatti et al. (in press) found that this percentage was two-thirds for the Texas coast as a whole. Such findings have obvious implications for defining these risk areas (using readily recognizable geographical features and political boundaries), but also underscore the importance of carefully assessing risk area residents' beliefs about even the seemingly most obvious aspects of emergency preparedness.

These and other studies have led to the development of practical guidance on the design of public education campaigns for earthquakes. Nathe (2000), for example, provided research-based advice for practitioners on such questions as what people need to know in order to actually change their behavior with respect to hazards, how to craft risk-related messages that address these informational needs, how best to convey scientific information to the lay public, and how to take advantage of the window of opportunity provided by a disaster. A recent report developed by social scientists affiliated with the three earthquake engineering research centers was designed specifically to provide guidance to earthquake safety advocates—including advice on risk communication and the design of strategies for educating the public (Alesch et al., 2004). Although derived from research on earthquakes, this guidance also incorporates findings from studies on many other types of hazards, and the principles outlined there can be applied to other natural, technological, and human-induced threats.

## RECOMMENDATIONS FOR RESEARCH ON PRE-IMPACT HAZARD MANAGEMENT

This section presents recommendations for future research that are organized in the order in which the corresponding topics were addressed in earlier sections of this chapter. The committee is cognizant of research in areas other than disaster research that addresses similar issues but has not been cited in this chapter. However, much of this relevant literature has been addressed by hazards and disasters researchers in the work that is cited here. For example, research on protective action decision making for environmental hazards and disasters has been linked to research on persuasive communications, social conformity, behavioral decision theory, attitude-behavior theory, information seeking, health behavior, and innovation processes (Lindell and Perry, 2004). Thus, the research recommendations that follow have been formulated in light of such research even though it is not explicitly referenced.

**Recommendation 3.1:** *Research should be conducted to assess the degree to which hazard event characteristics affect physical and social impacts of disasters and, thus, hazard mitigation and preparedness for disaster response and recovery.*

This very broad recommendation is essentially a call for comprehensive tests of the model described in Figure 1.2. The practical value of research on this topic is to resolve the apparent conflict between the results of previous disaster research, which support an all-hazards approach, and the increased focus on specific hazards that has emerged in recent approaches to homeland security. Expedient hazard mitigation is arguably specific to a single hazard or group of hazards with similar effects, and emergency assessment arguably also has hazard-specific aspects. However, most aspects of population protection and incident management appear to apply to a wide variety of hazards. Research is needed to determine if this assumption is correct.

Threat classifications will continue to play a significant role in the way researchers define events to study. However, few of the conclusions derived from crude threat classifications—the natural, technological, and willful classifications in particular—are based on empirical findings. It remains to be determined how human responses to intentional terrorist events differ from responses to natural or technological events. There has been much speculation that we cannot use past history to understand and predict how people will respond to events not previously experienced in this country. However, the likely responses to events such as suicide bombings, releases of biological agents, attacks with radiological dispersion devices, or releases of chemical warfare agents can be studied using careful empirical research before such disasters occur. Preliminary findings from the large number of post 9/11 investigations—not to mention studies of the 1993 World Trade Center and Oklahoma City bombings—suggest that some types of behavior are similar to those observed in other large-scale disasters. Thus, the absence of panic and the large amount of altruistic behavior should come as no surprise. Other types of behavior, such as changes in travel behavior and product purchases, have not been studied in connection with disasters but have been observed in connection with stigmatized products such as cyanide-contaminated bottles of Tylenol and Alar-tainted apples. It is critical that comparative research efforts be made to document and understand variations in human response to a wide range of hazards and social conditions.

**Recommendation 3.2:** *Research should be conducted to refine the concepts involved in all three components (hazard exposure, physical vulnerability, and social vulnerability) of hazard vulnerability analysis (HVA).*

Research is needed to understand the ways in which appointed (e.g., emergency managers, land-use planners, public health officers) and elected officials and risk area residents interpret information about hazard exposure. Research is also needed to assess the ways in which these stakeholder groups interpret the structural vulnerability of the buildings in which they live and work. In addition to assessing risk perceptions, these studies also should assess the degree to which users can and do make use of the work that physical scientists and engineers produce on hazard exposure and structural vulnerability, respectively.

Finally, research is needed to better understand the concept of social vulnerability. Following Cutter (2003a), Clark et al. (2000), Kasperson and Kasperson (2001), and the Heinz Center (2002), the first objective should be to understand the driving forces that determine the level of vulnerability and the scale (household, neighborhood, community, region) at which they are most pronounced. The second objective is to assess how current practices and public policies foster and transfer vulnerability both spatially and temporally. The third objective is to develop theoretical models and research methods that improve the prediction of future vulnerability. The fourth objective is to develop multihazard models that integrate hazard exposure and physical and social vulnerability. The fifth objective is to develop better metrics for comparing the relative levels of vulnerability from place to place and region to region, thus improving the linkage between the conceptualization of vulnerability and its measurement. The sixth objective is to improve visualizations of vulnerability and disseminate them to the practitioner and lay communities. The seventh objective is to develop a more robust understanding of the perception of vulnerability by various stakeholder groups (especially emergency managers, policy makers, and the public). The eighth objective is to develop rigorous and systematic methods for examining the similarities and differences in concepts, models, and exposure units of vulnerable groups, ecosystems, places, human-environment conditions, or coupled human-ecological systems.

**Recommendation 3.3:** *Research should be conducted to identify better mechanisms for intervening into the dynamics of hazard vulnerability.*

Recent examinations have revealed an exponentially increasing toll of disaster losses (Mileti, 1999a) that are exacerbated, if not caused, by existing federal hazard management policies (Burby et al., 1999). Hazard insurance has been identified as a promising alternative, but even subsidized flood insurance has had limited success—at best. An even broader issue concerns the ways in which there is escalating hazard vulnerability in specific population segments—especially the poor (Blaikie et al., 1994). Research is needed to assess the degree to which socially vulnerable population segments might be "pushed" into geographical areas of high hazard exposure and structures that were built under outdated building codes and are poorly maintained. However, it will also be important to assess the degree to which socially advantaged population segments are "pulled" into exposed areas and vulnerable structures. In the latter case, more affluent groups might choose high hazard areas for their normal amenities (views of rivers and coasts can carry the risk of flood and wind hazard; mountain views are associated with the risk of landslide and wildfire hazard). In addition, they might choose older houses for their historic and aesthetic qualities. Research is needed to assess the relative importance of these "push" and "pull" forces in determining vulnerability to different hazards in all regions of the country.

**Recommendation 3.4:** *Research should be conducted to identify the factors that promote the adoption of more effective community-level hazard mitigation measures.*

Specifically, most NEHRP-supported social science research on hazard mitigation has focused on intergovernmental issues in land-use regulation. Such research has substantially increased the scientific understanding of these issues, but this is only a portion of the problem. More research is needed on other mitigation measures—community protection works and building construction practices. In connection with the latter, the Earthquake Engineering Research Institute conducted a study of factors affecting building code compliance, but more research is needed on this topic (Hoover and Greene, 1996). In addition, more research is needed on strategies other than regulation. Such research should examine the joint effects of regulations, incentives, and risk communication on households and businesses, and should address new construction and retrofits to existing construction.

**Recommendation 3.5:** *Research should be conducted to assess the effectiveness of hazard mitigation programs.*

In particular, Project Impact was instituted during the 1990s but terminated immediately after a change in political administration. Project Impact was widely touted during the Clinton administration for its effectiveness in promoting hazard mitigation. Nevertheless, it was canceled by the Bush administration. Prater (2001) noted that it would be extremely difficult to evaluate the effectiveness of Project Impact since the cities that received the greatest financial support were selected specifically because they had already demonstrated support for hazard mitigation. However, a recently released study by the National Institute of Building Sciences Multihazard Mitigation Council that quantified the future savings from three FEMA mitigation programs, including Project Impact, found that they provided significant net benefits to society (Multihazard Mitigation Council, 2005). FEMA's Hazard Mitigation Grant Program and the Flood Mitigation Assistance Program were the other programs examined. The study, which focused on eight communities in depth during the period of 1993–2003, was requested by Congress and considered earthquake, wind, and flood hazards. The conclusion was that mitigation is sufficiently cost-effective to warrant significant federal funding. Many such studies are needed to examine the benefits and costs of mitigation efforts for all types of hazards. Research is also needed on methods for assessing the full costs and benefits of mitigations, comparing cost-effectiveness of different types of mitigations, and better incorporating such methods and knowledge into a decision-making process that reflects the needs of all stakeholders.

A principle intellectual tool relating public policy to social science research is benefit-cost analysis. In general, benefit-cost analysis of natural hazards policies has lagged. The need to adapt benefit-cost analysis to the study of catastrophic events has recently been highlighted by Posner (2004) and Sunstein (2002).

**Recommendation 3.6:** *Research should be conducted to identify the factors that promote the adoption of more effective emergency response preparedness measures.*

Previous studies have identified community hazard vulnerability, community resources, and especially, strategies and structures that emergency managers and other hazards professionals can adopt at low cost. Nonetheless, these studies have relied on very limited samples and need further work to replicate and extend their findings.

**Recommendation 3.7:** *Research should be conducted to assess the extent to which disaster research findings are being implemented in local emergency operations plans, procedures, and training.*

Anecdotal evidence suggests a very poor level of utilization, in part because of the lack of communication mechanisms between researchers (who customarily publish their findings in academic journals or present them at academic conferences) and practitioners (who customarily seek information from peers or at professional conferences).

**Recommendation 3.8:** *Research is needed to identify the factors that promote the adoption of more effective disaster recovery preparedness measures.*

The idea of recovery preparedness is intuitively appealing and initial research is promising, but there is little research on the extent to which local jurisdictions have adopted this practice and the ways in which it is being implemented. There is some evidence that pre-impact recovery planning is successful in accelerating housing recovery and integrating hazard mitigation into the recovery process (Wu and Lindell, 2004), but much more research needs to be conducted in this area.

**Recommendation 3.9:** *Research should be conducted to develop better models to guide protective action decision making in emergencies.*

Research on evacuation decision making is needed for a wide range of hazards such as hurricanes, floods, volcanic eruptions, and terrorist incidents. In addition, research is needed to choose between evacuation and sheltering in-place during tsunamis, hazardous materials releases, and wildfires. Such research will require social scientists to collaborate with transportation planners and engineers on evacuation modeling and with mechanical engineers on shelter-in-place modeling.

Specifically, research is needed to assess emergency managers' and responders' preparedness for protective action selection, warning, protective action implementation, impact zone access control and security, reception and care of victims, search and rescue, emergency medical care and morgues, and hazard exposure control. Research on preparedness for protective action selection should assess emergency managers' beliefs about the relative merits of evacuation and sheltering in-place—including compliance by the risk area population. Research on preparedness for warning should address the choice of warning sources, warning mechanisms, and warning content and the reasons for choosing them. In addition, research should examine the extent to which emergency managers systematically consider the time required to disseminate warnings and the role of informal warning networks in the dissemination process. Finally, research on preparedness for protective action implementation should address 11 behavioral parameters that affect the time required to complete an evacuation (see Box 3.3). These variables can have a significant influence on ETEs, but evacuation analysts appear to be making unfounded assumptions about them in the absence of reliable data.

---

**BOX 3.3**
**Evacutation Parameters**

1. Evacuation scope
2. Evacuation route system capacity
3. Hotel occupancy rate
4. Risk area resident population
5. Transit dependent resident population
6. Number of persons per household
7. Number of vehicles per household
8. Number of trailers per household
9. Evacuees' PAR compliance/spontaneous evacuation
10. Trip generation time distribution
11. Evacuees' utilization of the primary evacuation route system

---

In addition, many areas of research on preparedness for incident management are necessary. There is a major need to assess the extent to which the Incident Command System (ICS) successfully addresses problems identified by decades of research on emergency response (Drabek et al., 1981; Kreps, 1989a, 1991b; Tierney et al., 2001). One obvious disparity between the ICS framework and social science research findings is the absence of any explicit mention of population protection.

**Recommendation 3.10:** *Research is needed on training and exercising for disaster response.*

There has been some research on emergency response planning, but there appears to have been little or no research on training and exercising for disaster response. This is an unfortunate oversight because disaster response often requires the performance of tasks that are difficult, critical, and because of the rarity of such events, infrequently performed. There is an extensive literature on team training in organizational psychology that Ford and Schmidt (2000) found to be quite relevant to disaster response, but there is no evidence that this literature has been addressed by disaster researchers or utilized by practitioners. Analysis of the role of training and exercising before Hurricane Katrina should provide needed insight (see Box 3.4 for discussion on Hurricanes Katrina and Rita).

**Recommendation 3.11:** *Research should be conducted to develop better models of hazard adjustment adoption and implementation by community organizations.*

Specifically, these research needs can be organized in terms of methodological issues, as well as by the units of analysis discussed in the previous sections—households, businesses, government agencies, and neighborhood organizations.

**Recommendation 3.12:** *There is a continuing need for further research on hazard insurance.*

There must be some public constraints on private choices, but there is a delicate balance between the near term acceptability and the long-term effec-

---

**BOX 3.4**
**Research Implications of Hurricanes Katrina and Rita**

The impact of Hurricane Katrina underscores a number of the recommendations in this chapter. First, the failure to evacuate a significant number of transit dependent households during Katrina calls attention to the need for research to assess social vulnerability and its relation to hazard exposure and physical vulnerability. In addition, it also raises questions about the extent to which hazard/vulnerability analyses are conducted and used as a planning basis for developing local emergency operations plans. Second, the continued occupancy of areas below sea level that were protected only to the expected surge from a Category 3 storm raises questions about the dynamics of hazard vulnerability and the potential for more effective land-use practices and building construction practices to reduce this vulnerability. Future research should carefully examine the extent unfettered market forces reproduce previous vulnerability or, alternatively, whether new structural protection works, land-use practices, and building construction practices are integrated into the reconstruction process that will reduce this vulnerability. Third, Katrina revealed a conspicuous lack of coordination among agencies and levels of government during the emergency response. This suggests not only that planned multi-organizational networks (e.g., the National Incident Management System—NIMS) failed, but also that emergent multi-organizational networks failed to develop adequately. Research is needed to identify the organizational design and training problems that must be corrected to prevent future breakdowns.

Hurricane Rita provided yet another example of widespread traffic jams resulting from the evacuation of urbanized coastal areas. A survey by the Houston Chronicle found that approximately 2.5 million households (approximately 50 percent of the population) in the eight-county metropolitan Houston area evacuated. The large number of evacuating households, 46 percent of whom took more than one vehicle, grossly exceeded the capacity of the evacuation routes. This caused massive queues that resulted in 40 percent of the evacuees taking more than 12 hours to reach their destinations and 10 percent taking more than 24 hours—even though 95 percent of them were traveling to locations that are normally within a four hour drive. Although spontaneous evacuation was incorporated into evacua-

tiveness of any hazard insurance program. Some questions address institutional relationships such as the methods by which regulators can monitor insurers' catastrophe and insolvency risks and intervene to protect policyholders. Other questions address individual decision processes, such as how insurance premiums can be structured to encourage people to avoid hazard-prone areas where appropriate, to purchase insurance if they do decide to live there, and to implement hazard mitigation practices that reduce the likelihood of losses.

tion analyses conducted four years earlier, the over-response to Hurricane Rita greatly exceeded expectations because only 18 percent of the population of these counties is within officially designated hurricane risk areas. The excessive evacuation rate (327 percent of the projected rate) cannot be attributed solely to over warning because only 25 percent of the households reported receiving a mandatory evacuation order and another 12 percent reported receiving a voluntary evacuation order. Nor was it due only to misperception of risk because only 36 percent thought they were at either high or moderate risk from the hurricane. Thus, further research is needed to determine more clearly why so many households evacuated and if this over response is likely to occur in future hurricanes. In addition, the Hurricane Rita evacuation indicates a need for better methods of hurricane evacuation management. In particular, the evacuation analyses conducted for the state of Texas predicted that traffic queues could form in the hurricane surge zone south of Houston if a hurricane tracking directly west made a late change in direction to the north, as was the case for Hurricane Bret in 1999 and Hurricane Charley in 2004. Such a scenario could cause thousands of deaths if the evacuation were initiated less than 24 hours before landfall. During Hurricane Rita, the evacuation queues formed much earlier and about 20 miles farther inland than predicted in the Texas evacuation analyses because the storm tracked directly toward the Houston-Galveston area. Consequently, local officials initiated evacuations approximately 60 hours before landfall. Even though the late changing track scenario did not occur in Hurricane Rita, it might happen in a future hurricane. The likelihood of a major loss of life in this scenario could be reduced by better highway capacity management techniques such as contra flow. However, this technique is difficult to implement and can only increase capacity by 50-75 percent. Even greater safety can be provided by better evacuation demand management that uses more effective risk communication, improved structural protection works, better land-use practices, and better building construction practices to sharply reduce the number of evacuating vehicles. A significant amount of research will be needed to support the development of feasible hurricane hazard mitigation and emergency response preparedness plans.

# 4

# Research on Disaster
# Response and Recovery

T his chapter and the preceding one use the conceptual model
presented in Chapter 1 (see Figure 1.1) as a guide to understanding
societal response to hazards and disasters. As specified in that model,
Chapter 3 discusses three sets of pre-disaster activities that have the potential
to reduce disaster losses: hazard mitigation practices, emergency preparedness
practices, and pre-disaster planning for post-disaster recovery. This chapter
focuses on National Earthquake Hazards Reduction Program (NEHRP)
contributions to social science knowledge concerning those dimensions of
the model that are related to post-disaster response and recovery activities.
As in Chapter 3, discussions are organized around research findings regarding different units of analysis, including individuals, households, groups
and organizations, social networks, and communities. The chapter also
highlights trends, controversies, and issues that warrant further investigation. The contents of this chapter are linked to key themes discussed elsewhere in this report, including the conceptualization and measurement of
societal vulnerability and resilience, the importance of taking diversity into
account in understanding both response-related activities and recovery processes and outcomes, and linkages between hazard loss reduction and
sustainability. Although this review centers primarily on research on natural
disasters and to a lesser degree on technological disasters, research findings
are also discussed in terms of their implications for understanding and
managing emerging homeland security threats.

The discussions that follow seek to address several interrelated questions: What is currently known about post-disaster response and recovery,

and to what extent is that knowledge traceable to NEHRP-sponsored research activities? What gaps exist in that knowledge? What further research—both disciplinary and interdisciplinary—is needed to fill those gaps?

## RESEARCH ON DISASTER RESPONSE

Emergency response encompasses a range of measures aimed at protecting life and property and coping with the social disruption that disasters produce. As noted in Chapter 3, emergency response activities can be categorized usefully as *expedient mitigation actions* (e.g., clearing debris from channels when floods threaten, containing earthquake-induced fires and hazardous materials releases before they can cause additional harm) and *population protection actions* (e.g., warning, evacuation and other self-protective actions, search and rescue, the provision of emergency medical care and shelter; Tierney et al., 2001). Another common conceptual distinction in the literature on disaster response (Dynes et al., 1981) contrasts *agent-generated demands*, or the types of losses and forms of disruption that disasters create, and *response-generated demands*, such as the need for situation assessment, crisis communication and coordination, and response management. Paralleling preparedness measures, disaster response activities take place at various units of analysis, from individuals and households, to organizations, communities, and intergovernmental systems. This section does not attempt to deal exhaustively with the topic of emergency response activities, which is the most-studied of all phases of hazard and disaster management. Rather, it highlights key themes in the literature, with an emphasis on NEHRP-based findings that are especially relevant in light of newly recognized human-induced threats.

### Public Response: Warning Response, Evacuation, and Other Self-Protective Actions

The decision processes and behaviors involved in public responses to disaster warnings are among the best-studied topics in the research literature. Over nearly three decades, NEHRP has been a major sponsor of this body of research. As noted in Chapter 3, warning response research overlaps to some degree with more general risk communication research. For example, both literatures emphasize the importance of considering source, message, channel, and receiver effects on the warning process. While this discussion centers mainly on responses to official warning information, it should be noted that self-protective decision-making processes are also initiated in the absence of formal warnings—for example, in response to cues that people perceive as signaling impending danger and in disasters that occur without warning. Previous research suggests that the basic decision processes involved in self-protective action are similar across different types of disaster events, although the challenges posed and the problems that may develop can be agent specific.

As in other areas discussed here, empirical studies on warning response and self-protective behavior in different types of disasters and emergencies have led to the development of broadly generalizable explanatory models. One such model, the protective action decision model, developed by Perry, Lindell, and their colleagues (see, for example, Lindell and Perry, 2004), draws heavily on Turner and Killian's (1987) emergent norm theory of collective behavior. According to that theory, groups faced with the potential need to act under conditions of uncertainty (or potential danger) engage in interaction in an attempt to develop a collective definition of the situation they face and a set of new norms that can guide their subsequent action.[1] Thus, when warnings and protective instructions are disseminated, those who receive warnings interact with one another in an effort to determine collectively whether the warning is authentic, whether it applies to them, whether they are indeed personally in danger, whether they can reduce their vulnerability through action, whether action is possible, and when they should act. These collective determinations are shaped in turn by such factors as (1) the *characteristics of warning recipients*, including their prior experience with the hazard in question or with similar emergencies, as well as their prior preparedness efforts; (2) *situational factors*, including the presence of perceptual cues signaling danger; and (3) the *social contexts* in which decisions are made—for example, contacts among family members, coworkers, neighborhood residents, or others present in the setting, as well as the strength of preexisting social ties. Through interaction and under the influence of these kinds of factors, individuals and groups develop new norms that serve as guidelines for action.

Conceptualizing warning response as a form of collective behavior that is guided by emergent norms brings several issues to the fore. One is that far from being automatic or governed by official orders, behavior undertaken in response to warnings is the product of interaction and deliberation among members of affected groups—activities that are typically accompanied by a search for additional confirmatory information. Circumstances that complicate the deliberation process, such as conflicting warning information that individuals and groups may receive, difficulties in getting in touch with others whose views are considered important for the decision-making process, or disagreements among group members about any aspect of the

---

[1]Note that what is being discussed here are *group-level* deliberations and decisions, not individual ones. Actions under conditions of uncertainty and urgency such as those that accompany disaster warnings should not be conceptualized in individualistic terms.

threat situation, invariably lead to additional efforts to communicate and confirm the information and lengthen the period between when a warning is issued and when groups actually respond.

Another implication of the emergent norm approach to protective action decision making is the recognition that groups may collectively define an emergency situation in ways that are at variance from official views. This is essentially what occurs in the shadow evacuation phenomenon, which has been documented in several emergency situations, including the Three Mile Island nuclear plant accident (Zeigler et al., 1981). While authorities may not issue a warning for a particular geographic area or group of people, or may even tell them they are safe, groups may still collectively decide that they are at risk or that the situation is fluid and confusing enough that they should take self-protective action despite official pronouncements.

The behavior of occupants of the World Trade Center during the September 11, 2001 terrorist attack illustrates the importance of collectively developed definitions. Groups of people in Tower 2 of the World Trade Center decided that they should evacuate the building after seeing and hearing about what was happening in Tower 1 and after speaking with coworkers and loved ones, even when official announcements and other building occupants indicated that they should not do so. Others decided to remain in the tower or, perhaps more accurately, they decided to delay evacuating until receiving additional information clarifying the extent to which they were in danger. Journalistic accounts suggest that decisions were shaped in part by what people could see taking place in Tower 1, conversations with others outside the towers who had additional relevant information, and directives received from those in positions of authority in tenant firms. In that highly confusing and time-constrained situation, emergent norms guiding the behavior of occupants of the second tower meant the difference between life and death when the second plane struck (NIST, 2005).

The large body of research that exists regarding decision making under threat conditions points to the need to consider a wide range of individual, group, situational, and resource-related factors that facilitate and inhibit self-protective action. Qualitatively based decision-tree models developed by Gladwin et al. (2001) demonstrate the complexity of self-protective decisions. As illustrated by their work on hurricane evacuation, a number of different factors contribute to decisions on whether or not to evacuate. Such factors range from perceptions of risk and personal safety with respect to a threatened disaster, to the extent of knowledge about specific areas at risk, to constraining factors such as the presence of pets in the home that require care, lack of a suitable place to go, counterarguments by other family members, fears of looting (shown by the literature to be unjustified; see, for example, Fischer, 1998), and fear that the evacuation process may be more dangerous than staying home and riding out a hurricane. Warning recipients may decide that they should wait before evacuating, ultimately missing the opportunity to escape, or they may decide to shelter in-place after concluding that their homes are strong enough to resist hurricane forces despite what they are told by authorities.

In their research on Hurricane Andrew, Gladwin and Peacock describe some of the many factors that complicate the evacuation process for endangered populations (1997:54):

> Except under extreme circumstances, households cannot be compelled to evacuate or to remain where they are, much less to prepare themselves for the threat. Even under extraordinary conditions many households have to be individually located and assisted or forced to comply. Segments of a population may fail to receive, ignore, or discount official requests and orders. Still others may not have the resources or wherewithal to comply. Much will depend upon the source of the information, the consistency of the message received from multiple sources, the nature of the information conveyed, as well as the household's ability to perceive the danger, make decisions, and act accordingly. Disputes, competition, and the lack of coordination among local, state, and federal governmental agencies and between those agencies and privately controlled media can add confusion. Businesses and governmental agencies that refuse to release their employees and suspend normal activities can add still further to the confusion and noncompliance.

The normalcy bias adds other complications to the warning response process. While popular notions of crisis response behaviors seem to assume that people react automatically to messages signaling impending danger—for example, by fleeing in panic—the reality is quite different. People typically "normalize" unusual situations and persist in their everyday activities even when urged to act differently. As noted earlier, people will not act on threat information unless they perceive a personal risk to themselves. Simply knowing that a threat exists—even if that threat is described as imminent—is insufficient to motivate self-protective action. Nor can people be expected to act if warning-related guidance is not specific enough to provide them with a blueprint for what to do or if they do not believe they have the resources required to follow the guidance. One practical implication of research on warnings is that rather than being concerned about panicking the public with warning information, or about communicating too much information, authorities should instead be seeking better ways to penetrate the normalcy bias, persuade people that they should be concerned about an impending danger, provide directives that are detailed enough to follow during an emergency, and encourage pre-disaster response planning so that people have thought through what to do prior to being required to act.

## Other Important Findings Regarding the Evacuation Process

As noted earlier, evacuation behavior has long been recognized as the reflection of social-level factors and collective deliberation. Decades ago, Drabek (1983) established that households constitute the basic deliberative units for evacuation decision making in community-wide disasters and that the decisions that are ultimately made tend to be consistent with pre-disaster household authority patterns. For example, gender-related concerns often enter into evacuation decision making. Women tend to be more risk-averse and more inclined to want to follow evacuation orders, while males are less inclined to do so (for an extensive discussion of gender differences in vulnerability, risk perception, and responses to disasters, see Fothergill, 1998). In arriving at decisions regarding evacuation, households take official orders into account, but they weigh those orders in light of their own priorities, other information sources, and their past experiences. Information received from media sources and from family and friends, along with confirmatory data actively sought by those at risk, generally has a greater impact on evacuation decisions than information provided by public officials (Dow and Cutter, 1998, 2000).

Recent research also suggests that family evacuation patterns are undergoing change. For example, even though families decide together to evacuate and wish to stay together, they increasingly tend to use more than one vehicle to evacuate—perhaps because they want to take more of their possessions with them, make sure their valuable vehicles are protected, or return to their homes at different times (Dow and Cutter, 2002). Other social influences also play a role. Neighborhood residents may be more willing to evacuate or, conversely, more inclined to delay the decision to evacuate if they see their neighbors doing so. Rather than becoming more vigilant, communities that are struck repeatedly by disasters such as hurricanes and floods may develop "disaster subcultures," such as groups that see no reason to heed evacuation orders since sheltering in-place has been effective in previous events.

NEHRP-sponsored research has shown that different racial, ethnic, income, and special needs groups respond in different ways to warning information and evacuation orders, in part because of the unique characteristics of these groups, the manner in which they receive information during crises, and their varying responses to different information sources. For example, members of some minority groups tend to have large extended families, making contacting family members and deliberating on alternative courses of action a more complicated process. Lower-income groups, inner-city residents, and elderly persons are more likely to have to rely on public transportation, rather than personal vehicles, in order to evacuate. Lower-income and minority populations, who tend to have larger families, may also be reluctant to impose on friends and relatives for shelter. Lack of financial resources may leave less-well-off segments of the population less able to afford to take time off from work when disasters threaten, to travel long distances to avoid danger, or to pay for emergency lodging. Socially isolated individuals, such as elderly persons living alone, may lack the social support that is required to carry out self-protective actions. Members of minority groups may find majority spokespersons and official institutions less credible and believable than members of the white majority, turning instead to other sources, such as their informal social networks. Those who rely on non-English-speaking mass media for news may receive less complete warning information, or may receive warnings later than those who are tuned into mainstream media sources (Aguirre et al., 1991; Perry and Lindell, 1991; Lindell and Perry, 1992, 2004; Klinenberg, 2002; for more extensive discussions, see Tierney et al., 2001).

Hurricane Katrina vividly revealed the manner in which social factors such as those discussed above influence evacuation decisions and actions. In many respects, the Katrina experience validated what social science research had already shown with respect to evacuation behavior. Those who stayed behind did so for different reasons—all of which have been discussed in past research. Some at-risk residents lacked resources, such as automobiles and financial resources that would have enabled them to escape the city. Based on their past experiences with hurricanes like Betsey and Camille, others considered themselves not at risk and decided it was not necessary to evacuate. Still others, particularly elderly residents, felt so attached to their homes that they refused to leave even when transportation was offered.

This is not to imply that evacuation-related problems stemmed solely from individual decisions. Katrina also revealed the crucial significance of evacuation planning, effective warnings, and government leadership in facilitating evacuations. Planning efforts in New Orleans were rudimentary at best, clear evacuation orders were given too late, and the hurricane rendered evacuation resources useless once the city began to flood.

With respect to other patterns of evacuation behavior when they do evacuate, most people prefer to stay with relatives or friends, rather than using public shelters. Shelter use is generally limited to people who feel they have no other options—for example, those who have no close friends and relatives to take them in and cannot afford the price of lodging. Many people avoid public shelters or elect to stay in their homes because shelters do not allow pets. Following earthquakes, some victims, particularly Latinos in the United States who have experienced or learned about highly damaging earthquakes in their countries of origin, avoid indoor shelter of all types, preferring instead to sleep outdoors (Tierney, 1988; Phillips, 1993; Simile, 1995).

Disaster warnings involving "near misses," as well as concerns about the possible impact of elevated color-coded homeland security warnings, raise the question of whether warnings that do not materialize can induce a "cry-wolf" effect, resulting in lowered attention to and compliance with future warnings. The disaster literature shows little support for the cry-wolf hypothesis. For example, Dow and Cutter (1998) studied South Carolina residents who had been warned of impending hurricanes that ultimately struck North Carolina. Earlier false alarms did not influence residents' decisions on whether to evacuate; that is, there was little behavioral evidence for a cry-wolf effect. However, false alarms did result in a decrease in confidence in official warning sources, as opposed to other sources of information on which people relied in making evacuation decisions—certainly not the outcome officials would have intended. Studies also suggest that it is advisable to clarify for the public why forecasts and warnings were uncertain or incorrect. Based on an extensive review of the warning literature, Sorensen (2000:121) concluded that "[t]he likelihood of people responding to a warning is not diminished by what has come to be labeled the 'cry-wolf' syndrome *if the basis for the false alarm is understood* [emphasis added]." Along those same lines, Atwood and Major (1998) argue that if officials explain reasons for false alarms, that information can increase public awareness and make people more likely to respond to subsequent hazard advisories.

## PUBLIC RESPONSE

### Dispelling Myths About Crisis-Related Behavior: Panic and Social Breakdown

Numerous individual studies and research syntheses have contrasted commonsense ideas about how people respond during crises with empirical data on actual behavior. Among the most important myths addressed in these analyses is the notion that panic and social disorganization are common responses to imminent threats and to actual disaster events (Quarantelli and Dynes, 1972; Johnson, 1987; Clarke, 2002). True panic, defined as highly individualistic flight behavior that is nonsocial in nature, undertaken without regard to social norms and relationships, is extremely rare prior to and during extreme events of all types. Panic takes place under specific conditions that are almost never present in disaster situations. Panic only occurs when individuals feel completely isolated and when both social bonds and measures to promote safety break down to such a degree that individuals feel totally on their own in seeking safety. Panic results from a breakdown in the ongoing social order—a breakdown that Clarke (2003:128) describes as having moral, network, and cognitive dimensions:

There is a *moral* failure, so that people pursue their self interest regardless of rules of duty and obligation to others. There is a *network* failure, so that the resources that people can normally draw on in times of crisis are no longer there. There is a *cognitive* failure, in which someone's understanding of how they are connected to others is cast aside.

Failures on this scale almost never occur during disasters. Panic reactions are rare in part because social bonds remain intact and extremely resilient even under conditions of severe danger (Johnson, 1987; Johnson et al., 1994; Feinberg and Johnson, 2001).

Panic persists in public and media discourses on disasters, in part because those discourses conflate a wide range of other behaviors with panic. Often, people are described as panicking because they experience feelings of intense fear, even though fright and panic are conceptually and behaviorally distinct. Another behavioral pattern that is sometimes labeled panic involves intensified rumors and information seeking, which are common patterns among publics attempting to make sense of confusing and potentially dangerous situations. Under conditions of uncertainty, people make more frequent use of both informal ties and official information sources, as they seek to collectively define threats and decide what actions to take. Such activities are a normal extension of everyday information-seeking practices (Turner, 1994). They are not indicators of panic.

The phenomenon of shadow evacuation, discussed earlier, is also frequently confused with panic. Such evacuations take place because people who are not defined by authorities as in danger nevertheless determine that they are—perhaps because they have received conflicting or confusing information or because they are geographically close to areas considered at risk (Tierney et al., 2001). Collective demands for antibiotics by those considered not at risk for anthrax, "runs" on stores to obtain self-protective items, and the so-called worried-well phenomenon are other forms of collective behavior that reflect the same sociobehavioral processes that drive shadow evacuations: emergent norms that define certain individuals and groups as in danger, even though authorities do not consider them at risk; confusion about the magnitude of the risk; a collectively defined need to act; and in some cases, an unwillingness to rely on official sources for self-protective advice. These types of behaviors, which constitute interesting subjects for research in their own right, are not examples of panic.

Research also indicates that panic and other problematic behaviors are linked in important ways to the manner in which institutions manage risk and disaster. Such behaviors are more likely to emerge when those who are in danger come to believe that crisis management measures are ineffective, suggesting that enhancing public understanding of and trust in preparedness measures and in organizations charged with managing disasters can lessen the likelihood of panic. With respect to homeland security threats, some researchers have argued that the best way to "vaccinate" the public against the emergence of panic in situations involving weapons of mass destruction is to provide timely and accurate information about impending threats and to actively include the public in pre-crisis preparedness efforts (Glass and Shoch-Spana, 2002).

Blaming the public for panicking during emergencies serves to diffuse responsibility from professionals whose duty it is to protect the public, such as emergency managers, fire and public safety officials, and those responsible for the design, construction, and safe operation of buildings and other structures (Sime, 1999). The empirical record bears out the fact that to the extent panic does occur during emergencies, such behavior can be traced in large measure to environmental factors such as overcrowding, failure to provide adequate egress routes, and breakdowns in communications, rather than to some inherent human impulse to stampede with complete disregard for others. Any potential for panic and other problematic behaviors that may exist can, in other words, be mitigated through appropriate design, regulatory, management, and communications strategies.

As discussed elsewhere in this report, looting and violence are also exceedingly rare in disaster situations. Here again, empirical evidence of what people actually do during and following disasters contradicts what many officials and much of the public believe. Beliefs concerning looting are based not on evidence but rather on assumptions—for example, that social control breaks down during disasters and that lawlessness and violence inevitably result when the social order is disrupted. Such beliefs fail to take into account the fact that powerful norms emerge during disasters that foster prosocial behavior—so much so that lawless behavior actually declines in disaster situations. Signs erected following disasters saying, "We shoot to kill looters" are not so much evidence that looting is occurring as they are evidence that community consensus condemns looting.

The myth of disaster looting can be contrasted with the reality of looting during episodes of civil disorder such as the riots of the 1960s and the 1992 Los Angeles unrest. During episodes of civil unrest, looting is done publicly, in groups, quite often in plain sight of law enforcement officials. Taking goods and damaging businesses are the hallmarks of modern "commodity riots." New norms also emerge during these types of crises, but unlike the prosocial norms that develop in disasters, norms governing behavior during civil unrest permit and actually encourage lawbreaking. Under these circumstances, otherwise law-abiding citizens allow themselves to take part in looting behavior (Dynes and Quarantelli, 1968; Quarantelli and Dynes, 1970).

Looting and damaging property can also become normative in situations that do not involve civil unrest—for example, in victory celebrations following sports events. Once again, in such cases, norms and traditions governing behavior in crowd celebrations encourage destructive activities

(Rosenfeld, 1997). The behavior of participants in these destructive crowd celebrations again bears no resemblance to that of disaster victims.

In the aftermath of Hurricane Katrina, social scientists had no problem understanding why episodes of looting might have been more widespread in that event than in the vast majority of U.S. disasters. Looting has occurred on a widespread basis following other disasters, although such cases have been rare. Residents of St. Croix engaged in extensive looting behavior following Hurricane Hugo, and this particular episode sheds light on why some Katrina victims might have felt justified in looting. Hurricane Hugo produced massive damage on St. Croix, and government agencies were rendered helpless. Essentially trapped on the island, residents had no idea when help would arrive. Instead, they felt entirely on their own following Hugo. The tourist-based St. Croix economy was characterized by stark social class differences, and crime and corruption had been high prior to the hurricane. Under these circumstances, looting for survival was seen as justified, and patterns of collective behavior developed that were not unlike those seen during episodes of civil unrest. Even law enforcement personnel joined in the looting (Quarantelli, 2006; Rodriguez et al., forthcoming).

Despite their similarities, the parallels between New Orleans and St. Croix should not be overstated. It is now clear that looting and violent behavior were far less common than initially reported and that rumors concerning shootings, rapes, and murders were groundless. The media employed the "looting frame" extensively while downplaying far more numerous examples of selflessness and altruism. In hindsight, it now appears that many reports involving looting and social breakdown were based on stereotyped images of poor minority community residents (Tierney et al., forthcoming).

Extensive research also indicates that despite longstanding evidence, beliefs about disaster-related looting and lawlessness remain quite common, and these beliefs can influence the behavior of both community residents and authorities. For example, those who are at risk may decide not to evacuate and instead stay in their homes to protect their property from looters (Fischer, 1998). Concern regarding looting and lawlessness may cause government officials to make highly questionable and even counter-productive decisions. Following Hurricane Katrina, for example, based largely on rumors and exaggerated media reports, rescue efforts were halted because of fears for the safety of rescue workers, and Louisiana's governor issued a "shoot-to-kill" order to quash looting. These decisions likely resulted in additional loss of life and also interfered with citizen efforts to aid one another. Interestingly, recent historical accounts indicate that similar decisions were made following other large-scale disasters, such as the 1871 Chicago fire, the 1900 Galveston hurricane, and the 1906 San Francisco earthquake and firestorm. In all three cases, armed force was used to stop looting, and immigrant groups and the poor were scapegoated for their putative "crimes" (Fradkin, 2005). Along with Katrina, these events caution against making decisions on the basis of mythical beliefs and rumors.

As is the case with the panic myth, attributing the causes of looting behavior to individual motivations and impulses serves to deflect attention from the ways in which institutional failures can create insurmountable problems for disaster victims. When disasters occur, communications, disaster management, and service delivery systems should remain sufficiently robust that victims will not feel isolated and afraid or conclude that needed assistance will never arrive. More to the point, victims of disasters should not be scapegoated when institutions show themselves to be entirely incapable of providing even rudimentary forms of assistance—which was exactly what occurred with respect to Hurricane Katrina.

### Patterns of Collective Mobilization in Disaster-Stricken Areas: Prosocial and Helping Behavior

In contrast to the panicky and lawless behavior that is often attributed to disaster-stricken populations, public behavior during earthquakes and other major community emergencies is overwhelmingly adaptive, prosocial, and aimed at promoting the safety of others and the restoration of ongoing community life. The predominance of prosocial behavior (and, conversely, a decline in antisocial behavior) in disaster situations is one of the most longstanding and robust research findings in the disaster literature. Research conducted with NEHRP sponsorship has provided an even better understanding of the processes involved in adaptive collective mobilization during disasters.

**Helping Behavior and Disaster Volunteers.** Helping behavior in disasters takes various forms, ranging from spontaneous and informal efforts to provide assistance to more organized emergent group activity, and finally to more formalized organizational arrangements. With respect to spontaneously developing and informal helping networks, disaster victims are assisted first by others in the immediate vicinity and surrounding area and only later by official public safety personnel. In a discussion on search and rescue activities following earthquakes, for example, Noji observes (1997:162)

> In Southern Italy in 1980, 90 percent of the survivors of an earthquake were extricated by untrained, uninjured survivors who used their bare hands and simple tools such as shovels and axes. . . . Following the 1976 Tangshan earthquake, about 200,000 to 300,000 entrapped people crawled out of the debris on their own and went on to rescue others. . . . They became the backbone of the rescue teams, and it was to their credit that more than 80 percent of those buried under the debris were rescued.

Thus, lifesaving efforts in a stricken community rely heavily on the capabilities of relatively uninjured survivors, including untrained volunteers, as well as those of local firefighters and other relevant personnel.

The spontaneous provision of assistance is facilitated by the fact that when crises occur, they take place in the context of ongoing community life and daily routines—that is, they affect not isolated individuals but rather people who are embedded in networks of social relationships. When a massive gasoline explosion destroyed a neighborhood in Guadalajara, Mexico, in 1992, for example, survivors searched for and rescued their loved ones and neighbors. Indeed, they were best suited to do so, because they were the ones who knew who lived in different households and where those individuals probably were at the time of the disaster (Aguirre et al., 1995). Similarly, crowds and gatherings of all types are typically comprised of smaller groupings—couples, families, groups of friends—that become a source of support and aid when emergencies occur.

As the emergency period following a disaster lengthens, unofficial helping behavior begins to take on a more structured form with the development of emergent groups—newly formed entities that become involved in crisis-related activities (Stallings and Quarantelli, 1985; Saunders and Kreps, 1987). Emergent groups perform many different types of activities in disasters, from sandbagging to prevent flooding, to searching for and rescuing victims and providing for other basic needs, to post-disaster cleanup and the informal provision of recovery assistance to victims. Such groupings form both because of the strength of altruistic norms that develop during disasters and because of emerging collective definitions that victims' needs are not being met—whether official agencies share those views or not. While emergent groups are in many ways essential for the effectiveness of crisis response activities, their activities may be seen as unnecessary or even disruptive by formal crisis response agencies. In the aftermath of the attack on the World Trade Center, for example, numerous groups emerged to offer every conceivable type of assistance to victims and emergency responders. Some were incorporated into official crisis management activities, while others were labeled "rogue volunteers" by official agencies (Halford and Nolan, 2002; Kendra and Wachtendorf, 2002).[2]

Disaster-related volunteering also takes place within more formalized organizational structures, both in existing organizations that mobilize in response to disasters and through organizations such as the Red Cross,

---

[2]Indeed, many individuals persisted in literally demanding to be allowed to serve as volunteers, even after being repeatedly turned away. Some of those who were intent on serving as volunteers managed to talk their way into settings that were off-limits in order to offer their services.

which has a federal mandate to respond in presidentially declared disasters and relies primarily on volunteers in its provision of disaster services. Some forms of volunteering have been institutionalized in the United States through the development of the National Voluntary Organizations Active in Disaster (NVOAD) organization. NVOAD, a large federation of religious, public service, and other groups, has organizational affiliates in 49 states, the District of Columbia, Puerto Rico, and U.S. territories. National-level NVOAD affiliates include organizations such as the Salvation Army, Church World Service, Church of the Brethren Disaster Response, and dozens of others that provide disaster services. Organizations such as the Red Cross and the NVOAD federation thus provide an infrastructure that can support very extensive volunteer mobilization. That infrastructure will likely form the basis for organized volunteering in future homeland security emergencies, just as it does in major disasters.

Helping behavior is very widespread after disasters, particularly large and damaging ones. For example, NEHRP-sponsored research indicates that in the three weeks following the 1985 earthquake in Mexico City, an estimated 1.7 to 2.1 million residents of that city were involved in providing volunteer aid. Activities in which volunteers engaged after that disaster included searching for and rescuing victims trapped under rubble, donating blood and supplies, inspecting building damage, collecting funds, providing medical care and psychological counseling, and providing food and shelter to victims (Wenger and James, 1994). In other research on post-earthquake volunteering, also funded by NEHRP, O'Brien and Mileti (1992) found that more than half of the population in San Francisco and Santa Cruz counties provided assistance to their fellow victims after the 1989 Loma Prieta earthquake—help that ranged from assisting with search and rescue and debris removal activities to offering food, water, and shelter to those in need. Thus, the volunteer sector responding to disasters typically constitutes a very large proportion of the population of affected regions, as well as volunteers converging from other locations.

Social science research, much of it conducted under NEHRP auspices, highlights a number of other points regarding post-disaster helping behavior. One such insight is that helping behavior in many ways mirrors roles and responsibilities people assume during nondisaster times. For example, when people provide assistance during disasters and other emergencies, their involvement is typically consistent with gender role expectations (Wenger and James, 1994; Feinberg and Johnson, 2001). Research also indicates that mass convergence of volunteers and donations can create significant management problems and undue burdens on disaster-stricken communities. In their eagerness to provide assistance, people may "overrespond" to disaster sites, creating congestion and putting themselves and others at risk or insisting on providing resources that are in fact not needed. After disasters, communities typically experience major difficulties in dealing with unwanted and unneeded donations (Neal, 1990).

Research on public behavior during disasters has major implications for homeland security policies and practices. The research literature provides support for the inclusion of the voluntary sector and community-based organizations in preparedness and response efforts. Initiatives that aim at encouraging public involvement in homeland security efforts of all types are clearly needed. The literature also provides extensive evidence that members of the public are in fact the true "first responders" in major disasters. In using that term to refer to fire, police, and other public safety organizations, current homeland security discourse fails to recognize that community residents themselves constitute the front-line responders in any major emergency

One implication of this line of research is that planning and management models that fail to recognize the role of victims and volunteers in responding to all types of extreme events will leave responders unprepared for what will actually occur during disasters—for example, that, as research consistently shows, community residents will be the first to search for victims, provide emergency aid, and transport victims to health care facilities in emergencies of all types.[3] Such plans will also fail to take advantage of the public's crucial skills, resources, and expertise. For this reason, experts on human-induced threats such as bioterrorism stress the value of public engagement and involvement in planning for homeland security emergencies (Working Group on "Governance Dilemmas" in Bioterrorism Response, 2004).

These research findings have significant policy implications. To date, Department of Homeland Security initiatives have focused almost exclusively on providing equipment and training for uniformed responders, as opposed to community residents. Recently, however, DHS has begun placing more emphasis on its Citizen Corps component, which is designed to mobilize the skills and talents of the public when disasters strike. Public involvement in Citizen Corps and Community Emergency Response Team (CERT) activities have expanded considerably since the terrorist attacks of

---

[3]In one illustrative case, nearly half of those killed in the Northridge earthquake died as a consequence of damage in one of the buildings in the Northridge Meadows apartment complex, which was located not far from the earthquake's epicenter. Fire department personnel dispatched in vehicles to the damaged area following the earthquake mistook the structure, a three-story building that had pancaked on the first floor, for a two-story building, and they did not stop to inspect the structure or look for victims. The fact that fire personnel failed to recognize the severity of the earthquake's impact at the Northridge Meadows location made little difference in this case, because by that time, survivors had already escaped on their own or had been rescued by their fellow tenants.

9/11—a sign that many community residents around the nation wish to play an active role in responding to future disasters. The need for community-based preparedness and response initiatives is more evident than ever following the Katrina disaster.

**Organizational, Governmental, and Network Responses.** The importance of observing disaster response operations while they are ongoing or as soon as possible after disaster impact has long been a hallmark of the disaster research field. The quick-response tradition in disaster research, which has been a part of the field since its inception, developed out of a recognition that data on disaster response activities are perishable and that information collected from organizations after the passage of time is likely to be distorted and incomplete (Quarantelli, 1987, 2002). NEHRP funds, provided through grant supplements, Small Grants for Exploratory Research (SGER) awards, Earthquake Engineering Research Institute (EERI) reconnaissance missions, earthquake center reconnaissance funding, and small grants such as those provided by the Natural Hazards Research and Applications Information Center, have supported the collection of perishable data and enabled social science researchers to mobilize rapidly following major earthquakes and other disasters.

NEHRP provided substantial support for the collection of data on organizational and community responses in a number of earthquake events, including the 1987 Whittier Narrows, 1989 Loma Prieta, and 1994 Northridge earthquakes (see, for example, Tierney, 1988, 1994; EERI, 1995), as well as major earthquakes outside the United States such as the 1985 Mexico City, 1986 San Salvador, and 1988 Armenia events. More recently, NEHRP funds were used to support rapid-response research on the September 11, 2001 terrorist attacks and Hurricanes Katrina and Rita. Many of those studies focused on organizational issues in both the public and private sectors. (For a compilation of NEHRP-sponsored quick-response findings on the events of September 11, see Natural Hazards Research and Applications Information Center, 2003).

In many cases, quick-response research on disaster impacts and organizational and governmental response has led to subsequent in-depth studies on response-related issues identified during the post-impact reconnaissance phase. Following major events such as Loma Prieta, Northridge, and Kobe, insights from initial reconnaissance studies have formed the basis for broader research initiatives. Recent efforts have focused on ways to better take advantage of reconnaissance opportunities and to identify topics for longer-term study. A new plan has been developed to better coordinate and integrate both reconnaissance and longer-term research activities carried out with NEHRP support. That planning activity, outlined in the report *The Plan to Coordinate NEHRP Post-earthquake Investigations* (Holzer et al., 2003), encompasses both reconnaissance and more systematic research activities in the earth sciences, engineering, and social sciences.

Through both initial quick-response activities and longer-term studies, NEHRP research has added to the knowledge base on how organizations cope with crises. Studies have focused on a variety of topics. A partial list of those topics includes organizational and group activities associated with the post-disaster search and rescue process (Aguirre et al., 1995); intergovernmental coordination during the response period following major disaster events (Nigg, 1998); expected and improvised organizational forms that characterize the disaster response milieu (Kreps, 1985, 1989b); strategies used by local government organizations to enhance interorganizational coordination following disasters (Drabek, 2003); and response activities undertaken by specific types of organizations, such as those in the volunteer and nonprofit sector (Neal, 1990) and tourism-oriented enterprises (Drabek, 1994).

Focusing specifically at the interorganizational level of analysis, NEHRP research has also highlighted the significance and mix of planned and improvised networks in disaster response. It has long been recognized that post-disaster response activities involve the formation of new (or emergent) networks of organizations. Indeed, one distinguishing feature of major crisis events is the prominence and proliferation of network forms of organization during the response period. Emergent multiorganizational networks (EMON) constitute new organizational interrelationships that reflect collective efforts to manage crisis events. Such networks are typically heterogeneous, consisting of existing organizations with pre-designated crisis management responsibilities, other organizations that may not have been included in prior planning but become involved in crisis response activities because those involved believe they have some contribution to make, and emergent groups. EMONs tend to be very large in major disaster events, encompassing hundreds and even thousands of interacting entities. As crisis conditions change and additional resources converge, EMON structures evolve, new organizations join the network, and new relationships form. What is often incorrectly described as disaster-generated "chaos" is more accurately seen as the understandable confusion that results when mobilization takes place on such a massive scale and when organizations and groups that may be unfamiliar with one another attempt to communicate, negotiate, and coordinate their activities under extreme pressure. (For more detailed discussions on EMONs in disasters, including the 2001 World Trade Center attack, see Drabek, 1985, 2003; Tierney, 2003; Tierney and Trainor, 2004.)

This is not to say that response activities always go smoothly. The disaster literature, organizational after-action reports, and official investigations contain numerous examples of problems that develop as interorganizational and intergovernmental networks attempt to address disaster-related challenges. Such problems include the following: failure to recognize the magnitude and seriousness of an event; delayed and insufficient responses; confusion regarding authorities and responsibilities, often resulting in major "turf battles;" resource shortages and misdirection of existing resources; poor organizational, interorganizational, and public communications; failures in intergovernmental coordination; failures in leadership and vision; inequities in the provision of disaster assistance; and organizational practices and cultures that permit and even encourage risky behavior. Hurricane Katrina became a national scandal because of the sheer scale on which these organizational pathologies manifested. However, Katrina was by no means atypical. In one form or another and at varying levels of severity, such pathologies are ever-present in the landscape of disaster response (for examples, see U.S. President's Commission on the Accident at Three Mile Island, 1979; Perrow, 1984; Shrivastava, 1987; Sagan, 1993; National Academy of Public Administration, 1993; Vaughan, 1996, 1999; Peacock et al., 1997; Klinenberg, 2002; Select Bipartisan Committee to Investigate the Preparations for and Response to Hurricane Katrina, 2006; White House, 2006).

## Management Considerations in Disaster Response

U.S. disaster researchers have identified two contrasting approaches to disaster response management, commonly termed the "command-and-control" and the "emergent human resources," or "problem-solving," models. The command-and-control model equates preparedness and response activities with military exercises. It assumes that (1) government agencies and other responders must be prepared to take over management and control in disaster situations, both because they are uniquely qualified to do so and because members of the public will be overwhelmed and will likely engage in various types of problematic behavior, such as panic; (2) disaster response activities are best carried out through centralized direction, control, and decision making; and (3) for response activities to be effective, a single person is ideally in charge, and relations among responding entities are arranged hierarchically.

In contrast, the emergent human resources, or problem-solving, model is based on the assumption that communities and societies are resilient and resourceful and that even in areas that are very hard hit by disasters, considerable local response capacity is likely to remain. Another underlying assumption is that preparedness strategies should build on existing community institutions and support systems—for example by pre-identifying existing groups, organizations, and institutions that are capable of assuming leadership when a disaster strikes. Again, this approach argues against highly specialized approaches that tend to result in "stovepiped" rather than well-integrated preparedness and response efforts. The model also recognizes that when a disaster occurs, responding entities must be flexible if they are to be effective and that flexibility is best achieved through a decentralized response structure that seeks to solve problems as they arise, as opposed to top-down decision making. (For more extensive discussions of these two models and their implications, see Dynes, 1993, 1994; Kreps and Bosworth, forthcoming.)

Empirical research, much of which has been carried out with NEHRP support, finds essentially no support for the command-and-control model either as a heuristic device for conceptualizing the disaster management process or as a strategy employed in actual disasters. Instead, as suggested in the discussion above on EMONs, disaster response activities in the United States correspond much more closely to the emergent resources or problem-solving model. More specifically, such responses are characterized by decentralized, rather than centralized, decision making; by collaborative relationships among organizations and levels of government, rather than hierarchical ones; and, perhaps most important, by considerable emergence—that is, the often rapid appearance of novel and unplanned-for activities, roles, groups, and relationships. Other hallmarks of disaster responses include their fluidity and hence the fast pace at which decisions must be made; the predominance of the EMON as the organizational form most involved in carrying out response activities; the wide array of improvisational strategies that are employed to deal with problems as they manifest themselves; and the importance of local knowledge and situation-specific information in gauging appropriate response strategies. (For empirical research supporting these points, see Drabek et al., 1982; Stallings and Quarantelli, 1985; Kreps, 1985, 1989b; Bosworth and Kreps, 1986; Kreps and Bosworth, 1993; Aguirre et al., 1995; Drabek and McEntire, 2002; Waugh and Sylves, 2002; Webb, 2002; Drabek, 2003; Tierney, 2003; Tierney and Trainor, 2004; Wachtendorf, 2004.)

## NEW WAYS OF FRAMING DISASTER MANAGEMENT CHALLENGES: DEALING WITH COMPLEXITY AND ACCOMMODATING EMERGENCE

Advancements brought about through NEHRP research include new frameworks for conceptualizing responses to extreme events. In *Shared Risk: Complex Systems in Seismic Response*, a NEHRP-supported comparative study of organized responses to 11 different earthquake events, Comfort argues that the major challenge facing response systems is to use information in ways that enhance organizational and interorganizational learning and develop ways of "integrating both technical and organizational components in a socio-technical system to support timely, informed collective action" (Comfort, 1999:14). Accordingly, effective responses depend on the ability of organizations to simultaneously sustain structure and allow for flexibility in the face of rapidly changing disaster conditions and unexpected demands. Response networks must also be able to accommodate processes of *self-organization*—that is, organized action by volunteers and emergent groups. This approach again contrasts with command-and-control notions of how major crises are managed (Comfort, 1999:263-264):

> A socio-technical approach requires a shift in the conception of response systems as reactive, command-and-control driven systems to one of *inquiring systems*, activated by processes of inquiry, validation, and creative self-organization. . . . Combining technical with organizational systems appropriately enables communities to face complex events more effectively by monitoring changing conditions and adapting its performance accordingly, increasing the efficiency of its use of limited resources. *It links human capacity to learn with the technical means to support that capacity in complex, dynamic environments* [emphasis added].

Similarly, research stressing the importance of EMONs as the predominant organizational form during crisis response periods points to the importance of improving strategies for network management and of developing better methods to take advantage of emergent structures and activities during disasters. Planning and management approaches must, in other words, support rather than interfere with the open and dynamic qualities of disaster response activities. Indicators of improved capacity to manage emergent networks could include the diversity of organizations and community sectors involved in pre-crisis planning; plans and agreements facilitating the incorporation of the voluntary sector and emergent citizen groups into response activities; plans and tools enabling the rapid expansion of crisis communication and information-sharing networks during disasters to include new organizations; and protocols, such as mutual aid agreements, making it possible for new actors to more easily join response networks (Tierney and Trainor, 2004).

In the wake of the Katrina disaster, the need for disaster management by command-and-control-oriented entities has once again achieved prominence. For example, calls have increased for greater involvement on the part of the military in domestic disaster management. Such recommendations are not new. Giving a larger role in disaster management to the military was an idea that was considered—and rejected—following Hurricane Andrew (National Academy of Public Administration, 1993). Post-Katrina debates on needed policy and programmatic changes will likely continue to focus on how to most effectively deploy military assets while ensuring that disaster management remains the responsibility of civilian institutions.

## Additional Considerations: Do Responses to Natural, Technological, and Human-Induced Events Differ?

One issue that has come to the fore with the emergence of terrorism as a major threat involves the extent to which findings from the field of disaster research can predict responses to human-induced extreme events. Although some take the position that terrorism and bioterrorism constitute such unique threats that behavioral and organizational responses in such events will differ from what has been documented for other types of extreme events, others contend that this assumption is not borne out by social science disaster research.

The preponderance of evidence seems to suggest that there is more similarity than difference in response behaviors across different types of disaster agents. Regarding the potential for panic, for example, there is no empirical evidence that panic was a problem during the influenza pandemic of 1918, among populations under attack during World War II (Janis, 1951), in catastrophic structure fires and crowd crushes (Johnson, 1987; Johnson et al., 1994; Feinberg and Johnson, 2001), or in the Chernobyl nuclear disaster (Medvedev, 1990). Nor was panic a factor in the 1993 bombing of the World Trade Center (Aguirre et al., 1998), the 1995 Tokyo subway sarin attack (Murakami, 2000), or the terrorist attacks of September 11, 2001 (NIST, 2005; National Commission on Terrorist Attacks upon the United States, 2004). The failure to find significant evidence of panic across a wide range of crisis events is a testimony to the resilience of social relationships and normative practices, even under conditions of extreme peril.

Similarly, as noted earlier, research findings on challenges related to risk communication and warning the public of impending extreme events are also quite consistent across different types of disaster events. For individuals and groups, there are invariably challenges associated with understanding what self-protective actions are required for different types of emergencies, regardless of their origin.

In all types of disasters, organizations must likewise face a common set of challenges associated with situation assessment, the management of primary and secondary impacts, communicating with one another and with the public, and dealing with response-related demands. The need for more effective communication, coordination, planning, and training transcends hazard type. Although recent government initiatives such as the National Response Plan will result in the incorporation of new organizational actors into response systems for extreme events, most of the same local-, state-, and federal-level organizations will still be involved in managing extreme events of all types, employing common management frameworks such as the Incident Command System and now the National Incident Management System (NIMS).

Social scientific studies on disasters have long shown that general features of extreme events, such as geographic scope and scale, impact severity, and speed of onset, combined with the overall quality of pre-disaster preparedness, have a greater influence on response patterns than do the specific hazard agents that trigger response activities. Regardless of their origins, very large, near-catastrophic, and catastrophic events all place high levels of stress on response systems.

In sum, social science disaster research finds little justification for the notion that individual, group, and community responses to human-induced extreme events, including those triggered by weapons of mass terror, will differ in important ways from those that have been documented in natural and technological disasters. Instead, research highlights the importance of a variety of general factors that affect the quality and effectiveness of responses to disasters, irrespective of the hazard in question. With respect to warning the public and encouraging self-protective action, for example, warning systems must be well designed and warning messages must meet certain criteria for effectiveness, regardless of what type of warning is issued. Members of the public must receive, understand, and personalize warning information; must understand what actions they need to take in order to protect themselves; and must be able to carry out those actions, again regardless of the peril in question. Community residents must feel that they can trust their leaders and community institutions during crises of all types. For organizations, training and exercises and effective mechanisms for interorganizational communication and coordination are critical for community-wide emergencies of all types. When such criteria are not met, response-related problems can be expected regardless of whether the emergency stems from a naturally occurring event, a technological accident, or an intentional act.

Individual and group responses, as well as organizational response challenges, are thus likely to be consistent across different types of crises. At the same time, however, it is clear that there are significant variations in the behavior of responding institutions (as opposed to individuals, groups, and first responders) according to event type. In most technological disasters, along with the need to help those affected, questions of negligence and liability typically come to the fore, and efforts are made to assign blame and make responsible parties accountable. In terrorist events, damaged areas are always treated as crime scenes, and the response involves intense efforts both to care for victims and to identify and capture the perpetrators. Further, although as noted earlier, scapegoating can occur in disasters of all types, the tendency for both institutions and the public to assign blame to particular groups may be greater in technological and terrorism-related crises than in natural disasters.[4]

Finally, with respect to responses on the part of the public, even though evidence to the contrary is strong, the idea that some future homeland security emergencies could engender responses different from those observed in past natural, technological, and intentional disasters cannot be ruled out entirely. The concluding section of this chapter highlights the need for further research in this area.

## Research on Disaster Recovery

Like hazards and disaster research generally, NERHRP-sponsored research has tended to focus much more on preparedness and response than on either mitigation or disaster recovery. This is especially the case with respect to long-term recovery, a topic that despite its importance has received very little emphasis in the literature. However, even though the topic has not been well studied, NEHRP-funded projects have done a great deal to advance social science understanding of disaster recovery. As discussed later in this section, they have also led to the development of decision tools and guidance that can be used to facilitate the recovery process for affected social units.

It is not an exaggeration to say that prior to NEHRP, relatively little was known about disaster recovery processes and outcomes at different levels of analysis. Researchers had concentrated to some degree on analyzing the impacts of a few earthquakes, such as the 1964 Alaska and 1971 San Fernando events, as well as earthquakes and other major disasters outside the United States. Generally speaking, however, research on recovery was quite sparse. Equally important, earlier research oversimplified the recovery process in a variety of ways. First, there was a tendency to equate recovery, which is a social process, with reconstruction, which involves restoration and replacement of the built environment. Second, there was an assumption that disasters and their impacts proceed in a temporal, stage-like fashion, with "recovery" following once "response" activities have

---

[4]At the same time, consistent with positions taken elsewhere in this report, it is important to recognize that in crises of all kinds, blame and responsibility are socially constructed. For example, although triggered by a natural disaster, the levee failures during Hurricane Katrina are increasingly being defined as the result of human error. The disaster itself is also framed as resulting from catastrophic failures in decision making at all levels of government (Select Bipartisan Committee to Investigate the Preparation for and Response to Hurricane Katrina, 2006). While the connections are obviously clearer in crisis caused by willful attacks, it is now widely recognized that human agency is involved in disastrous events of all types—including not only terrorist events but also technological and natural disasters.

been concluded.[5] Earlier research also underemphasized the extent to which recovery may be experienced differently by different sectors and subpopulations within society. Some of these problems were related to the fact that at a more abstract level, earlier work had not sufficiently explored the concept of recovery itself—for example, whether recovery should be equated with a return to pre-disaster circumstances and social and economic activities, with the creation of a "new normal" that involves some degree of social transformation, or with improvements in community sustainability and long-term disaster loss reduction. Since the inception of NEHRP and in large measure because of NEHRP sponsorship, research has moved in the direction of a more nuanced understanding of recovery processes and outcomes that has not entirely resolved but at least acknowledges many of these issues.

The sections that follow discuss significant contributions to knowledge and practice that have resulted primarily from NEHRP-sponsored work. Those contributions can be seen (somewhat arbitrarily) as falling into four categories: (1) refinements in definitions and conceptions of disaster recovery, along with a critique and reformulation of stage-like models; (2) contributions to the literature on recovery processes and outcomes across different social units; (3) the development of empirically based models to estimate losses, anticipate recovery challenges, and guide decision making; and (4) efforts to link disaster recovery with broader ideas concerning long-term sustainability and environmental management.

**Conceptual Clarification.** Owing in large measure to NEHRP-sponsored efforts, the disaster field has moved beyond equating recovery with reconstruction or the restoration of the built environment. More usefully, research has moved in the direction of making analytic distinctions among different types of *disaster impacts, recovery activities* undertaken by and affecting different social units, and *recovery outcomes.* Although disaster *impacts* can be positive or negative, research generally tends to focus on various negative impacts occurring at different levels of analysis. As outlined in Chapter 3, these impacts include effects on the physical and built environment, including residential, commercial, and infrastructure damage as well as disaster-induced damage to the environment; other property losses; deaths and injuries; impacts on social and economic activity; effects at the community level, such as impacts on community cohesiveness and urban

---

[5]For example, Drabek's *Human System Responses to Disaster* (1986), which is organized according to disaster "stages," discusses short-term recovery in a chapter entitled "Restoration" and longer-term recovery in a chapter called "Reconstruction." Those two chapters address topics ranging from sheltering, looting, and emergent groups to mental health impacts, conflict during the recovery period, and organizational and community change.

form; and psychological, psychosocial, and political impacts. Such impacts can vary in severity and duration, as well as in the extent to which they are addressed effectively during the recovery process. An emphasis on recovery as a multidimensional concept calls attention to the fact that physical and social impacts, recovery trajectories, and short- and longer-term outcomes in chronological and social time can vary considerably across social units.

Recovery *activities* constitute measures that are intended to remedy negative disaster impacts, restore social units as much as possible to their pre-disaster levels of functioning, enhance resilience, and ideally, realize other objectives such as the mitigation of future disaster losses and improvements in the built environment, quality of life, and long-term sustainability.[6] Recovery activities include the provision of temporary and replacement housing; the provision of resources (government aid, insurance payment, private donations) to assist households and businesses with replacement of lost goods and with reconstruction; the provision of various forms of aid and assistance to affected government units; the development and implementation of reconstruction and recovery plans in the aftermath of disasters; coping mechanisms developed by households, businesses, and other affected social units; the provision of mental health and other human services to victims; and other activities designed to overcome negative disaster impacts. In some circumstances, recovery activities can also include the adoption of new policies, legislation, and practices designed to reduce the impacts of future disasters.

Recovery processes are significantly influenced by differential societal and group vulnerability; by variations in the range of recovery aid and support that is available; and by the quality and effectiveness of the help that is provided. The available "mix" of recovery activities and post-disaster coping strategies varies across groups, societies, and different types of disasters. For example, insurance is an important component in the reconstruction and recovery process for some societies, some groups within society, and some types of disasters, but not for others.

Recovery *outcomes*—or the extent to which the recovery activities are judged, either objectively or subjectively, as "complete" or "successful"— also show wide variation across societies, communities, social units, and disaster events. Outcomes can be assessed in both the short and the longer terms, although, as noted earlier, the literature is weak with respect to empirical studies on the outcomes of longer-term disasters. Additionally,

---

[6]The word "intended" is used here purposely, to highlight the point that the recovery process involves decisions made and actions carried out to remedy the problems that disasters create. Such decisions and actions can be made by governments, private sector entities, groups, households, and individuals.

outcomes consist not only of the intended effects of recovery programs and activities, but also of their unintended consequences. For example, the provision of government assistance or insurance payments to homeowners may make it possible for them to rebuild and continue to live in hazardous areas, even though such an outcome was never intended.

Keeping in mind the multidimensional nature of recovery, post-disaster outcomes can be judged as satisfactory along some dimensions, or at particular points in time, but unsatisfactory along others. Outcomes are perceived and experienced differently, when such factors as level of analysis and specific recovery activities of interest are taken into account. With respect to units of aggregation, for example, while a given disaster may have few discernible long-term effects when analyzed at the community level, the same disaster may well be economically, socially, and psychologically catastrophic for hard-hit households and businesses. A community may be considered "recovered" on the basis of objective social or economic indicators, while constituent social units may not be faring as well, in either objective or subjective terms. The degree to which recovery has taken place is thus very much a matter of perspective and social position.

In a related vein, research has also led to a reconsideration of linear conceptions of the recovery process. Past research tended to see disaster events as progressing from the pre-impact period through post-impact emergency response, and later recovery. In a classic work in this genre—*Reconstruction Following Disaster* (Haas et al., 1977:xxvi), for example—the authors argued that disaster recovery is "ordered, knowable, and predictable." Recovery was characterized as consisting of four sequential stages that may overlap to some degree: the emergency period; the restoration period; the replacement reconstruction period; and the commemorative, betterment, and developmental reconstruction period. In this and other studies, the beginning of the recovery phase was generally demarcated by the cessation of immediate life saving and emergency care measures, the resumption of activities of daily life (e.g., opening of schools), and the initiation of rebuilding plans and activities. After a period of time, early recovery activities, such as the provision of temporary housing, would give way to longer-term measures that were meant to be permanent. Kates and Pijawka's (1977) frequently cited four-phase model begins with the *emergency* period, lasting for a few days up to a few weeks, and encompassing the period when the emergency operations plan (EOP) is put into operation. Next comes the *restoration* period—when repairs to utilities are made; debris is removed; evacuees return; and commercial, industrial, and residential structures are repaired. The third phase, the *reconstruction replacement* period, involves rebuilding capital stocks and getting the economy back to pre-disaster levels. This period can take some years. Finally, there is the *development* phase, when commemorative structures are built, memorial dates are institutionalized in social time, and attempts are made to improve the community.

In another stage-like model focusing on the community level, Alexander (1993) identified three stages in the process of disaster recovery. First, the *rehabilitation* stage involves the continuing care of victims and frequently is accompanied by the reemergence of preexisting problems at the household or community level. During the *temporary reconstruction* stage, prefabricated housing or other temporary structures go up, and temporary bracing may be installed for buildings and bridges. Finally, the *permanent reconstruction* stage was seen as requiring good administration and management to achieve full community recovery.

Later work sees delineations among disaster phases as much less clear, showing, for example, that decisions and actions that affect recovery may be undertaken as early as the first days or even hours after the disaster's impact—and, importantly, even before a disaster occurs. The idea that recovery proceeds in an orderly, stage-like, and unitary manner has been replaced by a view that recognizes that the path to recovery is often quite uneven. While the concept of disaster phases may be a useful heuristic device for researchers and practitioners, the concept may also mask both how phases overlap and how recovery proceeds differently for different social groups (Neal, 1997). Recovery does not occur at the same pace for all who are affected by disasters or for all types of impacts. With respect to housing, for example, owing to differences in the availability of services and financing as well as other factors, some groups within a disaster-stricken population may remain in "temporary housing" for a very long time—so long, in fact, that those housing arrangements become permanent—while others may move rapidly into replacement housing (Bolin, 1993a). Put another way, as indicated in Chapters 1 and 3, while stage-like approaches to disasters are framed in terms of chronological time, for those who experience them, disasters unfold in social time.

Researchers studying recovery continue to contend with a legacy of conceptual and measurement difficulties. One such difficulty centers on the question of how the dependent variable should be measured. This problem itself is multifaceted. Should recovery be defined as a return to pre-disaster levels of psychological, social, and economic well-being? As a return to where a community, business, or household would have been were it not for the occurrence of the disaster? The study of disaster recovery also tends to overlap with research on broader processes of social change. Thus, in addition to focusing on what was lost or affected as a consequence of disaster events and on outcomes relative to those impacts, recovery research also focuses on more general post-disaster issues, such as the extent to which disasters influence and interact with ongoing processes of social change, whether disaster impacts can be distinguished from those resulting from broader social and economic trends, whether disasters simply magnify and accelerate those trends or exert an independent influence, and the extent to which the post-disaster recovery period represents continuity or discontinuity with the past. Seen in this light, the study of recovery can become indistinguishable from the study of longer-term social change affecting communities and societies. While these distinctions are often blurred, it is nevertheless important to differentiate conceptually and empirically between the recovery process, specific recovery outcomes of interest, and the wide range of other changes that might take place following (or as a consequence of) disasters.

**Analyzing Impacts and Recovery Across Different Social Units.** Following from the discussions above, it is useful to keep in mind several points about research on disaster recovery. First, studies differ in the extent to which they emphasize the objective, physical aspects of recovery—restoration and reconstruction of the built environment—or subjective, psychosocial, and experiential ones. Second, studies generally focus on particular units of analysis and outcomes, such as household, business, economic, or community recovery, rather than on how these different aspects of recovery are interrelated. This is due partly to the fact that researchers tend to specialize in particular types of disaster impacts and aspects of recovery, which has both advantages and disadvantages. While allowing for the development of in-depth research expertise, such specialization has also made it more difficult to formulate more general theories of recovery. Third, the literature is quite uneven. Some aspects of recovery are well understood, while there are others about which very little is known.

Even with these limitations, more general theoretical insights about recovery processes and outcomes have begun to emerge. Key among these is the idea that disaster impacts and recovery can be conceptualized in terms of *vulnerability and resilience.* As noted in Chapters 2 and 3, *vulnerability* is a consequence not only of physical location and the "hazardousness of place," but also of social location and of societal processes that advantage some groups and individuals while marginalizing others. The notion of vulnerability applies both to the likelihood of experiencing negative impacts from disasters, such as being killed or injured or losing one's home or job, and to the likelihood of experiencing recovery-related difficulties, such as problems with access to services and other forms of support. Social vulnerability is linked to broader trends within society, such as demographic trends (migration to more hazardous areas, the aging of the U.S. population) and population diversity (race, class, income, and linguistic diversity). Similarly, *resilience*, or the ability to survive and cope with disaster impacts and rebound after those events, is also determined in large measure by social factors. According to Rose (2004), resilience can be conceptualized as both *inherent* and *adaptive,* where the former term refers to resilience that is based on resources and options for action that are typically available during nondisaster times, and the latter refers to the ability to mobilize resources and create new options following disasters.[7] As discussed in Chapter 6, resilience stems in part from factors commonly associated with the concept of social capital, such as the extensiveness of social networks, civic engagement, and interpersonal, interorganizational, and institutional trust. (For an influential formulation setting out the vulnerability perspective, see Blaikie et al., 1994). As subsequent discussions show, the concepts of vulnerability and resilience are applicable to individuals, households, groups, organizations, economies, and entire societies affected by disasters. The sections that follow, which are organized according to unit of analysis, discuss psychosocial impacts and recovery; impacts and recovery processes for housing and businesses; economic recovery; and community-level and societal recovery.

**Psychological Impacts and Recovery.** There is no disagreement among researchers that disasters cause genuine pain and suffering and that they can be deeply distressing for those who experience them. Apart from that consensus, however, there have been many debates and disputes regarding the psychological and psychosocial impacts of disasters. One such debate centers on the extent to which disasters produce clinically significant symptoms of psychological distress and, if so, how long such symptoms last. Researchers have also struggled with the questions of etiology, or the causes of disaster-related psychological reactions. Are such problems the direct result of trauma experienced during disaster, the result of disaster-induced stresses, a reflection of a lack of coping capacity or weak social support networks, a function of preexisting vulnerabilities, or a combination of all these factors? Related concerns center on what constitute appropriate forms of intervention and service delivery strategies for disaster-related psychological problems. Do people who experience problems generally recover on their own, without the need for formally provided assistance, or does such assistance facilitate more rapid and complete recovery? What types of assistance are likely to be most efficacious and for what types of problems?

Research has yielded a wide array of findings on questions involving disaster-related psychological and psychosocial impacts and recovery. Findings tend to differ depending upon disaster type and severity, how disaster victimization is defined and measured, how mental health outcomes are measured, the research methodologies and strategies used (e.g., sampling,

---

[7]Rose was referring specifically to economic resilience, but the concepts of inherent and adaptive resilience can be (and indeed have been) applied much more broadly.

timing, variables of interest), and not inconsequentially, the discipline-based theoretical perspectives employed (Tierney, 2000). With respect to the controversial topic of post-traumatic stress disorder (PTSD), for example, well-designed epidemiological studies have estimated the lifetime prevalence of PTSD at around 5.4 percent in the U.S. population. An important epidemiologic study on the incidence of trauma and the subsequent risk of developing PTSD after various types of traumatic events estimates the risk at about 3.8 percent for natural disasters (Breslau et al., 1998; Kessler and Zhao, 1999). NEHRP-sponsored surveys following recent earthquakes in California found PTSD to be extremely rare among affected populations and not significantly associated with earthquake impacts (Seigel et al., 2000). Other studies show immense variation, with estimates of post-disaster PTSD ranging from very low to greater than 50 percent. Such variations could reflect real differences in the traumatic effects of different events, but it is equally likely that they are the result of methodological, measurement, and theoretical differences among investigators.

One key debate centers on the clinical significance of post-disaster emotional and mental health problems. Research is clear on the point that it is not unusual for disaster victims to experience a series of problems, such as headaches, problems with sleeping and eating, and heightened levels of concern and anxiety, that can vary in severity and duration (Rubonis and Bickman, 1991; Freedy et al., 1994). Perspectives begin to diverge, however, on the extent to which these and other disaster-induced symptoms constitute mental health problems in the clinical sense. In other words, would disaster victims, presenting their symptoms, be considered candidates for mental health counseling or medication if those symptoms were present in a nondisaster context? Do their symptoms correspond to survey based or clinically based measures of what constitutes a "case" for psychiatric diagnostic purposes? Again, as with PTSD, findings differ. While noting that many studies do document a rise in psychological distress following disasters, Shoaf et al. (2004:320) conclude that "those impacts are not of a nature that would significantly increase the rates of diagnosable mental illness." With respect to severe psychological impacts, these researchers found that suicide rates declined in Los Angeles County following the Northridge earthquake—a continuation of a trend that had already begun before that event. They also note that these findings are consistent with research on suicide following the Kobe earthquake, which showed that the suicide rate in the year following that quake was less than the average rate for the previous 10 years (Shoaf et al., 2004). Yet many researchers and practitioners rightly contend that psychosocial interventions are necessary following disasters, both to address clinically significant symptoms and to prevent more serious psychological sequelae.

There is also the question of whether some types of disasters are more likely than others to cause negative psychological impacts. Some researchers argue that certain types of technological hazards, such as nuclear threats and chronic exposures to toxic substances, are more pernicious in their effects than natural disasters because they persist longer and create more anxiety among potential victims, and especially because they tend to result in community conflict, causing "corrosive" rather than "therapeutic" communities to develop (Erikson, 1994). Events such as the Oklahoma City bombing, the Columbine school shootings, and the events of September 11, 2001 lead to questions about whether intentional attacks engender psychological reactions that are distinctive and different from those that follow other types of community crisis events. Some studies have suggested that the psychological impacts of terrorist attacks are profound, at least in the short term (North et al., 1999). Other research, focusing specifically on the short-term impacts of the September 11, 2001 terrorist attacks, indicates that the psychological impacts resulting from the events of 9/11 "are consistent with prior estimates of the impact of natural disasters and other terrorist events" (Miller and Heldring, 2004:21). Again, drawing conclusions about the relative influence of agent characteristics—as opposed to other factors—is difficult because studies vary so much in their timing, research designs, methodological approaches, and procedures for defining disaster victimization.

Another set of issues concerns factors associated with risk for poor psychological outcomes. Perilla et al. (2002) suggest that such outcomes can vary as a consequence of both *differential exposure* and *differential vulnerability* to extreme events. With respect to differential exposure, factors such as ethnicity and social class can be associated with living in substandard and vulnerable housing, subsequently exposing minorities and poor people to greater losses and disaster-related trauma. Regarding differential vulnerability, minorities and the poor, who are more vulnerable to psychosocial stress during nondisaster times, may also have fewer coping resources upon which to draw following disasters.

In a comprehensive and rigorous review of research on the psychological sequelae of disasters, Fran H. Norris and her colleagues (Norris et al., 2002a,b) carried out a meta-analysis of 20 years of research, based on 160 samples containing more than 60,000 individuals who had experienced 102 different disaster events. These data sets included a range of different types of surveys on both U.S. disaster victims and individuals in other countries, on various subpopulations, and on disasters that differed widely in type and severity. Impacts documented in these studies included symptoms of post-traumatic stress, depression, and anxiety; other forms of nonspecific distress not easily related to specific syndromes such as PTSD; health problems and somatic complaints; problems in living, including secondary stressors such as work-related and financial problems; and "psychosocial resource loss," a term that refers to negative effects on coping capacity, self-esteem, feelings of self-efficacy, and other attributes that buffer the effects of stress. According to their interpretation, which was based on accepted methods for rating indicators of psychological distress, the symptoms reported by as many as 39 percent of those studied reached clinically significant levels. However—and this is an important caveat—they found negative psychological effects to be much more prevalent in disasters occurring outside the United States. Generally, symptoms were most severe in the year following disaster events and declined over time.

Norris et al. (2002a, 2002b) classified U.S. disasters as low, moderate, and high in their psychosocial impacts, based on empirical data on postdisaster distress. The Loma Prieta and Northridge earthquakes were seen as having relatively few adverse impacts, and Hurricane Hugo and Three Mile Island were classified as moderate in their effects. Hurricane Andrew, the *Exxon* oil spill, and the Oklahoma City bombing were classified as severe with respect to their psychological impacts. As these examples suggest, the researchers found no evidence that natural, technological, and human-induced disasters necessarily differ in their effects.

This research review uncovered a number of vulnerability and protective factors that were associated with differential psychological outcomes following disasters. Broadly categorized, those risk factors most consistently shown to be negatively associated with post-disaster psychological well-being include severity of disaster exposure at both the individual and the community levels; being female; being a member of an ethnic minority; low socioeconomic status; experiencing other stressors or chronic stress; having had other mental health problems prior to the disaster; employing inappropriate coping strategies (e.g., withdrawal, avoidance); and reporting problems with both perceived and actual social support.

Overall, these findings are very consistent with perspectives in disaster research that emphasize the relationship between systemically induced vulnerability, negative disaster impacts, lower resilience, and poor recovery outcomes. Recent research situates disasters within the context of other types of stressful events (e.g., death of a loved one or other painful losses) that disproportionately affect those who are most vulnerable and least able to cope. At the same time, studies—many conducted under NEHRP auspices—show how social inequality and vulnerability both amplify the stress that results directly from disasters and complicate the recovery process over the longer term. For example, Fothergill (1996, 1998, 2004) and Enarson and Morrow (1998) have documented the ways in which gender is associated both with the likelihood of becoming a disaster victim and with a variety of subsequent post-disaster stressors. Peacock et al. (1997) and Bolin and Stanford (1998) have shown how pre-disaster conditions such as income disparities and racial and ethnic discrimination contribute both to disaster losses and to subsequent psychosocial stress and make recovery more difficult for vulnerable groups. Perilla et al. (2002), who studied ethnic differences in post-traumatic stress following Hurricane Andrew, also note that ethnicity can be associated with variations in personality characteristics such as fatalism, which tends to be associated with poor psychosocial outcomes resulting from stressful events, as well as with additional stresses associated with acculturation.[8]

Hurricane Katrina represents a critical test case for theories and research on psychosocial vulnerability and resilience. If, as Norris and her collaborators indicate, Hurricane Andrew resulted in relatively high levels of psychosocial distress, what will researchers find with respect to Katrina? For many victims, Katrina appears to contain all of the ingredients necessary to produce negative mental health outcomes: massive, catastrophic impacts; high property losses resulting in financial distress; exposure to traumas such as prolonged physical stress and contact with dead and dying victims; disruption of social networks; massive failures in service delivery systems; continual uncertainty about the future; and residential dislocation on a scale never seen in a U.S. disaster. Over time, research will result in important insights regarding the psychosocial dimensions of truly catastrophic disaster events.

**Household Impacts and Recovery.** Within the disaster recovery area, households and household recovery have been studied most often, with a significant proportion of that work focusing on post-earthquake recovery issues. Although this line of research predates NEHRP, many later studies have been undertaken with NEHRP support. Studies conducted prior to NEHRP include Bolin's research on household recovery processes following the Managua earthquake and the Rapid City flood, both of which occurred in 1972 (Bolin, 1976). Drabek and Key and their collaborators had also examined disaster impacts on families and the household recover process (Drabek et al., 1975; Drabek and Key, 1976, 1984). With NEHRP support, Bolin and Bolton studied household recovery following tornadoes in Wichita Falls, Vernon, and Paris, Texas; a hurricane in Hawaii; flooding in Salt Lake City; and the Coalinga earthquake (Bolin, 1982; Bolin and Bolton, 1986). Bolin's monograph *Household and Community Recovery after*

---

[8]This study found significant differences in post-disaster psychological well being among Caucasians, Latinos, and African Americans, with minority group members experiencing poorer outcomes. Interestingly, differences were seen between Latinos whose preferred language was English and those who preferred to speak Spanish. The latter experienced more overall psychological distress, while the reactions of the former more closely resembled those of their Caucasian counterparts.

*Earthquakes* was based on research on the 1987 Whittier Narrows and 1989 Loma Prieta events (Bolin, 1993b). Households have also been the focus of more recent studies on the impacts of Hurricane Andrew (Peacock et al., 1997) and the 1994 Northridge earthquake (Bolin and Stanford, 1998). Other NEHRP-sponsored work has focused more specifically on issues that are important for household recovery, such as post-disaster sheltering processes (Phillips, 1993, 1998) and housing impacts and recovery (Comerio, 1997, 1998). As Bolin (1993a:13) observes

> [d]isasters can have a multiplicity of effects on a household, including physical losses to property, injury and/or death, loss of job or livelihood, disruption of social and personal relations, relocation of some or all members of a family, physical disruption or transformation of community and neighborhood, and increased household indebtedness.

Accordingly, the literature has explored various dimensions of household impacts and recovery, including direct impacts such as those highlighted by Bolin; changes in the quality and cohesiveness of relationships among household members; post-disaster problems such as conflict and domestic violence; stressors that affect households during the recovery process; and coping strategies employed by households, including the use of both formal and informal sources of post-disaster support and recovery aid.

The literature also points to a number of factors that are associated with differences in short- and longer-term household recovery outcomes. Housing supply is one such factor—as indicated, for example, by housing costs, other real estate market characteristics, and rental vacancy rates Temporary housing options are affected by such factors as the proximity of friends and relatives with whom to stay, although use of this housing option is generally only a short-term strategy. Extended family members may not be able to help if they also are victims (Morrow, 1997). Such problems may be more prevalent in lower-income groups that have few alternative resources and when most members of an extended family live in the same affected community.

Availability of temporary and permanent housing generally is limited by their pre-impact supply in and near the impact area. In the U.S., in situations in which there is an insufficient supply of housing for displaced disaster victims, FEMA provides mobile homes, but even this expedient method of expanding the housing stock takes time. Even when houses are only moderately damaged, loss of housing functionality may be a problem if there is massive disruption of infrastructure. In such cases, tent cities may be necessary if undamaged housing is beyond commuting range (e.g., Homestead, Florida after Hurricane Andrew, as discussed in Peacock et al., 1997).

In the longer term, household recovery is influenced by such factors as household financial resources, the ability to obtain assistance from friends and relatives, insurance coverage, and the mix of housing assistance programs available to households. Typically, access to and adequacy of recovery resources are inversely related to socioeconomic status. Those with higher incomes are more likely to own their own homes, to be adequately insured, and to have savings and other financial resources on which to draw in order to recover—although disasters can also cause even better-off households to take on additional debt. With respect to formal sources of aid, the assistance process generally favors those who are adept at responding to bureaucratic requirements and who are able to invest time and effort to seek out sources of aid. The aid process also favors those living in more conventional, nuclear family living arrangements, as opposed to extended families or multiple households occupying the same dwelling unit (Morrow, 1997). Recovery may be particularly difficult for single-parent households, especially those headed by women (Enarson and Morrow, 1998; Fothergill, 2004).

The picture that emerges from research on household recovery is not that of a predictable and stage-like process that is common to all households, but rather of a multiplicity of recovery trajectories that are shaped not only by the physical impacts of disaster but also by axes of stratification that include income, race, and ethnicity, as well as such factors as the availability of and access to different forms of monetary aid, other types of assistance, and informal social support—which are themselves associated with stratification and diversity. Disaster severity matters, both because disasters that produce major and widespread impacts can limit recovery options for households and because they tend to be more damaging to the social fabric of the community. As Comerio's extensive research on housing impacts and issues following earthquakes and other disasters in different societal contexts illustrates, household recovery processes are also shaped by societal-level policy and institutional factors—which themselves have differential impacts (Comerio, 1998).[9]

**Large-Scale Comparative Research on Household Recovery.** Although there is clearly a need for such research, few studies exist that compare household recovery processes and outcomes across communities and disaster events. With NEHRP funding, Frederick Bates and his colleagues carried out what may well be the largest research efforts of this kind: a multicommunity

---

[9]Importantly, Comerio's work also highlights how policies themselves change and evolve in response to disasters and how these changes affect recovery options and outcomes in subsequent events. She shows, for example, that experience with deficiencies in housing programs after the Loma Prieta earthquake influenced the way in which programs were financed and managed in other major disasters, notably Hurricane Andrew.

longitudinal study on household and community impacts and recovery after the 1976 Guatemala earthquake and a cross-national comparative study on household recovery following six different disaster events. The Guatemala study, designed as a quasi-experiment, included households in 26 communities that were carefully selected to reflect differences in the severity of earthquake impacts, size, population composition, and region of the country. That study focused on a broad spectrum of topics, including changes over time in household composition and characteristics; household economic activity; housing characteristics and standards of living; household experiences with relief and reconstruction assistance; and fertility, health, and nutrition. Never replicated for any other type of disaster, the study provided detailed information on these topics, focusing in particular on how different forms of aid provision either facilitated or hampered household recovery (for detailed discussions, see Bates, 1982; Hoover and Bates, 1985; Bates et al., 1979).

The second study carried out by Bates and his colleagues extended methods developed to assess household recovery following the Guatemala earthquake to measure household recovery in disaster-stricken communities in six different countries. The tool used to measure disaster impacts and household recovery across different events and societies, the Domestic Assets Scale, made possible systematic comparisons with respect to one dimension of household recovery—the restoration of household possessions, tools, and technologies (Bates and Peacock, 1992, 1993).

**Vulnerability, Resilience, and Household Recovery.** Like the other aspects of recovery discussed here, what happens to households during and after disasters can be conceptualized in terms of vulnerability and resilience. With respect to vulnerability, social location is associated with the severity of disaster impacts for households. Poverty often forces people to live in substandard or highly vulnerable housing—manufactured housing is one example—leaving them more vulnerable to death, injury, and homelessness. As discussed in Chapter 3 with respect to disaster preparedness, factors such as income, education, and homeownership influence the ability of households to mitigate and prepare for disasters. Social-structural factors also affect the extent to which families can accumulate assets in order to achieve higher levels of safety, as well as their recovery options and access to resources after disasters strike—for example the forms of recovery assistance for which they are eligible. Households are thus differentially exposed to disasters, differentially vulnerable during the recovery period, and diverse in terms of both inherent and adaptive resilience.

## ECONOMIC AND BUSINESS IMPACTS AND RECOVERY: THE CHALLENGE OF ASSESSING DISASTER LOSSES

As discussed in Chapter 3, assessing how much disasters cost the nation and its communities has proven to be a major challenge. A National Research Council (NRC, 1999c) study concluded that such calculations are difficult in part because different agencies and entities calculate costs and losses differently. Moreover, no universally accepted standards exist for calculating economic impacts resulting from disasters, and there is no single agency responsible for keeping track of disaster losses. For any given disaster event, assessments of economic impacts may vary widely depending on which statistics are used—for example, direct or insured losses versus total losses.

NEHRP-sponsored research has addressed these problems to some degree. For example, as part of the NEHRP-sponsored "Second Assessment of Research on Natural Hazards," researchers attempted to estimates losses, costs, and other impacts from a wide array of natural and technological hazards.[10] For the 20 year period 1975–1994, they estimated that dollar losses from disasters amounted to $.5 billion per week, with climatological hazards accounting for about 80 percent of those losses; since 1989, losses have totaled $1 billion per week (Mileti, 1999a). Through work undertaken as part of the Second Assessment, data on losses from natural hazard events from the mid-1970s to 2000 are now available at the county level in geocoded form for the entire United States through the Spatial Hazard Events and Losses Database for the United States (SHELDUS). This data collection and database development effort has made it possible to analyze different types of losses, at different scales, using different metrics, and to assess locations in terms of their hazard proneness and loss histories. (For discussions of the data used in the SHELDUS database and associated challenges see Cutter, 2001.) What is still lacking is a national program to continue systematically collecting and analyzing impact and loss data.

Studies on economic impacts and recovery from earthquakes and other disasters can be classified according to the units of analysis on which they focus. Most research concerns economic losses and recovery at the community or, more frequently, the regional level. A smaller set of studies has analyzed economic impacts and recovery at the firm or facility level. There is even less research documenting national-level and macroeconomic impacts.

---

[10]However, it should be noted that, once again, those estimates were based on statistics from widely varied sources.

## Community-Level and Regional Studies

Studies on the economics of natural disasters at the community and regional levels of analysis differ significantly in methods, topics of interest, and conclusions. Some researchers, such as Rossi et al. (1978) and Friesema et al. (1979) have argued that at least in the United States, natural disasters have no discernible social or economic effects at the community level and that nondisaster-related trends have a far more significant influence on long-term outcomes than disasters themselves. This position has also been argued at the macroeconomic level, with respect to other developed and developing countries (Albala-Bertrand, 1993).[11] Dacy and Kunreuther (1969:168) even argued (although more than 30 years ago) that "a disaster may actually turn out to be a blessing in disguise" because disasters create reconstruction booms and allow community improvements to be made rapidly, rather than gradually. However, most research contradicts the idea that disasters constitute economic windfalls, emphasizing instead that economic gains that may be realized at one level (e.g., the community, particular economic sectors) typically constitute losses at another (e.g., the national tax base). One analyst has called the idea that disasters are beneficial economically "one of the most widely held misbeliefs in economics" (DeVoe, 1997:188).

Other researchers take the position that post-disaster economic and social conditions are generally consistent with pre-disaster trends, although disasters may amplify those changes (Bates and Peacock, 1993). Disasters may further marginalize firms and sectors of the economy that were already in decline, or they may speed up processes that were already under way prior to their occurrence. For example, Homestead Air Force Base was already slated for closure before Hurricane Andrew despite ongoing efforts to keep the base opened. When Andrew occurred, the base sustained damage and was closed for good. The closure affected businesses that had depended on the base and helped lead to the exodus of many middle-class families from the area, which in turn affected tax revenues in the impact region. These changes would have taken place eventually, but they were accelerated by Hurricane Andrew.

Related research has analyzed the distributive effects of earthquakes and other disasters. In an early formulation, Cochrane (1975) observed that lower-income groups consistently bear a disproportionate share of disaster losses, relative to higher-income groups. This theme continues to be promi-

---

[11]These findings refer to the impacts of disasters on societal-level economic indicators. Albala-Bertrand did document many instances in which disasters had both short- and longer-term political and economic impacts.

nent in the disaster literature; the notion that disasters create economic "winners and losers" has been borne out for both households and businesses (Peacock et al., 1997:Chapter 11; Tierney and Webb, forthcoming).

Another prominent research emphasis at the community and regional levels of analysis has grown out of the need to characterize and quantify the economic impacts of disasters (as well as other impacts) in order to be better able to plan for and mitigate those impacts. A considerable amount of NEHRP research on economic impacts and recovery has been driven by concern about the potentially severe economic consequences of major earthquakes, particularly those that could occur in highly populated urban areas. That concern is reflected in a number of NRC reports (1989, 1992, 1999c) on projected losses and potential economic impacts. Within the private sector, the insurance industry has also committed significant resources in an effort to better anticipate the magnitude of insured losses in future disaster events. (For new developments in research on the management of catastrophic insurance risk, see Grossi et al., 2004.)

Stimulated in large measure by NEHRP funding, new tools have been developed for both pre-disaster estimation of potential losses and post-disaster impact assessments, particularly for earthquakes. HAZUS, the national loss estimation methodology, which was originally developed for earthquakes and which has now been extended to flood and wind hazards, was formulated under FEMA's supervision with NEHRP funding. NEHRP funds have also supported the development of newer and more sophisticated modeling approaches through research undertaken at earthquake centers sponsored by the National Science Foundation (NSF).

The framework for estimating losses from natural hazards was initially laid out more than 20 years ago in publications such as Petak and Atkisson's *Natural Hazard Risk Assessment and Public Policy* (1982) and in applied studies such as the PEPPER (Pre-Earthquake Planning for Post-Earthquake Rebuilding) project (Spangle, 1987), which analyzed potential earthquake impacts and post-disaster recovery strategies for Los Angeles. According to the logic developed in these and other early studies (see, for example, NRC, 1989) and later through extensive NEHRP research, loss estimation consists of the analysis of scenario or probabilistic models that include data on *hazards; exposures*, or characteristics of the built environment at risk, including buildings and infrastructural systems; *fragilities*, or estimates of damage likelihood as a function of one or more parameters, such as earthquake shaking intensity; *direct losses*, such as deaths, injuries, and costs associated with damage; *and indirect losses and ripple effects* that result from disasters. Within this framework, recent research has focused on further refining loss models and reducing uncertainties associated with both the components of loss estimation models and their interrelationships (for representative work, see theme issue in *Earthquake Spectra*, 1997; Tierney et al., 1999; Okuyama and Chang, 2004).

This line of research has led both to advances in basic science knowledge and to a wide range of research applications. At the basic science level, loss modeling research—particularly studies supported through NEHRP—has helped distinguish and clarify relationships among such factors as physical damage, direct economic loss, business interruption effects, and indirect losses and ripple effects. For example, it is now more possible than ever before to disaggregate and analyze separately different types of economic effects and to understand how particular types of damage (e.g., damage to electrical power or transportation systems) contribute to overall economic losses. This research has shed light on factors that contribute to the resilience of regional economies, both during normal times and in response to sudden shocks. It has also shown how the application of newer economic modeling techniques, such as computable general equilibrium modeling and agent-based modeling, constitute improvements over more traditional input-output modeling, particularly for the study of extreme events (for discussions, see Rose et al., 2004; Chang, 2005; Rose and Liao, 2005). Econometric modeling provides another promising approach at both the micro and the regional levels (see West and Lenze, 1994), but this potential remains largely untapped.

At the applications level, loss estimation tools and products have proven useful for raising public awareness of the likely impacts of disaster events and for enhancing community preparedness efforts and mitigation programs. They have also made it possible to assess mitigation alternatives, not only in light of the extent to which those measures reduce damage, but also in terms of their economic costs and benefits. When applied in the disaster context, rapid economic loss estimates have also formed the basis for requests for federal disaster assistance. For the insurance industry, loss models provide important tools to improve risk management decision making, particularly with regard to catastrophic risks.

As noted earlier, loss modeling originally was driven by the need to better understand the economic impacts of earthquakes. In addition to economic losses, earthquake loss models are increasingly taking into account other societal impacts such as deaths, injuries, and residential displacement, as well as secondary effects such as earthquake-induced fires. The methodological approach developed to study earthquakes was first extended to other natural hazards and is now being used increasingly to assess potential impacts from terrorism. The nation is now better able to address the issue of terrorism-related losses because of the investments that had been made earlier for earthquakes and other natural hazards. Significantly, when the Department of Homeland Security decided in 2003 to begin funding university-based "centers of excellence" for terrorism research, the first topic that was selected for funding was risk and economic modeling for terrorist attacks in the United States.[12] Many of the investigators associated with that center had previously worked on loss modeling for earthquakes.

**Business and Facility-Level Impacts and Recovery.** Most research on recovery processes and outcomes has focused on households and communities. Prior to the 1990s, most research on the economic aspects of disasters focused not on individual businesses but rather on community-wide and regional impacts. Almost nothing was known about how private sector organizations are affected by and recover from disasters. Since then, a small number of studies have focused on business firms or, in some cases, commercial facilities, as units of analysis. Much of this work, including studies on large, representative samples of businesses, has been carried out with NEHRP support. Business impacts and recovery have been assessed following the Whittier Narrows, Loma Prieta, Northridge, and Kobe earthquakes; the 1993 Midwest floods; Hurricane Andrew; and other flood and hurricane events (for representative studies and findings, see Dahlhamer, 1998; Chang, 2000; Webb et al., 2000; Alesch et al., 2001). Long-term business recovery has been studied in the context of only two disaster events—the Loma Prieta earthquake and Hurricane Andrew (Webb et al., 2003).

These studies have shown that disasters disrupt business operations through a variety of mechanisms. Direct physical damage to buildings, equipment, vehicles, and inventories has obvious effects on business operation. It might be less obvious that disruption of infrastructure such as water/sewer, electric power, fuel (i.e., natural gas), transportation, and telecommunications frequently forces businesses to shut down in the aftermath of a disaster (Alesch et al., 1993; Tierney and Nigg, 1995; Tierney, 1997a, b; Webb et al., 2000). For example, Tierney (1997b) reported that extensive electrical power service interruption after the 1993 Midwest floods caused a large number of business closures in Des Moines, Iowa, even though the physical damage was confined to a relatively small area.

Other negative disaster effects include population dislocation, losses in discretionary income among those victims who remain in the impact area—which can weaken market demand for many products and services—and competitive pressure from large outside businesses. These kinds of impacts can cause small local businesses to experience major difficulties recovering from the aftermath of a disaster (Alesch et al., 2001). Indeed, such factors

---

[12]This research is being carried out by a consortium of universities, led by the University of Southern California. That consortium is called the Center for Risk and Economic Analysis of Terrorist Events (CREATE).

can produce business failures long after the precipitating event, especially if the community was already in economic decline before the disaster occurred (Bates and Peacock, 1993; Webb et al., 2003).

It is difficult to generalize on the basis of so few studies, particularly when the issues involved and the methodological challenges are so complex. However, studies to date have uncovered a few consistent patterns with respect to business impacts and recovery. First, studies show that most businesses do recover, and do so relatively quickly. In other words, typical businesses affected by disasters show a good deal of resilience in the face of major disruption.

Second, some businesses do tend to fare worse than others in the aftermath of disasters; clearly, not all businesses are equally vulnerable or equally resilient. Although findings from individual studies differ, the factors that seem to contribute most to vulnerability include small size; poor pre-disaster financial condition; business type, with wholesale and retail trade appearing to be especially vulnerable, while manufacturing and construction businesses stand to benefit most from disasters; and severity of disaster impacts. With regard to this last-mentioned factor, studies show that negative impacts on businesses include not only direct physical damage, lifeline-related problems, and business interruption, but also more long-lasting operational problems that businesses may experience following disasters, such as employee absenteeism and loss of productivity, earthquake-induced declines in demands for goods and services, and difficulties with shipping or receiving products and supplies.

Third, business recovery is affected by many factors that are outside the control of the individual business owner. For example, businesses located in highly damaged areas may experience recovery difficulties independent of whether or not they experience losses. In this case, recovery is complicated by the fact that disasters disrupt local ecologies on which individual businesses depend. Business recovery processes and outcomes are also linked to community-level decision making. After the Loma Prieta earthquake, for example, the City of Santa Cruz offered extensive support to businesses and used the earthquake as an opportunity to reinvent itself and to revitalize a business district that had fallen short of realizing its potential prior to the disaster (Arnold, 1998). Actions that communities take with respect to land-use, structural mitigation, infrastructure protection, community education, and emergency response planning also affect how businesses and business districts fare during and after disasters.

Fourth, recovery outcomes following disasters are linked to pre-disaster trends and broader market forces. For example, focusing on an important transport facility, the Port of Kobe, Chang (2000) showed that the port's inability to recover fully after the 1995 earthquake was due in part to losses in one part of the port's business—trans-shipment cargo—that had already been declining before the earthquake owing to severe competition from other ports in the region. Similarly, Dahlhamer (1998) found that businesses in the wholesale and retail trade sectors were more vulnerable to experiencing negative economic outcomes following the Northridge earthquake, perhaps because they constitute crowded and highly competitive economic niches and because turnover is high in those sectors during normal times. He also found that firms in industries that had been experiencing growth in the two-year period just before the earthquake were less likely than firms in declining industries to report being worse off following the Northridge event. Such findings are consistent with a more general theme in recovery research discussed earlier—that disasters do not generate change in and of themselves, but rather intensify or accelerate preexisting patterns.

**Community Recovery.** Although the topic of community recovery is still not well studied, significant progress has been made in understanding both recovery processes and factors that are associated with recovery outcomes for communities. Earlier research indicated that communities rebound well from disasters and that, at the aggregate level and net of other factors, the impacts of disasters are negligible (Friesema et al., 1979; Wright et al., 1979). However, other more recent research suggests that such findings paint an overly simplified and perhaps overly optimistic picture of post-disaster recovery. This may have been due to methodological shortcomings—for example, the tendency to aggregate data and to group together both more damaging disasters and those that did comparatively little damage—or because such studies were based on "typical" disasters in the United States, rather than catastrophic or near-catastrophic ones.[13] In contrast, in a methodologically sophisticated study focusing on a much more severe disaster, the 1995 Kobe event, Chang (2001) analyzed a number of recovery indicators, including measures of economic activity, employment in manufacturing, changes in the spatial distribution of work activities, and differences in recovery indicators among different districts within the city. She found that the earthquake did have lasting and significant negative effects on the City of Kobe. Equally important, poor recovery outcomes were more pronounced in some parts of the city than in others—specifically those areas that had already been experiencing declines. This study provides yet another illustration of how disasters exploit existing vulnerabilities. It also cautions against making blanket statements about disaster impacts and recovery.

---

[13]Additionally, recall that U.S. disasters began becoming more "disastrous" in the late 1980s. Both recent events (e.g., the 2004 hurricanes in Florida and Hurricanes Katrina and Rita) and scientific projections suggest that this trend will continue. It would thus be imprudent to overgeneralize from earlier work.

Another limitation of earlier work on community recovery was that it provided too little information on what actually happens in communities during the recovery process or what communities can do to ensure more rapid and satisfactory recovery outcomes. Later research, much of which has been undertaken with NEHRP support, has addressed these issues. For example, in *Community Recovery from a Major Natural Disaster*, Rubin et al. (1985) developed a set of propositions regarding factors that affect community recovery outcomes. That monograph, which was based on case study analyses of recovery following 14 disasters that occurred in the early 1980s, emphasized the importance of three general constructs—personal leadership, knowledge of appropriate recovery actions, and ability to act—as well as the influence of intergovernmental (state and federal) policies and programs. This work highlighted the effects of both government decision making and broader societal policies on community recovery.

Some more recent research has more explicitly incorporated community and population vulnerability as factors affecting community-level recovery. Bolin and Stanford (1998) traced how the post-Northridge recovery experiences of Los Angeles and smaller outlying towns differed as a function of such factors as political expertise and influence, preexisting plans, institutional capacity, involvement of community organizations, and interest group competition. In these diverse communities, the needs of more vulnerable and marginalized groups were sometimes addressed during the recovery process. However, recovery programs ultimately did little to improve the safety of those groups, because they failed to address the root causes of vulnerability (Bolin and Stanford, 1998:216):

> [s]ince vulnerability derives from political, economic, and social processes that deny certain people and groups access or entitlements to incomes, housing, health care, political rights, and, in some cases, even food, then post-disaster rebuilding by itself will have little effect on vulnerability.

**Societal-Level and Comparative Research on Disaster Recovery.** International research on disasters is discussed in greater detail in Chapter 6. This chapter focuses in a more limited way on what little research exists on disaster impacts and post-disaster change at the societal level. Regarding long-term societal impacts, researchers have generally found that disasters, even very large ones, typically do not in and of themselves result in significant change in the societies they affect. Instead, the broad consensus has been that to the extent disasters do have lasting effects, it is because they interact with other factors to accelerate changes that were already under way. Albala-Bertrand, for example has argued that while disasters can highlight preexisting political conflicts, whether such effects are sustained over time "has little to do with the disaster itself, but with preexisting economic and sociopolitical conditions" (1993:197). This research found that the potential for such changes was generally greater in developing countries than developed ones, although not great in any case.

With respect to the political impacts of disasters at the societal level, comparing very large disasters that occurred between 1966 and 1980, political scientist Richard Olson found that that major disasters can result in higher levels of political unrest, particularly in developing countries that are already politically unstable (Olson and Drury, 1997). In other research, Olson argues that under certain (and rare) circumstances, disasters can constitute "critical junctures," or crises that leave distinctive legacies within those societies. The 1972 earthquake in Managua, Nicaragua, was one such case. Following that devastating event, the corrupt and dictatorial Somoza regime took a large share of post-disaster aid for itself and mismanaged the recovery, in the process alienating Nicaraguan elites, the business establishment, and finally the middle class, and paving the way for the Sandinistas to assume power in 1979. The 1985 Mexico City earthquake also affected the political system of that nation by, among other things, helping to weaken the hegemony of the Institutional Revolutionary Party. However, rather than having a direct and independent influence on subsequent political changes, that earthquake interacted with factors and trends that were already beginning to affect Mexican society before it occurred. That disaster, which was not well managed by the ruling government, provided the Mexican people with a sharp contrast between the vibrancy and the capability of civil society and the government's lack of preparedness. Grass-roots response and recovery efforts also facilitated broader mobilization by groups that had been pressing for change. Although not a "critical juncture" in its own right, the earthquake did play a role in moving the political system in the direction of greater pluralism and strengthened the power of civil society institutions vis-à-vis the state (Olson and Gawronski, 2003).

Such findings assume particular significance in the aftermath of the December 2004 Indian Ocean earthquake and tsunami. The impacts of that catastrophe span at least 12 different nations and a number of semi-autonomous subnational units, each with its own distinctive history, mode of political organization, internal cleavages, and preexisting problems. Research is needed to better understand both recovery processes and outcomes and the longer-term societal effects of this devastating event.

## OTHER DISASTER RECOVERY-RELATED ISSUES

### Disaster Experience and the Mitigation of Future Hazards

Social science research has also focused in various ways on the question of whether the positive informational effects of disasters constitute learning experiences for affected social units by encouraging the adoption of mitigation measures and stimulating preparedness activity. While this idea seems intuitively appealing, the literature is in fact quite equivocal with regard to the extent to which disasters actually promote higher levels of safety. On the one hand, at the community and societal levels, there is considerable evidence to suggest that disasters constitute "windows of opportunity" for those seeking to enact loss reduction programs, making it possible to achieve policy victories that would not have been possible prior to those events (Alesch and Petak, 1986). Disasters have the potential to become "focusing events" (Birkland, 1997) that can alter policy agendas through highlighting areas in which current policy has failed, energizing advocates, and raising public awareness. On the other hand, many disasters fail to become focusing events and have no discernible impacts on the adoption and implementation of loss-reduction measures. For example, Burby et al., (1997), who studied communities in five different states, found no relationship between disaster experience and adoption of mitigation measures. Birkland (1997) suggests that these differences are related in part to the extent to which advocacy coalitions exist, are able to turn disaster events to their advantage, and are able to formulate appropriate policy responses.

Further complicating matters, policies adopted in the aftermath of disasters, like other policies, may meet with resistance and be only partially implemented—or implemented in ways that were never intended. While it is possible to point to examples of successful policy adoption and implementation in the aftermath of disasters, such outcomes are by no means inevitable, and when they do occur, they are typically traceable to other factors, not just to disaster events themselves.

Research does suggest that households, businesses, and other entities affected by disasters learn from their experiences and take action to protect themselves from future events. Those who have experienced disasters may, for example, step up their preparedness for future events or be more likely to heed subsequent disaster warnings. At the same time, it is also clear that there is considerable variability in the relationship between experience and behavioral change. While some studies document the positive informational effects of experience, others show no significant impact, and some research even indicates that repeated experiences engender complacency and lack of action (for a review of the literature, see Tierney et al., 2001).

## Role of Prices and Markets

Mainstream economic theory, models, and analytical tools (e.g., benefit-cost analysis) assume that markets generally function efficiently and equilibrate. Barring various situations of market failure, prices serve a key role as signals of resource scarcity. In this context, two broad areas of research needs can be identified. One is the role of prices and markets in pre-disaster mitigation (see also Chapter 3). Market-based approaches to reducing disaster risk involve such questions as how prices can serve as better signals of risk taking and risk protection, and the potential for new approaches to risk sharing (e.g., catastrophe bonds). At the same time, better understanding is also needed of market failures in mitigation (e.g., externalities in risk taking and risk protection). The second broad research need concerns markets in post-disaster loss and recovery. Little empirical research has been conducted on the degree to which assumptions of efficient markets actually hold in disasters, especially those having catastrophic impacts, and the degree to which markets are resilient in the face of disasters. Research is also needed on how economic models can capture the adjustment processes and disequilibria that are important as economies recover from disasters, and how economic recovery policies can influence recovery trajectories.

### Disaster Recovery and Sustainability

As discussed in more detail in Chapter 6, which focuses on international research, disaster theory and research have increasingly emphasized the extent to which vulnerability to disasters can be linked to unsustainable development practices. Indeed, the connection between disaster loss reduction and sustainability was a key organizing principle of the NEHRP-sponsored Second Assessment of Research on Natural Hazards. The title of the summary volume for the Second Assessment, *Disasters by Design* (Mileti, 1999b), was chosen to emphasize the idea that the impacts produced by disasters are the consequence of prior decisions that put people and property at risk. A key organizing assumption for the Second Assessment was the notion that societies and communities "design" the disasters of the future by failing to take hazards into account in development decisions; pursuing other values, such as rapid economic growth, at the expense of safety; failing to take decisive action to mitigate risks to the built environment; and ignoring opportunities to enhance social and economic resilience in the face of disasters. Conversely, communities and societies also have the ability to design safer futures by better integrating hazard reduction into their ongoing policies and practices in areas such as land-use and development planning, building codes and code enforcement, and quality-of-life initiatives.

Just as disasters dramatically highlight failures to address sources of vulnerability, the post-disaster recovery period gives affected communities and societies an opportunity to reassess pre-disaster plans, policies, and programs, remedy their shortcomings, and design a safer future (Berke et al., 1993). The federal government seeks to promote post-disaster mitigation through FEMA's Hazard Mitigation Grant Program, as well as programs that seek to reduce repetitive flood losses through relocating flood-prone properties. The need to weave a concern with disaster loss reduction into the fabric of ongoing community life has also guided federal initiatives such as Project Impact, FEMA's Disaster Resistant Communities program.

Yet the research record suggests that those opportunities are often missed. While it is clear that some disaster-stricken communities do act decisively to reduce future losses, for others the recovery period brings about a return to the status quo ante, marked at most by gains in safety afforded by reconstruction to more stringent building codes. The section above noted that disasters create "windows of opportunity" for loss reduction advocates, in part by highlighting policy failures and temporarily silencing opponents. At the same time, however, research evidence suggests that even under those circumstances, it is extremely difficult to advance sustainability goals in the aftermath of disasters. Changes in land use are particularly difficult to enact, both during nondisaster times and after disasters, despite the fact that such changes can significantly reduce vulnerability. Land use decision making generally occurs at the local level, but local jurisdictions have great difficulty enacting controls on development in the absence of enabling legislation from higher levels of government. Even when land-use and zoning changes and other mitigation measures are seen as desirable following disasters, community leaders may lack the political will to promote such efforts over the long term, allowing opponents to regroup and old patterns to reassert themselves (see, for example, Reddy, 2000; for more detailed discussions on land-use and hazards, see Burby, 1998). Assessing reconstruction following recent U.S. disasters, Platt (1998:51) observed that "[d]espite all the emphasis on mitigation of multiple hazards in recent years, political, social and economic forces conspire to promote rebuilding patterns that set the stage for future catastrophe." Overall, the research record suggests that while the recovery period should ideally be a time when communities take stock of their loss reduction policies and enact new ones, post-disaster change tends to be incremental at best and post-disaster efforts to promote sustainability are rare.

## RESEARCH RECOMMENDATIONS

This chapter closes by making recommendations for future research on disaster response and recovery. As the foregoing discussions have indicated, existing research has raised numerous questions that need to be addressed through future research. This concluding section highlights general areas in which new research is clearly needed, both to test the limits of current social science knowledge and to take into account broad societal changes and issues of disaster severity and scale.

Recommendation 4.1: *Future research should focus on further empirical explorations of societal vulnerability and resilience to natural, technological, and willfully caused hazards and disasters.*

Discussions of factors associated with differential vulnerability and resilience in the face of disasters appear in many places in this report. What these discussions reveal is that researchers have only begun to explore these two concepts and much work remains to be done. It is clear that vulnerability is produced by a constellation of psychological, attitudinal, physical, social, and economic factors. However, the manner in which these factors operate and interact in the context of disasters is only partially understood. For example, while sufficient evidence exists to indicate that race, gender, and ethnicity are important predictors of hazard vulnerability and disaster-related behavior, research has yet to fully explore such factors, their correlates, and their interactions across different hazard and disaster contexts. In many cases age is associated with vulnerability to disasters (see Ngo, 2001; Anderson, 2005), but other factors such as ethnicity and socioeconomic status have differential effects within particular age groups (Bolin and Klenow, 1988), and the vulnerability of elderly persons may be related not only to age but also to other factors that are correlated with age, such as social isolation, which can cut off older adults from sources of lifesaving aid under disaster conditions (Klinenberg, 2002).

Even less is known about how to conceptualize, measure, and enhance resilience in the face of disasters—whether that concept is applied to the psychological resilience of individuals or to the resilience of households, communities, local and regional economies, or other units of analysis. Resilience can be conceptualized as the ability to survive disasters without significant loss, disruption, and stress, combined with the ability to cope with the consequences of disasters, replace and restore what has been lost, and resume social and economic activity in a timely manner (Bruneau et al., 2003). Other dimensions of resilience include the ability to learn from disaster experience and change accordingly.

The large volume of literature on psychological resilience and coping offers insights into factors that facilitate resilient responses by individual disaster victims. Other work, such as research on "high-reliability organizations," organizational adaptation and learning under crisis conditions, and organizational effectiveness (Roberts, 1989; La Porte and Consolini, 1998; Comfort, 1999; Drabek, 2003) also offers insights into correlates of resilience at the organizational and interorganizational levels. As suggested in Chapter 6, the social capital construct and related concepts such as civic engagement and effective collective action are also related to resilience. The challenge is to continue research on the resilience concept while synthesizing theoretical insights from these disparate literatures, with the ultimate objective of developing an empirically grounded theory of resilience that is generalizable both across different social units and across different types of extreme events.

**Recommendation 4.2:** *Future research should focus on the special requirements associated with responding to and recovering from willful attacks and disease outbreaks.*

A better understanding is needed of likely individual, group, and public responses to intentional acts of terrorism, as well as disease outbreaks and epidemics. As indicated in this chapter, there appears to be no strong a priori reason for assuming that responses to natural, technological, or intentionally caused disasters and willful or naturally occurring disease outbreaks will differ. However, research on hazards and disasters also calls attention to factors that could well prove to be important predictors of responses to such occurrences, particularly those involving unique hazards such as chemical, biological, nuclear, and radiological agents. Research on individual and group responses to different types of disasters has high-lighted the importance of such factors as familiarity, experience, and per-ceptual cues; perceptions about the characteristics of hazards (e.g., their dread nature, lethality and other harms); the content, clarity, and consis-tency of crisis communications; knowledge of appropriate self-protective actions; and feelings of efficacy with respect to carrying out those measures (see, for example, classic work on risk perception, discussed in Slovic, 2000, as well as Lindell and Perry, 2004).

Recent research has also highlighted the importance of emotions in shaping perceptions of risk. Hazards that trigger vivid images of danger and strong emotions may be seen as more likely to occur, and more likely to produce harm, even if their probability is low (Slovic et al., 2004). If willful acts engender powerful emotions, they could potentially also engender unusual responses among threatened populations.

The potential for ambiguity and confusion with respect to public com-munications may also be greater for homeland security threats and public health hazards such as avian flu than for other hazards. For example, warning systems and protocols are more institutionalized and more widely understood for natural hazards than for homeland security and public health threats. While it is generally recognized that organizations such as the National Hurricane Center and the U.S. Geological Survey constitute reliable sources of information on hurricanes and earthquakes, respectively, members of the public may be less clear regarding responsibilities and authorities with respect to other risks, particularly since such threats and the expertise needed to assess them are so diverse.

These kinds of differences could translate into differences in public perceptions and subsequent responses. Research is needed on the manner in which the distinctive features of particular homeland security and public health threats, such as those highlighted here, as well as official plans and management strategies, could affect responses during homeland security emergencies.

> Recommendation 4.3: *Future research should focus on the societal consequences of changes in government organization and in emergency management legislation, authorities, policies, and plans that have occurred as a result of the terrorist attacks of September 11, 2001, as well as on changes that will almost certainly occur as a result of Hurricane Katrina.*

The period since the 2001 terrorist attacks has been marked by major changes in the nation's emergency management system and its plans and programs. Those changes include the massive government reorganization that accompanied the creation of the Department of Homeland Security (DHS); the transfer of FEMA, formerly an independent agency, into DHS; the shifting of many duties and responsibilities formerly undertaken by FEMA to DHS's Office of Domestic Preparedness, which was formerly a part of the Justice Department; the development of the National Response Plan, which supercedes the Federal Response Plan; Presidential Homeland Security Directives 5 and 8, which make the use of the National Incident Management System (NIMS) mandatory for all agencies and organizations involved in responding to disasters and also mandate the establishment of new national preparedness goals; and increases in funding for special homeland security-related initiatives, particularly those involving "first responders." Other changes include a greater emphasis on regionalized approaches to preparedness and response and the growth at the federal, state, and local levels of offices and departments focusing specifically on homeland security issues—entities that in many cases exist alongside "traditional" emergency management agencies. While officially stressing the need for an "all-hazards" approach, government initiatives are concentrating increasingly on preparedness, response, and recovery in the context of willful attacks. These changes, all of which have taken place within a relatively short period of time, represent the largest realignment of emergency management policies and programs in U.S. history.

What is not known at this time—and what warrants significant research—is how these changes will affect the manner in which organizations and government jurisdictions respond during future extreme events. Is the system that is evolving more centralized and more command-and-control oriented than before September 11? If so, what consequences will that have for the way organizations and governmental entities respond? What role will the general public and emergent groups play in such a system? How will NIMS be implemented in future disasters, and to what effect? What new forms will emergent multiorganizational networks assume in future disasters? Which agencies and levels of government will be most central, and how will shifts in authority and responsibility affect response and recovery efforts? Will the investment in homeland security preparedness translate into more rapid, appropriate, and effective responses to natural and technological disasters, or will the new focus on homeland security lead to an erosion in the competencies required to manage other types of emergencies? A major research initiative is needed to analyze the intended and unintended consequences in social time and space of the massive changes that have taken place in the nation's emergency management system since September 11, 2001.

These concerns loom even larger in the aftermath of Hurricane Katrina. That disaster revealed significant problems in virtually every aspect of intergovernmental preparedness and response. The inept management of the Katrina disaster was at least in part a consequence of the myopic institutional focus on terrorism that developed in the wake of the September 11, 2001 attacks—a focus that included marginalizing and underfunding FEMA and downplaying the challenges associated with responding to large-scale natural disasters (Tierney, 2006, forthcoming). Katrina is certain to bring about further efforts at reorganizing the nation's response system, particularly at the federal level. These reorganizations and their consequences merit special attention.

**Recommendation 4.4:** *Research is needed to update current theories and findings on disaster response and recovery in light of changing demographic, economic, technological, and social trends such as those highlighted in Chapter 2 and elsewhere in this report.*

It is essential to keep knowledge about disaster response and recovery current. The paragraphs above highlight the need for new research on homeland security threats and institutional responses to those threats. Research is also needed to update what is known about disaster response and recovery in light of other forms of social change and to reassess existing theories. Technological change is a case in point. Focusing on only one issue—disaster warnings—the bulk of the research that has been conducted on warning systems and warning responses was carried out prior to the information technology and communications revolutions. With the rise of the Internet and interactive Web-based communication, the proliferation of cellular and other wireless media, and the growing potential for ubiquitous communications, questions arise regarding the applicability of earlier research findings on how members of the public receive, interpret, and act on warnings. Changes in the mass media, including the rise of the 24-hour news cycle and the trend toward "narrowcasting" and now "podcasting" for increasingly specialized audiences, also have implications for the ways in which the public learns about hazards and receives warning-related information. In many respects, warning systems reflect a preference for "push-oriented" information dissemination approaches. However, current information collection practices are strongly "pull oriented." These and other trends in communications technology introduce additional complexity into already complex processes associated with issuing and receiving warnings, decision making under uncertainty, and crisis-related collective behavior. New research is needed both to improve theories and models and to serve as the basis for practical guidance.

Much the same can be said with respect to organizations charged with responding during disaster events. Along with being affected by policy and programmatic changes such as those discussed above, crisis-relevant agencies are also being influenced by the digital and communications revolution and by the diffusion of technology in areas such as remote sensing, geographic information science, data fusion, decision support systems, and visualization. In the more than 15 years since Drabek (1991b) wrote *Microcomputers and Emergency Management*, which focused on the ways in which computers were affecting the work of local emergency management agencies, technological change has been rapid and massive. How such changes are affecting organizational performance and effectiveness in disasters is not well understood and warrants extensive systematic study.

**Recommendation 4.5:** *More research is needed on response and recovery for near-catastrophic and catastrophic disaster events.*
Chapter 1 discusses issues of determining thresholds of disastrous conditions. NEHRP-sponsored social science research indicates that, in the main, U.S. communities have shown considerable resilience even in the face of major disasters. Similarly, at the individual level, U.S. disasters have produced a range of negative psychosocial impacts, but such impacts appear to have been neither severe nor long-lasting. While recognizing that disasters disproportionately affect the most vulnerable in U.S. society and acknowledging that recovery is extremely difficult for many, disasters have been less devastating in the United States and other developed societies than in the developing world. Disaster-related death tolls have also been lower by orders of magnitude, and economic losses, although often large in absolute terms, have also been lower relative to the size of the U.S. economy. At least that was the case until Hurricane Katrina, a catastrophic event that has more in common with disasters in the developing world than with the typical U.S. disaster.

The vast majority of empirical studies on which such generalizations are based have not focused on truly catastrophic disasters, and therefore research results may not be "scalable" to such events. Katrina clearly demonstrates that the nation is at risk for events that are so large that they overwhelm response systems and produce almost insurmountable post-disaster recovery challenges. What kinds of social and economic impacts and outcomes would result from a large earthquake under downtown Los Angeles, a 7.0 earthquake event on the Hayward Fault in the San Francisco Bay area, a repeat of Hurricane Andrew directly striking Miami, or another hurricane landfall in the already devastated Gulf Coast region? What about situations involving multiple disaster impacts, such as the 2004 hurricane season in Florida and multiple disaster events that produce protracted impacts over time, such as the large aftershocks that are now occurring after the Indian Ocean earthquake and tsunami? To move into the realm of worst cases, what about an attack involving weapons of mass destruction, or simultaneous terrorist attacks in different cities around the United States? Such events are not outside the realm of possibility. There is a need to envision the potential social and economic effects of very large disasters, to learn from catastrophic events such as Hurricane Katrina, and to analyze historical and comparative cases for the insights they can provide.

**Recommendation 4.6:** *More cross-societal research is needed on natural, technological, and willfully caused hazards and disasters.*

Most of the research discussed in this chapter has focused on studies conducted within the United States, but it is important to recognize that findings from U.S. research cannot be overgeneralized to other societies. Disaster response and recovery challenges are greater by many orders of magnitude in smaller and less developed societies than in larger and more developed ones.

Disaster impacts, disaster responses, and recovery processes and outcomes clearly vary across societies. Although the earthquakes that struck Los Angeles in 1994, Kobe in 1995, and Bam, Iran, in 2003 were roughly equivalent in size, they differed in almost every other way: lives lost, injuries, extent of physical damage, economic impacts, and subsequent response and recovery activities. Research suggests that such cross-societal differences are attributable to many factors, including differences in physical and social vulnerability; governmental and institutional capacity; government priorities with respect to loss reduction; and response and recovery policies and programs (see, for example, Davis and Seitz, 1982; Blaikie et al., 1994; Berke and Beatley, 1997; Olson and Gawronski, 2003). NEHRP has made significant contributions to cross-societal research through initiatives such as the U.S.-Japan research program on urban earthquake hazards, which was launched following the Northridge and Kobe earthquakes, as well as a similar initiative that was developed after the 1999 Turkey and Taiwan earthquakes. In some cases, these initiatives have led to longer-term research partnerships; Chapter 6 contains information on one such collaboration, involving the Texas A&M University Hazard Reduction and Recovery Center and the National Center for Hazards Mitigation at the National

Taiwan University. Significantly more cross-national and comparative research is needed to further document and explain cross-societal variations in response and recovery processes and outcomes across different scales and different disaster events. Disasters such as the Indian Ocean earthquake and tsunami merit intensive study because they allow for rich comparisons at various scales (individuals, households, communities, and institutional and societal levels), providing an opportunity to greatly expand existing social science knowledge.

> **Recommendation 4.7:** *Taking into account both existing research and future research needs, sustained efforts should be made with respect to data archiving, sharing, and dissemination.*

As noted in detail in Chapter 7, attention must be paid to issues related to data standardization, data archiving, and data sharing in hazards and disaster research. NEHRP has been a major driving force in the development of databases on response and recovery issues. However, vast proportions of these data have yet to be fully analyzed. For social scientists to be able to fully exploit the data that currently exist, let alone the volume of data that will be collected in the future, specific steps have to be taken to make available and systematically collect, preserve, and disseminate such data appropriately within the research community. As recommended in Chapter 7, information management strategies must be well coordinated, formally planned, and consistent with federal guidelines governing the protection of information on human subjects. Assuming that these foundations are established, the committee supports the creation of a *Disaster Data Archive* organized in ways that would encourage broader use of social science data on disaster response and recovery. Contents of this archive would include (but not be limited to) survey instruments; cleaned databases in common formats; code books, coding instructions and other forms of documentation; descriptions of samples and sampling methods; collections of papers containing analyses using those databases; photographs and Internet links (where applicable); and related research materials. Procedures for data archiving and sharing would build on existing protocols set out by organizations such as the Inter-University Consortium for Political and Social Research (e.g., ICPSR, 2005).

The distributed Disaster Data Archive would perform a number of important functions for social science hazards and disaster research and for the nation. The existence of the archive would make it much more likely that existing data sets will be used to their full potential by greatly improving accessibility. The archive would serve as an important tool for undergraduate and graduate education by making data more easily available for course projects, theses, and dissertations. By enabling researchers to access instruments used in previous research and incorporate past survey and interview items into their own research, the archive should help make social science research on disasters more cumulative and replicable. An archive would also make it easier for newcomers to the field of disaster research to become familiar with existing research and enable researchers to identify gaps in past research and avoid unnecessary duplication. The archive would also serve an important function in preserving data that might otherwise be lost. Finally, such an archive would enable social science disaster research to better respond to agency directives regarding the desirability of data sharing.

For an effort of this kind to succeed, a number of conditions must be met. Funds will be needed to support the development and maintenance of the archive, and researchers must be willing to make their data sets and all relevant documentation available. This second condition is crucial, because the committee is aware of a number of important data sets that are not currently being shared, and the archive cannot succeed without broad researcher support. Challenges related to human subjects review requirements, confidentiality protections, and disclosure risks must be fully explored and addressed. Other issues include challenges associated with the development and enforcement of quality control standards, rules and standards for data sharing, procedures to ensure that proper acknowledgment is given to project sponsors and principal investigators, and questions about long-term management of the archive.

Related to the need for better data archiving, sharing, and dissemination strategies, social scientists must be poised to take advantage of new capabilities for data integration and fusion. Strategies are needed to integrate social science data with other types of data collected by both pervasive in situ and mobile ad hoc sensor networks (Estrin et al., 2003), such as networks that collect data on environmental and ecological changes and disaster impacts. In light of the availability of such a wide array of data, the hazards and disasters research community must recognize that hazards and disaster informatics—the application of information science and technology to disaster research, education, and practice—is an emerging field.

To realize this potential, and with the foundation established through implementing recommendations in Chapter 7, the committee further supports the creation of a Data Center for Social Science Research on Hazards and Disasters. In addition to maintaining the Disaster Data Archive, this center would conduct research on automated information extraction from data, including the development of efficient and effective methods for storing, querying, and maintaining both qualitative and quantitative data from disparate and heterogeneous sources.

# 5

# Interdisciplinary Hazards and Disaster Research

This chapter addresses the committee's charge to examine challenges posed for the social science hazards and disaster research community due to the expectation that, like other relevant research community disciplines, it become a major partner in integrated research. Interdisciplinary research has been gaining prominence across all domains of science, engineering, and social sciences. The first section of this chapter draws from the literature on interdisciplinary research to discuss definitions, challenges, and factors in the success of interdisciplinary studies generally. The second section focuses on interdisciplinarity in hazards and disaster research, with particular reference to the social sciences. It emphasizes trends in research funding structures, the role of multidisciplinary research centers, and the importance of interdisciplinary research for addressing gaps in knowledge about hazards and disasters. The third section presents several exemplars of interdisciplinary research in this field and draws insights and lessons from them. The final section summarizes key findings and offers recommendations for supporting interdisciplinary research in the field.

## DEFINITIONS

Various terms have been used to describe research that crosses traditional disciplinary boundaries. These include "interdisciplinary," "multidisciplinary," "trans-disciplinary," and "cross-disciplinary." The terms have been used in multiple, confusing, and often conflicting ways. For example (Klein, 1990:55),

> The popular term *cross-disciplinary* . . . has been used for several different purposes: to view one discipline from the perspective of another, rigid axiomatic control by one discipline, the solution of a problem with no intention of generating a new science or paradigm, new fields that develop between two or more disciplines, a generic adjective for six different categories of discipline-crossing activities, and a generic adjective for all activities involving interaction across disciplines.

Emerging consensus suggests that research can generally be characterized by the degree of interaction among disciplines. In order of increasing interaction, the spectrum ranges from "multidisciplinary" to "interdisciplinary" to "trans-disciplinary" research.

In "multidisciplinary" research, investigators representing different disciplines often work in parallel, rather than collaboratively (Klein, 1990:56):

> "Multidisciplinarity" signifies the juxtaposition of disciplines. It is essentially *additive*, not *integrative*. Even in a common environment, educators, researchers, and practitioners still behave as disciplinarians with different perspectives their relationship may be mutual and cumulative but not interactive, for there is "no apparent connection," no real cooperation or "explicit" relationships, and even, perhaps, a "questionable eclecticism." The participating disciplines are neither changed nor enriched, and the lack of "a well-defined matrix" of interactions means disciplinary relationships are likely to be limited and "transitory."

Indeed, Klein (1990) finds that most activities purported to be "interdisciplinary" are in actuality "multidisciplinary," particularly research arising from problem-focused projects that intrinsically involve multiple disciplines. Multidisciplinary research in essence involves two or more disciplines, each making a separate contribution to the overall study (NRC, 2005).

"Interdisciplinary" research, in contrast, is often defined along the lines of referring to "integration of different methods and concepts through a cooperative effort by a team of investigators . . . [not referring simply to] the representation of different disciplines on a team nor to individuals who may 'themselves' incorporate different disciplines on a project themselves" (Rhoten, 2004:10). For example, a National Research Council (NRC) committee provided the following definition (Pellmar and Eisenberg, 2000:3):

> Interdisciplinary research is a cooperative effort by a team of investigators, each expert in the use of different methods and concepts, who have joined in an organized program to attack a challenging problem. Ongoing communication and reexamination of postulates among team members promote broadening of concepts and enrichment of understanding.

Although each member is primarily responsible for the efforts in his or her own discipline, all share responsibility for the final product.

Most recently, the NRC's Committee on Facilitating Interdisciplinary Research (NRC, 2005b:26) has conceptualized the term to refer not necessarily to the composition of a research team, but rather to the mode of investigation:

> Interdisciplinary research (IDR) is a mode of research by teams or individuals that integrates information, data, techniques, tools, perspectives, concepts, and/or theories from two or more disciplines or bodies of specialized knowledge to advance fundamental understanding or to solve problems whose solutions are beyond the scope of a single discipline or field of research practice.

This view emphasizes that true interdisciplinarity goes beyond involving two disciplines to create one product and is characterized by the synthesis of research ideas and methods. In some cases, particularly fruitful interdisciplinary efforts actually lead to the evolution of new disciplines, for example, neuroscience (Pellmar and Eisenberg, 2000:3).

The term "transdisciplinary" is distinct in referring to approaches that are "far more comprehensive in scope and vision [than interdisciplinary approaches]" (Klein, 1990:65). Examples of trans-disciplinary approaches include structuralism, Marxism, and policy sciences. Klein (1990) contrasts nondisciplinary versus disciplinary positions in the discourse:

> The nondisciplinary position is more scornful of the disciplines. Visible in the call to overturn disciplinary hegemony, it has figured in propositions of "transdisciplinarity," revisionist theories of "critical interdisciplinarity," and the "integrative"/"interdisciplinary" distinction that emerged in education and the social sciences. The disciplinary position holds that disciplinary work is essential to good interdisciplinary work (Klein, 1990:106).

For hazards and disaster studies, it is useful to make several other distinctions. First, collaborative research within the social sciences differs from collaborative efforts by social scientists with natural scientists and engineers. Both are important for addressing knowledge gaps. However, the challenges of the latter are particularly great, as discussed in the next section. Basic research can also be distinguished from more applied types of studies (e.g., problem-focused, evaluation, impact assessment) in which interdisciplinary research tends to be more prominent.

For purposes of this report, the committee adopts the following positions with regard to defining interdisciplinary research within the social science hazards and disaster research community:

- The term interdisciplinary is used as an umbrella term to represent efforts usually conducted by research teams that involve ideas and methods from more than one discipline.
- There exists a spectrum of degrees of interdisciplinarity. These range from parallel efforts with a research team comprising different disciplines, to sequentially linked efforts where outputs of one disciplinary research effort provide inputs to another, to fundamentally integrated research where multiple disciplines interact in mutually transforming ways from problem definition through to research design and execution.
- Research efforts across this spectrum are needed and appropriate to different types of problems.
- Interdisciplinary research is particularly challenging when it crosses boundaries between the social sciences and the natural sciences and engineering.

## CHALLENGES

Interdisciplinary research is challenging, and the potential of interdisciplinary research is often unrealized. "Across the spectrum of higher education, many initiatives deemed interdisciplinary are, in fact, merely reconfigurations of old studies—traditional modes of work patched together under a new label—rather than actual reconceptualizations and reorganizations or new research" (Rhoten, 2004:6). In the area of global environmental change, for example, there have been frequent calls for alliances between natural and social sciences but few successes (Stern et al., 1992).

The literature has identified numerous barriers to interdisciplinary research. These range from intellectual issues such as attitude and communication to organizational issues such as academic structure and funding mechanisms, for example (Pellmar and Eisenberg, 2000:4-5):

> Disciplinary jargon and cultural differences among disciplines are serious problems. Surveys show concerns among researchers about perceptions of interdisciplinary science as second-rate. . . . There are concerns that training in interdisciplinary fields will not prepare graduates for a career. The explosion of information within each scientific discipline raises concerns about how long it would take to attain expertise in one, let alone two or more, fields. . . . Because publications and successful grants are essential for promotion and tenure, the concern that interdisciplinary research will reduce the likelihood of first-authorship and of funding presents an additional obstacle.

Some of the most commonly cited barriers to interdisciplinary research include lack of funding, indifference or hostility on the part of researchers, and incompatibility with academic incentive and reward structures. In a recent study of interdisciplinary research centers and programs, Rhoten (2004:6) found that the latter may be most significant and perhaps even underestimated: "The transition to interdisciplinarity and consilience does not suffer from a lack of *extrinsic* attention at the 'top' or *intrinsic* motivation at the 'bottom,' but, rather, from a lack of *systemic* implementation in the 'middle'" as universities have implemented piecemeal and incoherent policies rather than systematic reforms.

While systemic barriers may be most significant for research centers and programs, in individual studies, difficulties typically relate to the failure of a research team to function collaboratively. This failure may derive from causes ranging from individual researchers devaluing the contributions of other team members to inability of the group to bridge culture gaps. As an example of the latter, the NRC (2005b:54) Committee on Facilitating Interdisciplinary Research points to the culture gap between mechanical engineers and software engineers in some early robotics research: "To the first group, a robot with adequate sensors had little need for software; to the second group, an abundance of mechanical sensors was a sign of inadequate software."

For hazards and disaster studies, the challenges of interdisciplinarity are compounded by additional hurdles. These relate to the marginal position of the social sciences relative to the natural science and engineering fields, perceptions of applied research, and attitudes toward mission-oriented research. Traditionally, hazard and disaster studies have been dominated by natural science and engineering fields. Public policy in the United States has emphasized scientific and technological "solutions" (e.g., earthquake prediction, earthquake engineering, flood control dams) to the hazards problem. Social science accounts for a small share of research funding, activity, and personnel in the hazards and disasters field generally; as noted in Chapter 9, there are approximately as many social scientists in the hazards and disasters field as there are volcanologists. This marginality means that when social scientists are involved in interdisciplinary research with scientists or engineers, their involvement typically resembles an afterthought or "add-on" to a primarily "scientific" or "technical" inquiry. This situation is changing, but it is still rare in collaborations with science or engineering for social science concerns and concepts to substantially shape the overarching research questions and approach (for an exception, see Box 5.1).

Additionally, hazards and disaster studies are commonly viewed as applied research aimed at "fixing problems" rather than basic science intended to advance knowledge. It is not uncommon for consultants to participate in these studies. The perception of applied research often marginalizes hazards and disaster research within the social sciences in relation to established academic disciplines, so that research is difficult to

---

**BOX 5.1**
**Social Scientists' Use of Engineers on**
**Public Policy Research Projects**

Local governments often adopt—or attempt to adopt—earthquake hazard mitigation policies that affect both future and existing buildings following events that damage their communities. These policy debates, decisions, or nondecisions usually pivot around highly technical engineering proposals that often have potentially significant impacts on building owners, especially those that own existing damaged buildings.

Social scientists must depend on earthquake engineering experts to interpret how the proposed policy measures could influence the ways that, at what cost, and over what period of time repair or retrofit measures for existing and standards for new buildings could affect owners, and through them, the adopters of such policies—locally elected officials.

Social scientists, in their studies of how such technically sound proposals can affect local politics, draw on earthquake engineers to characterize and interpret these proposals, which, when introduced into the local political system, may not go the way the engineering community desires. (See Olson and Olson, 1993; Olson et al., 1998, 1999.)

---

publish in mainstream disciplinary journals. The severity of this problem does vary across the social science disciplines. In geography, and to a lesser extent sociology and urban and regional planning, there are well-established traditions of hazards and disaster studies, and researchers in these areas have gained disciplinary prominence and intellectual influence. In other disciplines such as economics, psychology, anthropology, and political science, it is virtually impossible to publish hazards and disaster studies in mainstream journals. This constraint creates a substantial disincentive for researchers, particularly young scholars seeking tenure, to conduct research in this field. Consequently, the number of researchers in the field remains small (see Chapter 9).

Similarly, interdisciplinary journals, while widely read and influential within the hazards and disaster research community, are not well recognized by reviewers in mainstream disciplines. Consequently, they are given less weight than disciplinary journals by reviewers who make recommendations regarding tenure and promotion. Interdisciplinary journals include both traditional outlets such as the *International Journal of Mass Emergencies and Disasters* and *Risk Analysis*, as well as new interdisciplinary journals such as *Environmental Hazards* and the *Natural Hazards Review*.

Moreover, many hazards and disaster studies involve funding or collaboration with agencies and organizations other than the National Science Foundation (NSF). Historically, organizations such as the Federal Emergency Management Agency (FEMA), state emergency management agencies, the U.S. Army Corps of Engineers, the National Oceanic and Atmospheric Administration (NOAA), the National Aeronautics and Space Administration (NASA), and the insurance industry have supported social scientists conducting hazards and disaster research. As noted in Chapters 1 and 2 of this report, the Department of Homeland Security (DHS) has recently emerged as a major funding source for some types of disaster research. Studies not primarily supported by NSF are often viewed as "mission-oriented," responding to the interests of mission agencies (e.g., terrorism) rather than to intellectual curiosity or other motivations of basic science. This circumstance further impedes publication and acceptance by mainstream academic disciplines.

## FACTORS IN SUCCESS

These impediments can be overcome, and the accumulation of experience points to a number of factors that seem to be important in the success of interdisciplinary studies. These factors generally pertain to three dimensions of the research process: the research problem, the participants, and management. External support plays a role in each of these dimensions.

As noted by the NRC Committee on Facilitating Interdisciplinary Research, trends toward more interdisciplinary research are driven, in part, by the complexity of natural and social phenomena and the need to address societal problems. Accordingly, that committee found that interdisciplinary research "works best when it responds to a problem or process that exceeds the reach of any single discipline or investigator" (NRC, 2005:53). Problem-oriented research thus appears to be favorable for interdisciplinary collaboration in that the value of different disciplines' contributions and the need for integrative conceptualizations can be focused and driven by the complexity and demands of the societal problem itself.

Characteristics of the participants and research group also appear to be important. Experience from the National Laboratories, which routinely engage in interdisciplinary research, suggests that the first key to success is to "involve only people who find unraveling a complex transdisciplinary issue at least as important as their own discipline" (Wilbanks, in NRC, 2005:55-56). Interdisciplinary research and collaboration requires interpersonal skills beyond subject matter expertise in disciplinary methods (Pellmar and Eisenberg, 2000:43).

The size of the research team is also influential to some degree; studies have found that small groups (e.g., centers with less than 20 affiliates) that have stable membership tend to be most successful at interdisciplinary integration (Klein, 1990; Rhoten, 2004). As far as research centers are concerned (Rhoten, 2004:9):

> [I]nterdisciplinary centers need not only to be well-funded but to have an independent physical location and intellectual direction apart from traditional university departments. They should have clear and well-articulated organizing principles—be they problems, products, or projects—around which researchers can be chosen on the basis of their specific technical, methodological, or topical contributions, and to which the researchers are deeply committed. While a center should be established as a long-standing organizational body with continuity in management and leadership, its researchers should be appointed for flexible, intermittent but intensive short-term stays that are dictated by the scientific needs of projects rather than administrative mandates.

Much of the literature has focused on issues such as communication, leadership, rewards, and teamwork strategies that relate to project planning and management. For example, the NRC (2005b:18-19) found that "key conditions for effective [interdisciplinary research] . . . include sustained and intense communication, talented leadership, appropriate reward and incentive mechanisms (including career and financial rewards), adequate time, seed funding for initial exploration, and willingness to support risky research."

Communication appears to be critical for overcoming disciplinary preconceptions where they may hinder interdisciplinary collaboration. It has been noted that effective teamwork requires that team members have trust in one another's skills and expertise, which is difficult to evaluate when working with researchers from other disciplines. Good communication is thus essential for the process to succeed (Pellmar and Eisenberg, 2000:43). The experience of the National Laboratories suggests the importance of discouraging "disciplinary entitlements" wherein "something is accepted as truth because one discipline says so." It also stresses the need to overcome disciplinary stereotypes, replacing them with personal relationships that require substantial time to cultivate (Wilbanks, in NRC, 2005:55-56).

The literature on organizational psychology suggests that one of the inherent dilemmas in multidisciplinary groups concerns the contradictory consequences of member diversity. Diversity can exist in *underlying attributes* such as (task-related) knowledge, skills, and abilities, as well as (relations-oriented) values, needs, attitudes, and personality characteristics. Diversity can also exist in *readily detectable attributes* such as (task-related) educational level, disciplinary degree, and team tenure, as well as (relations-oriented) gender, age, and ethnicity (Jackson et al., 1995). Team members are selected to staff a project on the basis of readily detectable task-related attributes (e.g., educational level, disciplinary degree) because these are assumed to provide a valid indication of a person's underlying task-related attributes (e.g., knowledge, skills, abilities). However, some readily detectable task-related attributes are associated with underlying relations-oriented attributes (i.e., disciplines vary in the prevalence of people with certain values, needs, attitudes, and personality characteristics), and some readily detectable relations-oriented attributes are stereotypically associated with underlying task-related attributes (e.g., ethnicity and gender, are thought to be correlated with certain skills and abilities such as mathematics).

These incidental differences and stereotypic beliefs can have a significant effect on the performance of interdisciplinary projects where members of different disciplinary subgroups must work together. An obvious problem is that people's confidence in some stereotypes exceeds those stereotypes' predictive validity. A more fundamental dilemma is that diversity in underlying task-related attributes is an essential ingredient in innovation, adaptation, and performance (Jackson et al., 1995). However, diversity in underlying relations-oriented attributes can create friction, reduce normative consensus and cohesiveness, and cause members to leave the group. The positive effects of diversity can be attained and its negative effects minimized by promoting communication of information, cooperation in task performance, a positive work climate, and team cohesiveness.

Leadership is a second aspect of project management that is important in facilitating interdisciplinary research (Pellmar and Eisenberg, 2000:43):

> Interdisciplinary research teams need leaders who understand the challenges of group dynamics and who can establish and maintain an integrated program. Leaders need to have vision, creativity, and perseverance. . . . To coordinate the efforts of a diverse team requires credibility as a research scientist, skill in modulating strong personalities, the ability to draw out individual strengths, and skill in the use of group dynamics to blend individual strengths into a team.

Effective leaders should foster an organizational climate that is conducive to interdisciplinary research. Organizational climate affects organizational effectiveness by influencing the degree to which team members are motivated to contribute toward group goals. It includes dimensions of leadership climate (leader initiating structure, leader consideration, and leader communication), team climate (team coordination, team cohesion, team task orientation, and team pride), and role climate (role clarity, but not role conflict or role overload) (Lindell and Whitney, 1995; Lindell and Brandt, 2000).

Reward structures are also important in facilitating interdisciplinary research. In particular, research team members should "know that their reputations will be affected by the success or failure of the enterprise—that everybody's name will be on the product" (Wilbanks, in NRC, 2005:55-56). Rewarding performance at the group level, rather than at the individual level, is an effective means of promoting cooperative goals (Ellis and Fisher, 1994).

Finally, a number of teamwork strategies have been found to be effective in the context of interdisciplinary research. Clarity is important with respect to roles, expectations, and authority, especially in terms of sharing of data and resources (Pellmar and Eisenberg, 2000). Role clarification and role negotiation enable team members to assess their mutual needs and expectations while also clarifying differences in their methodologies and ideologies (Klein, 1990).

Iteration is another strategy that has proven especially useful. "Iteration allows authors to become readers and critics by going over each other's work in order to achieve a coherent, common assessment" (Klein, 1990:190). The team leader can facilitate the interaction by acting as a synthesizer.

More generally, cooperation can be enhanced by interdependence among subgroups' tasks. Task interdependence within a project can be characterized in one of three ways (Thompson, 1967). First, subgroups have *sequential interdependence* when the initiation of one subgroup's task is dependent on the completion of another subgroup's task. Second, subgroups have *reciprocal interdependence* when their outputs cycle iteratively until the team product reaches an acceptable state. Third, subgroups have *pooled interdependence* when both depend on the same resources. This last type of interdependence is important because organizational subgroups operating in parallel are usually assumed to be independent, but they actually have pooled interdependence because all depend indirectly on the success of the others for the continued survival of the project as a whole. Thus, the interdependence of organizational subgroups will be extremely obvious when it is reciprocal and also quite obvious when it is sequential. However, project managers may need to emphasize the existence of pooled interdependence when project members mistakenly assume that they are completely independent of others. One of the most important consequences of cooperation on the reciprocally and sequentially interdependent tasks characteristic of many interdisciplinary research projects is that sharing of information and ideas, especially constructive discussion of alternative views, leads to greater productivity (Tjosvold, 1995).

Another strategy is to collaboratively involve subject matter experts (SMEs) in project management. In multidisciplinary research projects, no single person or even small group of persons has all of the knowledge needed to plan and implement the project. Thus, setting project objectives, identifying and scheduling tasks, and estimating resource needs requires collaboration among SMEs who are knowledgeable about all of the distinct areas to be addressed by the project. Similarly, SMEs from all areas must collaborate in organizing project staff, monitoring task performance, and adjusting resources or objectives in response to deviations from plans.

Successful collaboration among SMEs from the different functional areas is sometimes accomplished by augmenting the project manager with a project management team that actively contributes to project decisions. On small projects, the project management team comprises all project members, whereas on very large projects the project management team might consist of representatives from each functional area. If the members of the project management team have not worked together previously, they must accomplish a number of social tasks at the same time they are attempting to plan and implement the project. That is, according to McIntyre and Salas (1995), members must perform *teamwork* in order to accomplish *task work*. Task work requires project staff to learn enough about each other's subject matter to develop a *shared mental model* of the project (Morgan and Bowers, 1995). This shared mental model must contain all of the elements needed for the project plan—project objectives, task schedules, and resource requirements. In a large project, it probably will not be possible for anyone other than the most interdisciplinary project personnel to develop a fully comprehensive mental model of the project; in an extremely large, complex project it probably will not be possible for anyone to develop a fully comprehensive mental model. Instead, project staff with the broadest scientific knowledge will have a detailed understanding of their own subject matter areas and the ways in which their areas interconnect with closely related areas. In addition, they would have a general understanding of other disciplines that do not link directly to their own. For example, a multidisciplinary earthquake center would be expected to have close linkages of earth scientists with structural engineers, structural engineers with planners, and planners with social scientists.

Finally, the literature on organizations suggests that group cohesiveness can be achieved in a number of ways (Ellis and Fisher, 1994). The first is through formulation of cooperative goals. The goals of individual team members are cooperative when they are positively linked, competitive when they are negatively linked and independent when they are unrelated. One of the easiest ways to establish cooperative goals is to reward performance at the group level, not at the individual level. A second method of achieving cohesiveness is to emphasize external threats. In the case of multidisciplinary projects, the threat of project failure raises the potential for mutual negative career consequences. A third way to achieve cohesiveness is for the group to rapidly achieve some visible goals. This can be accomplished if the team sets some easily attainable short-term goals that will provide early success experiences. Finally, cohesiveness can be enhanced by shared experiences, especially collaborative responses to difficult challenges such as preparing for external reviews.

## INTERDISCIPLINARY TRENDS IN SOCIAL SCIENCE HAZARDS AND DISASTER RESEARCH

The NRC Committee on Facilitating Interdisciplinary Research (2005b:2) identified four fundamental forces that are driving the growth of interdisciplinary research:

1. the inherent complexity of nature and society,
2. the desire to explore problems and questions that are not confined to a single discipline,
3. the need to solve societal problems, and
4. the power of new technologies.

While these forces have long been influential for social science hazards and disaster research, recent trends toward interdisciplinarity can be ascribed to more proximate drivers. It is especially important to recognize the influence of the National Science Foundation in terms of research funding criteria as well as the earthquake engineering research centers.

### Research Funding Structure

Interdisciplinary research has been gaining increasing emphasis from funding organizations, including NSF. Perceptions that there is little research funding for interdisciplinary studies appear to be unfounded, at least in recent years. One study (Rhoten, 2004:7) reported that

> [W]e have found substantial evidence of extrinsic attention to interdisciplinary research in the discourses and resources of government agencies, policymakers, scholarly associations, and university administrators. . . . Of the $4.11 billion that the NSF requested from Congress for research and related activities in 2004, $765 million—a 16.5percent increase over 2003—has been earmarked for four priority areas all designated as interdisciplinary [including] Human and Social Dynamics. . . . In addition, private dollars are also being poured into interdisciplinary endeavors at unprecedented levels.

Many observers believe that research funding for interdisciplinary studies is not only well established but also likely to continue growing. This trend derives from the juxtaposition of stable or declining national budgets for research with a political climate that demands research expenditures be justified on practical, tangible grounds (Hackett, 2000). Even observers who are ambivalent about the merits of interdisciplinary research advocate acknowledging and taking advantage of its growing prominence (Hackett, 2000:259):

Very little is known about such initiatives. . . . In light of all this ignorance and uncertainty, it is difficult to embrace interdisciplinary initiatives. Yet they may be a lasting instrument of science policy, and the one area of real growth and opportunity. Our best option, then, is to proceed boldly but reflectively, giving such investments our whole-hearted support while thinking critically and systematically about their performance and consequences.

Human-environment interactions broadly defined, including hazards and disasters, have been a focus of numerous interdisciplinary initiatives at NSF in recent years. Some examples include initiatives for studies on Long-Term Ecological Research (LTER), including urban LTER, several centers for climate study, biocomplexity, human dimensions of global change, and Human and Social Dynamics (HSD). The latter in particular identifies "Decision making and Risk" as one focal area, and requires that research teams include at least one social scientist. Some "centers of excellence" being established by DHS have been directed to conduct interdisciplinary research, sometimes requiring a fairly prominent role for the social sciences.

### Earthquake Engineering Research Centers

For social science hazards and disaster research, another important driver in interdisciplinary studies, particular with science and engineering, has been the NSF-supported earthquake engineering research centers, a major NSF contribution under the National Earthquake Hazards Reduction Program (NEHRP).

When the National Center for Earthquake Engineering Research (NCEER) was funded through NSF's Directorate for Engineering in 1986, the aim was to promote multidisciplinary team research that would provide a more comprehensive understanding of earthquake hazards and how to cope with them. This was quite a novel idea at the time because what NSF sought was nothing less than the creation of a program of integrated research that included the most talented investigators from the earth sciences, engineering, and the social sciences. The notion was that researchers from these different disciplines would not merely work in parallel, as NSF grantees sometimes did in research carried out in individual and small group projects, but collaboratively. It turned out that it was particularly difficult to get the social scientists integrated into the NCEER program in this fashion, at least until some of the researchers that made up the engineering leadership began accepting the NSF vision and actually became champions of it. It helped, too, that some outstanding social scientists became members of the NCEER team, demonstrating the importance of the social sciences in developing a truly comprehensive understanding of earthquake hazards—that it wasn't just about building design and performance.

NSF funding for NCEER lasted 10 years. With broad backing from the earthquake research community, NSF funded three new centers in 1997: the Pacific Earthquake Engineering Research Center (PEER), the Mid-America Earthquake Center (MAE), and the Multidisciplinary Center for Earthquake Engineering Research (MCEER), which replaced and built on NCEER. As before, NSF charged the three centers with the task of conducting interdisciplinary research that included the social sciences, something that the larger earthquake research community had come to accept at least in principle since it encouraged NSF to continue supporting center-based research. Implementing this charge has remained a challenge in many ways, however, even with a multidisciplinary NSF staff providing oversight for the centers program and with multidisciplinary review panels participating in the periodic assessments of progress. Based on presentations to the committee by center leaders, as well as social scientists knowledgeable about the centers (including funded participants and external reviewers), MCEER has come closest to meeting NSF's expectations for collaborative research. Perhaps a greater commitment on the part of its leadership is one factor, while another might be that as the successor to NCEER, it simply has had more time to make progress in that area.

It is evident that although the NSF mandate has been influential, it has not been sufficient to catalyze effective interdisciplinary research. Other factors, such as leadership and the duration of contact among researchers, have also been necessary for the development of trust and respect across the disciplines.

Table 5.1 provides some indication of the growth in interdisciplinary research involving the social sciences in the centers context. The table compares evidence of social science involvement in research activity at NCEER (1986–1994) and MCEER (2001–2004). Data pertain to the disciplinary

**TABLE 5.1** Percentage of Publications by Disciplines of Authors, NCEER and MCEER

| Disciplinary Affiliations of Authors | NCEER | MCEER |
|---|---|---|
| Science/engineering only | 92% | 48% |
| Social science only | 4% | 22% |
| Both science/engineering and social science | 4% | 30% |
| Total | 100% | 100% |
| Total number of publications represented | 26 | 27 |
| Research years represented | 1986-1994 | 2001-2004 |

Tabulated from NCEER *Research Accomplishments 1986-1994* and MCEER *Research Accomplishments 2001-2003* and *Research Accomplishments 2003-2004*. Available at http://mceer.buffalo.edu/publications/resaccom/default.asp.

affiliations of authors and coauthors on papers in the centers' Research Accomplishments publications covering the years indicated. The increase in social science research at the centers, particularly in interdisciplinary research, is evident. For example, social scientists were (co-)authors on only 8 percent of the NCEER papers, but 52 percent of the MCEER papers.

Funding for social science research has been similar at each of the three current centers (MCEER, MAE, and PEER). On average, the social sciences account for approximately 15 percent of research funding, varying from year to year in the range of 13 to 19 percent. At least one of the centers has indicated to the committee that it will continue to increase the social science share due to pressures from NSF site reviews (Bruneau, 2004; Dobson, 2005; May, 2005). The centers are supported with $2 million annual funding from NSF as well as substantial funds from industry and other sources. The centers thus represent a major source of interdisciplinary research funding and a locus of interdisciplinary research activity in recent years.

## Importance of Interdisciplinary Research

The increase in interdisciplinary studies in social science hazards and disaster research reflects a growing consensus within the field about the importance of research problems that cannot be addressed through disciplinary studies alone. There have been numerous calls for interdisciplinary research. For example, at a recent NSF workshop on Integrated Research in Risk Analysis and Decision Making in a Democratic Society, one of the major conclusions reached by participants was: "To advance the basic science and increase the utility of risk analysis and decision science, it is necessary to foster interdisciplinary and multidisciplinary research that includes engineering, information sciences, natural sciences, and social sciences" (NSF, 2002:7).

This importance has also been cited with reference to the George E. Brown, Jr. Network for Earthquake Engineering Simulation (NEES). At a cost of more than $80 million, NEES is a major new initiative in NSF's support for research on earthquake hazards. The focus of NEES is on laboratory experimentation and computer-based simulation. However, an NRC (2003b:125) committee charged with developing a research agenda for NEES cited the importance of collaboration between "scientists and engineers who will develop and test new theories on earthquakes, earthquake damage, and its mitigation" and "social and political scientists who will use the science and technology from NEES to develop better risk assessment tools, loss estimation models, and communication and teaching strategies to help enact and implement more enlightened policies on earthquake loss mitigation."

The need for interdisciplinary studies was also emphasized in the Earthquake Engineering Research Institute's (EERI) Research and Outreach Plan in earthquake engineering, a consensus document from the profession, which notes (EERI, 2003:29):

> Many of the engineering and earth science research programs will benefit directly from [breakthrough technologies and opportunities identified in the Plan], but these efforts, by themselves, will not assure protection from loss. Translating knowledge to action continues to frustrate loss reduction efforts in this and other hazard mitigation efforts. A significant groundbreaking effort is also required to understand the underlying societal factors that contribute to vulnerability and inhibit efforts intended to reduce this vulnerability.
>
> Recent advances in social science research hold particular promise in this regard. These include the challenging areas of risk perception and communication, societal inertia to change, decision making, effective fiscal instruments, and quantification of economic impacts. Consequently, a major component of this Plan is the complementary role of the social sciences, working in partnership with engineering and earth sciences, to achieve the goal of community resilience and protection from loss.

These calls for interdisciplinary research involving the natural sciences, engineering, and social sciences are not unique to the hazards and disaster field. Similar calls have been made in NRC and NSF reports on research agendas in environmental science. In environmental science, it is recognized that fundamental, disciplinary research in the natural sciences, social sciences, mathematics, and engineering is important and requires strengthening. At the same time, "present and future challenges include connecting across disciplines and scales, supporting synthesis studies and activities, more tightly linking science, technology, and decision making, and achieving predictive capability where possible" (Pfirman and AC-ERE, 2003:5). It has been noted that social science or human dimensions research, which supports as well as cuts across other scientific inquiry on global change, must involve both disciplinary and interdisciplinary approaches (NRC, 1999a). Environmental synthesis is needed "to frame integrated interdisciplinary research questions and activities and to merge data, approaches, and ideas across spatial, temporal, and societal scales" (Pfirman and AC-ERE, 2003:1).

Interdisciplinary research, moreover, requires new types of research groups, capabilities, educational frameworks, and forms of support. The NRC (2001) committee on Grand Challenges in Environmental Sciences found that no single discipline in environmental science could entirely capture how multiple driving variables affect environmental as well as human outcomes. It recommended new types of research teams and communities that can communicate and collaborate across the "gulf" that divides natural and social sciences. "These groups will require a large number of scientists with broad, interdisciplinary perspectives, as well as an increased capability for cross-disciplinary collaboration among environmental scientists, who may develop more interdisciplinary orientations as a result" (NRC, 2001:71).

To this end, one NSF report has stressed the importance of adaptability (Pfirman and AC-ERE, 2003:6):

> In this new era, imagination, diversity, and the capacity to adapt quickly are essential qualities for both institutions and individuals. This places a premium on the quality and evolutionary capacity of environmental research and education. In turn, the richness and complexity of interdisciplinary environmental research is creating the opportunity for more immediate and broad-based application of the results to human systems and problems.

These issues and recommendations resonate with the research needs in hazards and disaster research.

### Interdisciplinary Research Needs

In discussing needs for interdisciplinary research, it is useful to consider specific examples and how they transcend traditional disciplinary boundaries. To do this, the committee begins with priorities outlined in the EERI (2003) Research and Outreach Plan prepared with support from NSF. As a further illustration, a type of research that is important, but often overlooked—evaluation research—is considered. Finally, and most importantly, the committee discusses the role of interdisciplinarity in the research priorities identified from the overview of research progress, gaps, and opportunities in the field that are identified in Chapters 3 and 4.

As noted earlier, the EERI Research and Outreach Plan is a consensus document from the profession. It "provides a vision for the future of earthquake engineering research and outreach focused on security of the nation from the catastrophic effects of earthquakes" (EERI, 2003:5). Interdisciplinary collaboration is clearly needed to address several of the priority research tasks identified in the plan, including the following (EERI, 2003:38):

- *System level simulation and loss assessment tools*—e.g., validation studies to calibrate the accuracy of loss estimation models, incorporating the full range of physical and societal impacts and losses for earthquake and other hazards;
- *Assessment of cost effectiveness of loss mitigation*—e.g., definition of performance measures for lifelines and communities, comprehensive direct and indirect loss models, more in-depth demonstration studies (involving an integration of disciplinary approaches),

and examination of nonlinear adaptive behavior in complex organizations;

- *Financial instruments to transfer risk*—e.g., studies to assess the efficacy of alternative risk reduction or transfer methods, analysis of benefits and costs to various stakeholder groups, analysis of complementary roles of mitigation and insurance, and analysis of safeguards against insurance industry insolvency;

- *Advanced and emerging technologies for emergency response and effective recovery*—e.g., real-time loss estimation tools, remote-sensing technologies for damage assessment, advanced decision-support systems for response and recovery, and advanced communication and networking systems for response and recovery

- *Methodologies and measurement of progress in reducing vulnerability and enhancing community resilience to earthquakes*—e.g., risk management cost-effectiveness methodologies and analyses, investigation of societal impacts of catastrophic earthquakes, research on decision making and earthquake risk perceptions, and research on implementation of risk management and earthquake mitigation programs.

These research priorities call primarily for interdisciplinary collaboration between social scientists and researchers in the natural sciences and engineering, although a few (e.g., decision making, risk perception) would also involve interdisciplinary research within the social sciences.

Another example of needed research that is highly interdisciplinary is evaluation research. Evaluation is the systematic assessment of the value and worth of a program, policy, project, technology or some other object. The goal of evaluation research is to provide feedback on the appropriateness and effectiveness of the object of concern. Objects can be evaluated before they are implemented to aid in decision making about a range of alternatives or after they have been implemented to determine processes or impacts. Evaluation can help measure costs and benefits, implementation issues, outcomes, lessons learned, and effectiveness.

Most programs dealing with the reduction of losses from hazards are not evaluated on a systematic basis. For example, the nation's oldest and largest mitigation program is the National Flood Insurance Program (NFIP). In its 30 years of existence, the NFIP's effectiveness has never been evaluated (Mileti, 1999b). Most other major federal hazard reduction efforts have never been systematically evaluated, including the multiagency National Earthquake Hazards Reduction Program, FEMA's National Dam Safety Program, or the Urban Search and Rescue Program. In the absence of program evaluation, it is difficult to understand if risk reduction efforts are effective (see Box 5.2).

---

**BOX 5.2**
**Required Periodic Evaluation of the Implementation of**
**State and Local Plans Developed Pursuant to the**
**Disaster Mitigation Act of 2000**

To remain eligible for several federal hazard mitigation and disaster assistance programs, the Disaster Mitigation Act of 2000 (DMA 2000) requires all state and local governments, including special districts and tribal governments, to prepare and have approved multihazard mitigation plans.

One of the planning steps (10) requires that the adopted plan be maintained, reviewed, and updated periodically. This can be accomplished through a Mitigation Coordinating Committee that includes all original and new participants. The plan is to be reviewed annually and updated and resubmitted for state and federal approval every five years.

Changes in jurisdictions' vulnerabilities can be identified by noting (1) lessened vulnerability as a result of implementing mitigation actions, (2) increased vulnerability as a result of failed or ineffective mitigation actions, and (3) increased vulnerability as a result of new development (and/or annexations).

---

Of particular relevance to this committee's charge is the role of interdisciplinary research in addressing research gaps that the committee has identified in assessing knowledge in the field. The research recommendations presented in Chapters 3 and 4 of this report are repeated in Table 5.2. The table also indicates the types of studies required to address each of these research needs—whether they be disciplinary (social science) studies, interdisciplinary studies involving collaborations within the social sciences, interdisciplinary studies involving social science collaborations with natural sciences and engineering, or some combination of these types. For each recommendation, the table also indicates which type is the primary research need.

Table 5.2 shows that for all of the priority research recommendations identified in Chapters 3 and 4, some degree of interdisciplinary inquiry will be necessary. Disciplinary research needs remain in the vast majority of the recommendations. However, in only 4 of the 19 recommendations are the key research needs disciplinary. The primary research need in 11 of the cases is interdisciplinary within the social sciences. In the remaining four recommendations, it involves collaboration with the natural sciences and engineering. The committee therefore concludes that although important disciplinary research needs remain, the trend toward more interdisciplinary research appears to be consistent with major research needs in the field. Rather than leaving behind important unanswered disciplinary questions,

**TABLE 5.2**  Research Recommendations and Role of Interdisciplinary Studies[1,2]

| | Type(s) of Studies Needed | | |
| --- | --- | --- | --- |
| Recommendation | Disciplinary—Within a Single Social Science Discipline | Interdisciplinary—Within Social Sciences | Interdisciplinary—Beyond Social Sciences[3] |
| 3.1 Assess how event characteristics affect disaster impacts, mitigation, and preparedness | ○ | ○ | ● |
| 3.2 Refine concepts in hazard vulnerability analysis | | ○ | ● |
| 3.3 Identify mechanisms to reduce vulnerability | ○ | ● | |
| 3.4 Identify factors promoting mitigation adoption | | ● | ○ |
| 3.5 Assess effectiveness of mitigation programs | ○ | ● | ○ |
| 3.6 Identify factors promoting emergency preparedness | ○ | ● | |
| 3.7 Assess implementation of research findings | ○ | ● | |
| 3.8 Identify factors promoting recovery preparedness | ○ | ● | |
| 3.9 Develop models for decision making in emergencies | ● | ○ | |
| 3.10 Conduct research on training and exercising for disaster response | ● | ○ | |
| 3.11 Develop models of hazard adjustment adoption and implementation | ○ | ● | |
| 3.12 Research on hazard insurance | ● | ○ | |
| 4.1 Explore vulnerability and resilience | | ○ | ● |
| 4.2 Compare impacts and responses across natural, technological, and willful events | ○ | ● | ○ |
| 4.3 Homeland security and disaster response | | ● | |
| 4.4 Update knowledge in light of demographic and other societal changes | ○ | ○ | ● |

*continued*

**TABLE 5.2** Continued

| | Type(s) of Studies Needed | | |
| | --- | --- | --- |
| Recommendation | Disciplinary—Within a Single Social Science Discipline | Inter-disciplinary—Within Social Sciences | Inter-disciplinary—Beyond Social Sciences[3] |
| 4.5 Research on events of high magnitude and scope | ● | ○ | |
| 4.6 Cross-cultural research on hazards and disasters | ○ | ● | ○ |
| 4.7 Data archiving and data sharing (hazards and disaster informatics) | ○ | ● | ○ |

1. Recommendations on priority research needs identified in Chapters 3 and 4. 2. ● = type of research primarily needed, ○ = type of research also needed. 3. "Beyond social sciences" = between social sciences and natural sciences and/or engineering.

this trend is necessary and should be encouraged. Moreover, the greatest needs are for interdisciplinary research of a type that is often overlooked—studies that integrate knowledge across various disciplines within the social sciences.

## EXEMPLARS AND LESSONS

This section presents four successful examples of interdisciplinary research in hazards and disaster social science. These studies were selected by the committee to demonstrate various forms, advantages, and contributions of interdisciplinarity. They also provide specific insights into both the challenges faced by researchers engaging in interdisciplinary research and the key ingredients for overcoming these challenges.

The four studies summarized below represent different types of successful interdisciplinary research in social science hazards and disaster research. Three of these studies were supported to a large degree by NSF, primarily through NEHRP. In the first, a study of the economic impact of infrastructure failures in earthquake disasters, social scientists collaborated with engineers under the auspices of an NSF-supported earthquake engineering research center. In the second, a study of human casualties in earthquakes, social scientists and engineers collaborated through their own initiative, outside the context of centers. The third study, focusing on decision making for risk protection, represents a case of successful interdisciplinarity between different fields within the social sciences. In selecting the fourth example, sustainability science, the committee looked outside the traditional boundaries of hazards and disaster social science research to find an instructive example of successful interdisciplinary research.

## What Is Successful Interdisciplinary Research?

In the growing literature on interdisciplinary research, surprisingly little has been written on assessment. By what criteria should interdisciplinary research be judged to be successful? According to one former NSF program officer (Hackett, 2000:258),

> It is difficult to know whether interdisciplinary initiatives return fair value for the money invested, and it is difficult to measure their performance against that of traditional, disciplinary activities. Partly this is a problem of yardsticks and perspective, with metrics of any sort of science performance hard to come by and with sharp differences in perspectives on the fundamental merit of any sort of interdisciplinary effort.

According to the NRC Committee on Facilitating Interdisciplinary Research, a successful interdisciplinary research program will produce outcomes that influence multiple fields and feed back into disciplinary research. It would also enhance research personnel, creating "researchers and students with an expanded research vocabulary and abilities in more than one discipline and with an enhanced understanding of the interconnectedness inherent in complex problems" (NRC, 2005:150).

For purposes of this report, the committee has adopted the following general indicators of successful interdisciplinary research:

- It seeks to advance knowledge in ways not possible through traditional, disciplinary research (e.g., through questions addressed or methods used).
- It involves substantive collaboration among a team of researchers with diverse expertise and training (including at least one social scientist).
- It produces outcomes that are significant and influential.

While recognizing that interdisciplinary research can, in rare cases, be conducted by a single individual, the committee chooses to focus on the more typical cases involving a team of at least two researchers from different disciplines.

### Exemplar: Modeling How Infrastructure Failures
### Impact Urban Economies

Interdisciplinary research, particularly collaboration between social scientists and engineers, has been a continuing focus of the earthquake engineering research centers supported by NSF through NEHRP. As noted earlier, NCEER, particularly in the latter years of its 10-year existence, was a pioneer for involving social scientists as researchers and members of its leadership structure. The three earthquake engineering research centers (MCEER, MAE, and PEER) that succeeded it, which are approaching the end of their NSF-supported tenures, all engage social scientists in their research programs to varying degrees. Integrating research across the social sciences and engineering has proven challenging, however, even in the facilitating context of these centers. The learning curve has been steep, and success has been mixed.

One of the earliest cases of social scientists collaborating with engineers in the context of these centers has also been one of the most successful. Since the mid-1990s, researchers at NCEER and its direct successor MCEER have been developing increasingly sophisticated methods for assessing the social and economic losses caused by lifeline infrastructure failures in earthquakes. Lifelines such as electric power, water, and transportation systems provide critical services to every sector of society. Disasters have repeatedly shown that lifelines are highly vulnerable to damage and cause serious, wide-ranging impacts when they fail. Yet these broader impacts have not been considered in other loss estimation models, most notably FEMA's HAZUS™ model.

Assessing the societal impacts of lifeline outages is an intrinsically interdisciplinary research problem. It requires addressing many, if not all, of the elements and interactions of the conceptual model of societal response to disasters presented in Chapters 1, 3, and 4 (see Figure 1.2), including linkages between pre-impact conditions of hazard exposure, physical vulnerability, and social vulnerability with conditions of the specific earthquake, pre-impact interventions, and post-impact responses. In the case of economic impacts of a water delivery system, for example, researchers must assess not only the extent of physical damage, but also the spatial pattern of water outage, restoration plans, and outage duration, the spatial and sectoral distribution of impacted businesses; and the sensitivity of business activity to water outage. To address this problem, the NCEER-MCEER research team brought together expertise in structural and systems engineering, urban planning, sociology, and economics.

Some of the advances made in this research can be regarded as contributions by individual researchers to their home disciplines. For example, engineers developed new methods for conducting network flow analysis under conditions of earthquake damage (Hwang et al., 1998). Sociologists conducted business surveys that greatly advanced knowledge of how businesses prepare for, respond to, and are affected by disasters. Economists developed new models for capturing the way businesses and economic sectors respond resiliently to disasters (Rose et al., 1997; Rose and Liao, 2005).

Each of these individual advances contributed to the integrative core of the project, a model that estimates not just physical damage to lifelines in earthquakes but also the consequent outages and economic impacts (Shinozuka et al., 1997, 1998; Chang et al., 2002). Integration consisted largely of sequential linkages. For example, systems engineering models produced maps of initial outage patterns. These outage maps were used in a restoration model to assess utility restoration over time. Outage and restoration data were integrated in a geographic information system (GIS) with spatial data on business locations to estimate economic activity at risk. Survey-based information on the differential vulnerability of various types of businesses was used to translate these estimates into expected loss outcomes. This collaboration provides an example of a type of interdisciplinary research in which the contributions of different disciplines are effectively and productively linked, without fundamentally transforming the nature of the research in each area.

The integrated model was developed as a simulation tool. This allowed characterization of uncertainty in the outcomes. It also enabled policy analysis through "what-if" exploration and comparison of intervention options ranging from pre-disaster structural mitigations to strategies for post-disaster restoration. The NCEER-MCEER model has been applied to case studies of electric power and water systems in Memphis, Tennessee, and Los Angeles, California. The original loss estimation framework has been expanded and refined to address community resilience to disasters (Bruneau et al., 2003; Chang and Shinozuka, 2004).

The degree of interdisciplinarity in this effort is evident in the research outcomes. The integrated model has been documented in numerous publications coauthored by engineers and social scientists. These include journal articles, conference presentations, center research reports, and a center monograph.

The more intangible successes of this interdisciplinary research are also noteworthy. The NCEER-MCEER lifelines project produced research personnel experienced in and committed to interdisciplinary inquiry. Building on the experience of the centers, several of the key researchers—engineers as well as social scientists, established as well as young investigators—pursued further interdisciplinary research outside the context of the Centers. In some cases, the same investigators continued working together, and in others, they formed new collaborative teams. The project also paved the way for the recent blossoming of a distinct literature on modeling the spatial economic impacts of disasters (e.g., Okuyama and Chang, 2004).

How, in this research example, were barriers to interdisciplinarity discussed earlier in this chapter overcome? Indeed, the research team did encounter barriers ranging from miscommunication to mistrust to perceived lack of academic rewards. It took several years, perhaps the better part of a decade, for productive collaborations to form.

Three factors proved essential to success: problem-focused collaboration, certain characteristics of the research team, and the center environment. The collaboration focused initially on a demonstration study for the Memphis electric power system, where NCEER engineering researchers had already made headway and were eager to extend their engineering loss results to measures of societal impact. Social scientists (primarily economists) were able to communicate and collaborate with the engineers by focusing on the specifics of the problem—for example, by understanding that for a given earthquake scenario, the engineering models simulated electric power outage by the utility's service area and that these results could serve as linkages to economic impact models. The case study focus also provided a clear end goal for the collaboration: quantitative estimates of social and/or economic disruption resulting from power outage.

Characteristics of the research team, which evolved over time, were also crucial to success. The literature suggests that senior faculty may be best suited for interdisciplinary collaborations because they are able to "risk time out of the disciplinary mainstream" and, moreover, "often need new challenges" (Klein, 1990:182). This pattern applied in the NCEER-MCEER example, where a senior engineer (who served as Center director for a time) and a senior economist were mainstays of the effort. However, the project was only able to get off the ground through the efforts of other researchers who were able to serve as translators between engineering and economics. Key linkages were made by one junior researcher who had been formally trained in both areas.

Finally, the importance of the center environment in fostering this research cannot be underestimated. The literature on interdisciplinarity suggests the benefits of small groups with stable membership. In this case, the centers served as an incubator for an initially risky endeavor with little precedent in the field. It allowed the research team to evolve its membership and develop trust, interdisciplinary language, and collaborative practices over a period of several years.

The centers also provided important support for the endeavor. In addition to grant funding, this support included research infrastructure, such as the opportunity to publish in center technical reports, research accomplishment volumes, and monographs. Regular center meetings provided forums for researchers to meet and develop relationships with investigators from other disciplines. Frequent peer reviews of the centers by NSF panels, which consistently sought evidence of collaboration between engineers and social scientists, also provided immediate and important impetus to the interdisciplinary research. For example, the lifeline project was showcased in several NSF site reviews of the centers.

It is fair to say that this research would not have occurred or succeeded—certainly not in its initial stages—without the supportive and facilitative center environment.

## Exemplar: Analyzing Casualties Through a Standardized Framework

After the 1994 Northridge earthquake, a number of researchers in Southern California were funded to study injuries in that earthquake (Shoaf et al., 1998, 2001; Park et al., 2001; Seligson and Shoaf, 2002; Seligson et al., 2002; Peek-Asa et al., 2003). The group included researchers at University of California, Los Angeles, the Los Angeles County Department of Health Services (LAC-DHS), and the California Department of Health Services. Furthermore, the California Governor's Office of Emergency Services (OES) had funded a major risk-consulting firm in Southern California to gather data on many aspects of the earthquake, including fatalities and injuries. The funding for these studies came from numerous sources, including NSF, and had different requirements. Since the senior researchers from the public health sector were all well acquainted, they made a conscious decision to meet to ensure the consistency of their methodologies and definitions of injury. One of the senior researchers at UCLA was also involved in the hazards and disasters community and invited researchers from the consulting firm to attend the meeting. The meeting included two sets of researchers from UCLA, researchers from LAC-DHS, and researchers from the consulting firm. Each of the research teams consisted of a senior researcher and at least one advanced doctoral student or junior researcher who served as project manager.

As the research on the Northridge earthquake evolved, the teams agreed on consistent terminology and methodology for the collection of data. As research continued, it was carried out primarily by junior researchers. This group of junior researchers included two injury epidemiologists, a public health educator, and an earthquake engineer. As the data collection came to an end and analysis began, this multidisciplinary team of researchers began to look at analysis and the usefulness of the complete data set collected for improving casualty estimation. The earthquake engineer had special expertise in loss estimation modeling. A proposal was submitted to NSF to utilize this unique data set to improve casualty estimation modeling. The public health educator and earthquake engineer served as coprincipal investigators on the project.

The research project, funded by NSF, culminated in a standardized data classification scheme for earthquake-related casualties (Shoaf et al., 2002). This classification scheme was unique in that it attempted to include standards for data that were in the domain of the engineers and geosciences (hazard characteristics and building characteristics) as well as those in the domain of public health and the medical sciences (sociodemographic characteristics and injury characteristics). This process led the earthquake engineer and the public health educator to see themselves as translators. The earthquake engineer would translate technical information from the engineering and geoscience communities into language that the public health educator could translate back to the epidemiology and medical communities, and vice versa. Ongoing collaboration between the public health educator and the engineer has resulted in a number of studies on earthquake casualties, as well as the development of casualty models for other hazards including flooding and a number of terrorism scenarios (Peek-Asa et al., 2001).

Factors influencing the success of this interdisciplinary association included characteristics of the team, mentorship of senior researchers, and the focused nature of the research. While the literature suggests that senior researchers are more likely to be successful in interdisciplinary projects, the success of this team was primarily the result of the junior researchers. Perhaps because public health is in itself an interdisciplinary field, the public health educator has been able to succeed in the field of public health while conducting research almost exclusively on disasters in an interdisciplinary fashion. Furthermore, the engineer engaged on this team had worked extensively in loss estimation and not exclusively in structural engineering. As young researchers, this team developed a new discipline in which, over the decade, they have become leaders in the field.

The effect of the fact that researchers from a variety of fields had been calling for this type of research cannot be overlooked in the success of this team. Early in the research on the Northridge earthquake, members of this Northridge research team participated in a meeting of the U.S. Interdisciplinary Working Group on Earthquake Casualties. Participating in this working group, which had been meeting since the 1980s, lent credibility to the new research being done by this research team and encouraged it to continue the efforts begun by a number of other, more established researchers in the field.

### Exemplar: Understanding Decision Making for Risk Protection

Knowledge of hazards and disasters has also been advanced by research that crosses disciplinary boundaries within the social sciences. One of the most successful examples concerns a long-standing collaboration between an economist, Howard Kunreuther, and a psychologist, Paul Slovic.

Through work spanning three decades, they have individually and jointly made major contributions to understanding how individuals perceive risk, manage risk, and make decisions regarding insurance and other forms of risk protection, and the implications for public policy (Slovic et al., 1974; Kunreuther and Slovic, 1978, 1996; Kunreuther et al., 1978, 1998; Slovic, 2000; Flynn et al., 2001). Their initial collaboration was supported by NSF through the Directorate of Research Applied to National Needs; much of their later collaborative research was also supported by NSF, some of it through NEHRP.

This collaboration was initially catalyzed by regular meetings of the Natural Hazards Research and Applications Information Center (established in 1976 at the University of Colorado at Boulder) and the encouragement of its founder, Gilbert White. As Slovic (2000:xxi) recalls,

> In 1970, I was introduced to Gilbert White, who asked if the studies on decision making under risk that [another collaborator] and I had been doing could provide insights into some of the puzzling behaviors he had observed in the domain of human response to natural hazards. Much to our embarrassment, we realized that our laboratory studies had been too narrowly focused on choices among simple gambles to tell us much about risk taking behavior in the flood plain or on the earthquake fault.

Questions from White's pioneering work on risk perception of flood hazard, such as why people who live in dangerous areas always return to live there after a disaster, or whether it was true that people react differently to risk if consequences are immediate as opposed to delayed, intrigued the psychologist and induced him to begin working on applied research problems. Discussions with the economist, Kunreuther, led initially to an influential overview paper (Slovic et al., 1974) that introduced recent research in psychology, including the work of A. Tversky and D. Kahneman (who won the Nobel Prize in Economics in 2002), and made linkages to the hazards and disasters field.

A few years later, Kunreuther began an NSF project on individual decision making for insurance and invited Slovic to participate on the team. The project, documented in *Disaster Insurance Protection* (Kunreuther and Slovic, 1978), involved an unusual blend of laboratory experimental work with field study. The field study included an extensive telephone survey of more than 3,000 insured and uninsured homeowners in floodplains and earthquake zones across the United States. Collaboration occurred throughout the project; for example, the economist and psychologist worked together to design the survey and jointly pilot-tested the questionnaire in person in neighborhoods of San Francisco. The laboratory experiments, led by the psychologist and closely advised by the economist, were designed to complement the survey. For instance, the survey found that homeowners had poor knowledge of the hazard and generally took little action to mitigate their risk. Laboratory findings suggested an explanation: "that people refuse to attend to or worry about events whose probability is below some threshold," where the threshold could vary between individuals and between situations (Kunreuther et al., 1978:236). Results further showed that people did not perceive insurance in the ways that economists had assumed. The study found, for instance, that people insured not against low-probability, high-consequence events, but against high-probability, low-consequence events—in effect viewing insurance as a form of investment (Slovic, 2005).

This early interdisciplinary collaboration appears to have played an important role in the careers of these influential researchers. Slovic, in particular, credits his early experiences in natural hazards research with expanding his horizons beyond the "usual narrow path" of the experimental psychologist, in particular, sensitizing him to "risks in the real world." This led him to study technological risk, an issue of great currency in the 1970s, and to focus on issues of risk perception, whereas in his laboratory work, he had been more interested in issues of risk taking. This led to productive collaborations with a number of other researchers, work on risk and decision making in a societal context, and more than 50 papers on risk perception (Slovic, 2005).

A number of factors appear to have been significant in the success of this case. First, and arguably most important, is the involvement of researchers "who find unraveling a complex transdisciplinary issue at least as important as their own discipline" (Wilbanks in NRC, 2005). Curiosity and open-mindedness appear, along with a proclivity for intellectual collaboration, to have been important drivers. It may have helped that Slovic was working outside a university environment. A second factor was the problem-focused nature of the research. The complexity of the applied problem—that is, how people behave in the face of natural hazards and how they make decisions concerning insurance—demanded an interdisciplinary approach. It is also significant that the researchers placed high value on "integrating descriptive and prescriptive elements" in their research, insisting on both advancing knowledge and providing guidance for policy (Slovic, 2005). Third, the Natural Hazards Center at Boulder, the mentorship of Gilbert White, and grant support from NSF all appear to have provided crucial support in both tangible and intangible forms.

### Exemplar: Sustainability Science

Instructive experiences can also be found in fields allied with hazards and disaster studies. The case of "sustainability science" demonstrates the possibility, processes, and challenges associated with developing fundamentally interdisciplinary conceptual frameworks and research agendas that cross boundaries between the natural and social sciences. This example is especially apt because of the prominence of "sustainability" as a vision in the recent Second Assessment of research in the hazards and disaster field, wherein sustainability "means that a locality can tolerate—and overcome— damage, diminished productivity, and reduced quality of life from an extreme event without significant outside assistance" (Mileti, 1999b:4).

The idea of sustainable development emerged in the 1980s originating from multidisciplinary scientific perspectives on the interactions and inter-dependencies between society and the environment. The concept gained political traction and broader acceptance through two important and influential endeavors, both supported by the United Nations—the Brundtland Commission report (WCED, 1987), and the UN Conference on Environment and Development in Rio de Janeiro in 1992 and its Agenda 21 report (UNCED, 1992).

For the past two decades the international science plan for global environmental change was largely based on getting the correct scientific understanding of the interactions between the geosphere and biosphere as they influence climate change and other perturbations. The Intergovernmental Panel on Climate Change (IPCC) process initially focused on scientific questions, but within the past decade, the emphasis has shifted toward understanding the societal responses to climate change. One milestone in this transition from a purely natural science to an integrated natural science-social science perspective was the publication of the NRC report *Our Common Journey* (NRC, 1999b). Then-president of the National Academy of Sciences (NAS), Bruce Alberts, "saw in the idea of a sustainability transition the great challenge of the coming century and consistently urged the board to explore and articulate how the science and technology enterprise could provide the knowledge and know-how to help enable that transition" (NRC, 1999b:xiv). Funded with foundation support (rather than by federal agencies asking for advice), and a strong personal interest and leadership from the National Academies, this report lays out a research agenda for "sustainability science" (NRC 1999b:11):

1. Develop a research framework that integrates global and local perspectives to shape a "place-based" understanding of the inter-actions between environment and society.
2. Initiate focused research programs on a small set of understudied questions that are central to a deeper understanding of interactions between society and the environment.
3. Promote better utilization of existing tools and processes for linking knowledge to action in pursuit of a transition to sustainability.

Many of the members of the original Board on Sustainable Development (and others who participated in the workshops) had known one another for a long time and shared similar philosophical and intellectual predilections (Turner, 2005). Their work and interaction continued beyond the publication of the NRC report, especially in the promotion of scientific research in sustainability (Kates et al., 2001). When the U.S. Global Change Research Program wanted to explore some of the themes in more detail, they went to this group of scholars (William C. Clark, Robert Kates, Pamela Matson, Robert Corell, and Billie L. Turner, among others). With National Science Foundation support (with contributions from NOAA and NASA), an interdisciplinary group began meeting to discuss the conceptual and methodological development of sustainability science. The entire group met annually for a period of three years, with side conversations and work done at the participating institutions—Clark, Stanford, and Harvard universities and the Stockholm Environment Institute. The intensive summer annual workshops were a "must go." From these workshops, the initial result was a series of published articles in the *Proceedings of the National Academy of Sciences* in 2003 that articulated both the conceptualization of the field and exemplars of how to implement them at various scales (Cash et al., 2003; Clark and Dickson, 2003; Kates and Parris, 2003; Parris and Kates, 2003; Turner et al., 2003a,b).

The success of the interdisciplinary research collaboration has been fostered by personal relationships among key participants, a common scholarly view of the need for better understanding of nature-society interactions, keen personal interest from leaders of the scientific establishment (NAS and the American Association for the Advancement of Science [AAAS]) and outside political forces (societal needs identified by the United Nations and others). The most significant outcomes to date have been in the conceptual development of the field, but the actual implementation of the science agenda has not happened in any meaningful way. The barrier has and continues to be funding. When sympathetic program managers left the primary mission agencies (NASA and NOAA) that were funding such work, funding languished. Despite this, sustainability science (as an integrated and interdisciplinary field of study) continues to enjoy strong intellectual support from the leadership of the scientific community (Raven, 2002).

## Lessons for Successful Interdisciplinary Research

These four exemplars were selected to represent different types of interdisciplinary research. In reviewing factors leading to their success, however, a number of commonalities emerge:

- *Support from senior leaders.* All four exemplars cite this factor, whether in the form of personal commitment to the interdisciplinary inquiry from senior researchers who were themselves involved in the research, mentoring and encouraging of junior researchers to collaborate, or personal interest from leaders of the scientific community.
- *Financial support from granting agencies.* All four cases were supported by NSF, as well as other sources. In some cases, the grant funding catalyzed the collaboration, while, in others it enabled in-depth empirical studies to follow on conceptual discussions of interdisciplinary frameworks. Although difficult to verify, it appears that without this financial support, none of the collaborations would have flourished for long or, in some cases, materialized at all. In one case, as previously noted, research progress was impeded when funding was lost because sympathetic program officers left the supporting funding agencies.
- *Forum for continuous dialogue.* In three of the cases, an institutional meeting ground (either a multidisciplinary center or a series of formal meetings) appears to have been important for fostering, if not also initiating, the intellectual dialogue across disciplines. This seems to have been particularly important when collaborators did not already have long-standing personal relationships, particularly where social scientists needed to establish new collaborations with natural scientists and engineers.
- *Focus on an applied problem.* The three exemplars from the hazards and disaster field all noted that focus on an applied problem greatly facilitated interdisciplinary research. The complexity of the societal problem exceeded the bounds of any traditional discipline and required an interdisciplinary approach. Moreover, the problem focus provided clarity and specificity regarding the nature of the interdisciplinary knowledge needed.

A number of other common factors in success are also apparent, to a somewhat lesser degree, across the cases. Although each case cited "characteristics of the research team," somewhat different characteristics were noted for each. They included junior researchers who could serve as interdisciplinary links or translators, long-standing personal relationships between the collaborators, and open rather than discipline-bound intellectual perspectives. Three of the cases involved at least one key participant from outside a university setting, which may have reduced the academic institutional barriers to collaboration that are often cited in the literature. Two of the cases noted the importance of external calls from the scientific community for interdisciplinary research on the specific problem. These factors appear to be important in some circumstances, but they do not appear to be as robust explanations as the ones listed above. The cases profiled here corroborate many of the findings summarized earlier from the larger literature on interdisciplinary research.

## RECOMMENDATIONS

Interdisciplinarity in hazards and disaster research is growing. Interdisciplinary research, both within and beyond the social sciences, has made major contributions to the field. Interdisciplinarity figures prominently in the research needs of the field. While unanswered disciplinary questions remain, all of the priority research needs identified by the committee (see Chapters 3 and 4) involve multiple disciplines and are in part, if not fundamentally, interdisciplinary.

Research centers have proven to be very important in facilitating interdisciplinary research, as demonstrated in the hazards field and reinforced by recommendations in related fields. A workshop on integrated research in risk analysis and decision making yielded a consensus recommendation that "the most effective way to achieve program goals is to fund multidisciplinary centers." (NSF, 2002:7) In the area of human dimensions of global environmental change, centers have been advocated in order to strengthen key linkages between the natural and social sciences (Stern et al., 1992).

The committee makes three recommendations regarding interdisciplinary hazards and disaster research.

**Recommendation 5.1:** *As NSF funding for the three earthquake engineering research centers (EERCs) draws to a close, NSF should institute mechanisms to sustain the momentum that has been achieved in interdisciplinary hazards and disaster research.*

In 2007, the three EERCs will come to the end of their 10-year terms of NSF support. At the same time, the Network for Earthquake Engineering Simulation (NEES), at a cost of more than $80 million, will soon dominate the landscape of NSF-supported hazards research. Both of these changes threaten the momentum that has developed with regard to social science involvement in interdisciplinary hazards and disaster research. Within the EERCs, a necessary condition for the fostering of social science research and interdisciplinary collaborations was the sustained pressure from annual NSF site review teams. As the EERCs "graduate" to self-sustaining financing structures and seek support from the private sector, it is likely that the role of social science research will be diminished. At risk are the valuable lessons, experience, and momentum developed over the last two decades.

Within NEES, because of its emphasis on laboratory testing of physical structures, opportunities for social science involvement appear to be very limited. While the NEES research agenda cites the importance of interdisciplinary collaboration with the social sciences, none of the specific recommendations within that research agenda reflect this importance (NRC, 2003b). Within NEES, the Grand Challenges program, which funds research on "compelling national research" problems that require a "comprehensive systems approach" and "in-depth, cross-disciplinary, and multi-organizational investigation" (http://www.nsf.gov/pubs/2005/nsf05527/nsf05527.htm), provides the most likely context for social science involvement. However, the NEES program has not funded any Grand Challenges research projects to date.

Recommendation 5.2: *The hazards and disaster research community should take advantage of current, unique opportunities to study the conditions, conduct, and contributions of interdisciplinary research itself.*

Social science expertise on subjects ranging from individual decision making to organizational effectiveness and evaluation research should be utilized to study interdisciplinary research in the hazards and disaster field. One opportunity consists of research on NEES; for example, to investigate how a spatially distributed network structure influences the research enterprise and to evaluate the effectiveness of such a structure. A second opportunity is the impending "graduation" of the earthquake engineering research centers from NSF funding to industry and other forms of financial support: for example, to study how this change affects the role of interdisciplinary research generally and interdisciplinary research involving the social sciences, in particular; to study centers and how they do or do not work effectively; and to systematically investigate team building in hazards research. A third opportunity would be to make similar comparisons between research supported by NEHRP and that supported by the Department of Homeland Security.

Recommendation 5.3: *NSF should support the establishment of a National Center for Social Science Research on Hazards and Disasters.*

In such a center, the committee envisions a distributed consortium of researchers and research units across the United States, with affiliated members located across the world. Similarly to NEES, it would take advantage of telecommunications technology to link spatially distributed data repositories, facilities, and researchers. It would provide an institutionalized, integrative forum for social science research on hazards and disasters, much as the Southern California Earthquake Center (SCEC) does for the earthquake earth sciences community. The key charges of the center would include

- facilitating access to and use of disaster data;
- coordinating post-disaster reconnaissance efforts of social scientists;
- providing consensus statements from the research community to inform public policy;
- providing educational materials (i.e., integrating existing materials, developing new ones, and disseminating both), such as Web-based short courses, that can help disseminate social science research findings to a broad range of audiences, including students, investigators new to the field, potential collaborators in other disciplines, and researchers in developing countries;
- supporting researchers in developing the expertise they need to successfully engage in interdisciplinary research—for example, through doctoral and post-doctoral opportunities, sabbaticals, career development awards, or formal training (see Pellmar and Eisenberg, 2000:11; for an example, see www.nianet.org); and
- catalyzing interdisciplinary collaborations, both within the social sciences and between the social sciences and natural sciences and/or engineering; for example, through convening workshops and symposia.

Core nodes of the network would include existing university-based research centers that are focused on hazards and disaster research (see Chapter 8), those DHS centers of excellence that involve social science research (e.g., the National Center for the Study of Terrorism and Responses to Terrorism), and the new centers recommended by this committee—the Data Center for Social Science Research on Hazards and Disasters (see Chapter 4) and the Center for Modeling, Simulation, and Visualization of Hazards and Disasters (see Chapter 7). However, individual researchers not associated with these existing centers would also have access to this distributed network.

The center would receive core funding from NSF and mission agencies such as DHS, NOAA, and NASA. It would leverage these funds to attract support from state and local governments, as well as international agencies and the private and not-for-profit sectors.

Such a center arrangement would provide several important benefits for social science research on hazards and disasters. First, it would provide a "critical mass" research network. The field is small, characterized by a modest number of core researchers, spread over many disciplines and many institutions, and bolstered by others who are only intermittently involved in hazards research (see Chapter 9). Achieving a critical mass is important for attracting and retaining researchers, as well as catalyzing interdisciplinary collaborations (see, for instance the first and third exemplars above).

Second, such a center would elevate the stature of the field. This would enable social scientists to negotiate interdisciplinary, collaborative research agendas with their natural science and engineering counterparts on coequal footing. This could lead, for example, to interdisciplinary collaborations on hazards and disasters that address fundamental dynamics of social change (see Chapter 2). Such research has not been possible in the context of the EERCs, where social scientists comprise a small minority and research agendas have been set predominantly by engineers. The envisioned center would allow social science insights and concerns to influence, rather than simply extend, priorities in natural science and engineering research for the ultimate goal of making society safer.

Third, such a center would provide needed international leadership. The benefits of critical mass and stature noted above could be especially important in other countries, where social science research on hazards and disasters is often poorly established. Moreover, the benefits of an international network also extend to U.S. researchers, particularly in promoting collaborative research on the linkages between disasters and development (see Chapter 6).

# 6

# International Research: Confronting the Challenges of Disaster Risk Reduction and Development

Worldwide, natural disasters cause catastrophic losses. Average annual economic losses caused by disasters were $75 billion in the 1960s, $138 billion in the 1970s, $213 billion in the 1980s, and more than $659 billion in the 1990s (Munich Re, 2002).[1] While most losses are in developed countries, these estimates fail to capture the impact of disasters on poor countries that often bear the brunt of losses in terms of lives and livelihoods. Compared to developing countries, the absorptive capacities of developed countries are greater, the impact ratios on economies are smaller, and the recovery rates are more rapid. Further, 85 percent of people exposed to natural disasters reside in countries of medium or low economic development (Munich Re, 2002).

The process of development has a major impact on disaster risk (ISDR, 2004). In some countries, development means greater ability to afford the investments needed to build more disaster-resilient communities. In other countries, growth is accompanied by haphazard development decisions that place more people and property at risk. In the wake of these patterns, rebuilding from disasters has been devastating to poor countries, as losses consume vast amounts of limited available capital, significantly reducing resources for new investment. The adverse effects on employment, balance of trade, and foreign indebtedness can be felt for years.

---

[1] All loss estimates based on 2002 dollars.

This chapter is concerned with assessing the current state of knowledge about disasters and development. It begins by reviewing global patterns in disaster risk and development and introduces the concept of sustainable development as a vision for creating disaster-resilient places. Next, the major institutional obstacles to the advancement of disaster resiliency are discussed. The committee then offers a definition of success in terms of disaster resiliency and reviews influences on achievement of this goal premised on theories of governance and social capital. Next, collaborative international research efforts are reviewed that can potentially offer robust opportunities for comparative analyses of these influences on disaster resiliency. Finally, the committee develops research-based recommendations that offer guidance for confronting the challenges posed by disasters to development and outlines future research needs.

## GLOBAL PATTERNS IN DISASTER RISK AND VULNERABILITY

Understanding global patterns of disaster risk entails a review of key concepts of development that address the vulnerability of human communities. Use of these concepts to model relationships between risk and development requires reliable data across disaster events and cultures. Although there are significant limitations in data, preliminary studies have begun to explore the links between disaster risk and development at various spatial scales.

### Concepts of Risk and Vulnerability

The relationship between disaster risk and development is complex and multifaceted. Risk refers to potential for loss of life and property damage. As noted in Chapter 1, *disaster risks* are products of the *disaster event* and the degree of *vulnerability* of human communities that sustain losses from the event. The destructive power of the *disaster event* is influenced by several physical characteristics (e.g., magnitude and scope of impact, length of forewarning) as well as the degree of exposure to impacts. The physical force of a disaster, however, is insufficient to explain risk. Areas that experience equivalent levels of physical force of a given *disaster event* have widely varying levels of risk. *Vulnerability* is the concept that explains why, with the equivalent force of disaster, people and property are at different levels of risk.

*Vulnerability* consists of various social, economic, and natural and built environmental indicators of societal development that represent the capability of a human community to cope with a disaster event (Kasperson et al., 2001). Sen (1981) has demonstrated that given equivalent availability of food, food crises may occur in some areas but not others due to unequal vulnerabilities in human communities. The difference is rooted in social and economic entitlements, but not in the severity of the physical characteristics of natural disaster events such as drought or floods. Other lines of research focus on the importance of natural systems and the effects of system change on disaster risk. Extensive research on the impacts of climate change which were assessed by the Intergovernmental Panel on Climate Change (IPCC, 2001), revealed that system change has direct effects on risk and indirect effects through the vulnerability of human populations. A direct effect is increased risk to flooding in coastal settlements, especially those settlements in low-lying coastal areas, in deltas, and on small islands. An indirect effect is the decline of life support functions of coastal ecosystems (e.g., coral reefs and estuaries that support fisheries, recreation, and wildlife habitat) causing increased vulnerability of coastal populations, especially in developing countries, that depend on these ecosystems, which in turn would decrease their capability to cope with future risk (Adger et al., 2005).

## Reliability of Data

A major constraint to conducting risk assessments is the absence of reliable data. Comparative assessments of losses at various spatial scales have been made through the use of a wide variety of sources from government compilations, scientific publications, and census information (LA RED, 2002; IPCC, 2001; UNDP, 2004).[2] However, systematic record keeping on losses and associated vulnerability indicators is sketchy at best in developed countries and almost nonexistent in developing countries.

Loss of life is the most quantifiable measure, and the most consistently recorded type of disaster loss throughout the world, and frequently constitutes the only loss data available after disasters. Because mortality is considered more reliable than other types of data, it is often viewed as the best indicator for comparative assessments, especially between disasters in developed and developing countries (UNDP, 2004). However, use of deaths as a proxy for disaster risk limits its analysis in relationship to societal development. As noted in Chapter 1, disasters affect people's lives and livelihoods in many ways other than loss of life. Mortality data do not capture a broader range of other development losses linked to disaster risk trends, and can only point to comparative orders of magnitude in vulnerability and loss. Thus, social, economic, and environmental (built and natural) losses linked to disaster risk should complement analyses based on life losses.

---

[2]The University of Leuven, Belgium, maintains a central repository of disaster loss data (see www.em-dat.net, accessed March 24, 2005).

## Relationships Between Disasters and Development

While death may not be the best indicator, it offers some insight about the relationships between disaster losses and development. A study by the United Nations Development Programme (UNDP, 2004) found that millions of people in the world suffer from disasters each year, with a disproportionate share in less developed countries, and that these disasters claim increasingly high tolls in loss of life. The study indicates that low- and medium-development countries have similar loss patterns due to close relationship between deaths and level of development. For example, Guinea Bissau (low development) and Bulgaria (medium development) experience low levels of death, but Venezuela (medium development) and Sudan (low development) experience high levels of death. However, high-development countries consistently experience low levels of deaths associated with disasters. Specifically, no high-development countries experienced more than an average of 10 deaths per million and more than an average of 500 deaths per year. Both of these figures are exceeded by numerous medium- and low-development countries. Further, countries classified as high development represent 15 percent of the exposed population, but only 1.8 percent of the deaths.

From an economic loss perspective, a study by Munich Re (2002)—a German reinsurance company—estimates that losses are rapidly increasing in developed and developing countries and that between 1992 and 2002 global losses from disasters were 7.3 times greater than in the 1960s. Based on single cases of disaster events, evidence suggests that while total economic losses are greater in developed countries, they are disproportionately greater for developing countries. The economic costs of disasters in poor countries often exceed 3 to 4 percent of the gross domestic product (GDP), and in some extremely economically vulnerable countries in Africa the cost can exceed 20 percent or more of GDP. In instances of geographically small, poor countries, the impacts can be devastating to national economies. The $330 million loss from Hurricane Lewis sustained by the island state of Antigua in 1995 was equivalent to 66 percent of GDP (UNDP, 2004).

In contrast, the $30 billion in losses caused by Hurricane Andrew in South Florida in 1992 (the most costly hurricane ever to strike the United States until Katrina in 2005) represents an almost undetectable percentage of the country's $6 trillion economy in 1992. Losses thus have less to do with the scope of the physical impact than the relative proportion of the population and economy involved. Thus a key research and public policy issue is the link between poverty and vulnerability to disasters. The committee addresses this issue by placing disasters in the context of sustainable development.

## SUSTAINABLE DEVELOPMENT AND DISASTERS

Sustainable development offers a promising public policy perspective for guiding decisions to create more disaster-resilient societies and communities. The committee's intent is to define sustainable development and discuss how this concept can be applied in ways that integrate disaster and development issues.

### What Is Sustainable Development?

By the turn of this century, hazards and disaster management had become energized by the challenges of achieving the goal of sustainable development. The concept of sustainable development seeks to focus attention on integrating often competing normative visions about ecological limits, economic development, and intergenerational equity, as reflected in the familiar definition of the report *Our Common Future* by the World Commission on Economic Development (WCED, 1987:8): "Sustainable development is development that meets the needs of the present generation without compromising the ability of future generations to meet their own needs." As its United Nations origin attests, sustainable development is a global vision. It has been taken up by multinational development institutions such as the World Bank and UN organizations, national government groups in developed and developing countries designing conservation strategies, and NGOs (nongovernmental organizations) active in the worldwide environmental movement.

Since the WCED report was published in 1987, increased attention has been given to the role of the sciences in fostering societal transition to sustainability. The National Research Council (NRC) report *Our Common Journey: A Transition Toward Sustainability* (1999) indicated a need for "significant advances in basic knowledge, in the social capacity and technological capabilities to utilize it, and in the political will to turn this knowledge and know how into action" (NRC, 1999b:7). The major recommendation in that report was to outline a research agenda for "sustainability science" that includes the development of an interdisciplinary research framework. This framework builds on the intellectual foundations of the geophysical, biological, social, and technological sciences. Hazards and disaster research is a major component of this agenda as evidenced by the 2003 *Proceedings of the National Academy of Sciences* that focuses on vulnerability analysis and sustainability science (see www.pnas.org/cgi/content/start/100/14/8080; accessed March 14, 2005). A key conclusion of these proceedings is that vulnerability is explained not by exposure to hazards alone, but also by the resilience of the system experiencing such hazards. The proceedings include a recommendation for revising and enlarging the basic design of current vulnerability assessment models to account for the capacity of human-environment systems, including social structures, institutions, and level of economic development.[3]

## How Does Sustainable Development Apply to Disasters?

The vision of sustainability has influenced the formulation of a generation of international initiatives and also the thinking of hazards and disaster researchers and policy makers that followed the 1987 WCED report. The importance of natural disasters in devising sustainable development strategies was recognized by the United Nations resolution declaring the 1990s as the International Decade for Natural Disaster Reduction (IDNDR). This resolution helped galvanize support for incorporating disasters into development initiatives by stipulating that member nations establish a national program for a decade of disaster loss reduction. The successor to IDNDR, the International Strategy for Disaster Reduction (ISDR), which was created by the United Nations in 2000, promoted the sustainability agenda by focusing on the integration of citizen participation, awareness building, and consensus with technical disaster risk assessment.

The United Nations Commission on Sustainable Development (2001:2) recently stated the linkage between disasters and sustainability succinctly:

> Can sustainable development, along with the international instruments aiming at poverty reduction and environmental protection, be successful without taking into account the risks of natural hazards and their impacts? Can the planet take the increasing costs and losses due to natural disasters? The short answer is no.

Hazards and disaster researchers have proposed various conceptualizations of the links between disasters and sustainability premised on *disaster resiliency*. The concept of resilience has long been a tradition in ecology (see, for example, Holling, 1973). Resilience, whether for individual organisms or communities, is based on accommodation and ability to adapt to a disturbance from a change agent, such as vector-borne diseases, overharvesting, pollution, fires, and hurricanes. The idea of resilience is increasingly present in social science analysis, and in developing a theory for linked social-ecological processes. In the context of disasters, the concept of resil-

---

[3]Various units within the National Academies have undertaken programs to advance "sustainability science." The Science and Technology for Sustainability Program in the Division of Policy and Global Affairs was created in 2002 to focus on cross-cutting thematic issues (e.g., pollution prevention, biodiversity, water and sanitation) that emphasize how principles of science are an integral part of societal decision making (http://nationalacademies.org/ sustainability; accessed March 14, 2005).

iency denotes strength, flexibility, and the ability to deal with a loss or misfortune and recover quickly. Mileti (1999b:5) defines disaster resiliency as the capacity to "withstand an extreme natural event with a tolerable level of losses" and taking "mitigation actions consistent with achieving that level of protection." Bruneau et al. (2003:735) define community disaster resiliency as "the ability of social units (e.g., organizations, communities) to mitigate hazards, contain the effects of disasters when they occur, and carry out recovery activities in ways that minimize social disruption and mitigate the effects of future [disasters]." Chang and Shinozuka (2004) extend this definition by conceptualizing, measuring, and evaluating resiliency of a community to earthquakes along four interrelated dimensions: technical, organization, social and economic. Other definitions stress the role of city and regional planning in creating resilient natural and built environmental systems (Godschalk, 2003) and cultural values related to historic meanings of resilience and urban trauma (Vale and Campanella, 2005).

It is possible to understand how several of the underlying principles of sustainable development outlined by the United Nation's *Agenda 21*, the first United Nation's agenda for action on sustainability, can be applied to disasters (Sitarz, 1993). These principles can be referred to as the four "E"s of sustainable development for disaster resiliency:

1.  *Ecological limits*: Recognize that disasters are limiting environmental factors to development to ensure that basic health and safety needs essential to human development are met.
2.  *Equity*:
    *   Intergenerational—Account for disasters to ensure efficiency in use of development funds that might otherwise not be available for future investment.
    *   Intragenerational—Improve equity within generations by providing for sufficient low-cost, low-risk development opportunities for the least advantaged.
3.  *Economic development*: Sustainability means that living standards in the future will be higher than in the present and higher levels of development will be associated with greater mitigation and emergency preparedness.
4.  *Engagement*: Development actions that address disaster reduction (and other significant issues) must be formulated through a fair and equitable process that provides an opportunity for all affected parties to participate.

Spatial and social scale is an important factor in translating these principles into practice. Local issues may be quite different, but are often inextricably linked to global processes. For example, global warming may increase the spread of infectious diseases and threaten food production systems at the regional and local scales. At the same time, global processes may be affected by local land-use decisions that support greater dependence on automobiles and increased $CO_2$ emissions, which contribute to global climate change. As scale changes, the disaster mitigation tools change. For example, urban infrastructure investments and land-use plans can shape urban forms and reduce the dependence on cars, while individual countries will be less likely to enact more stringent emission standards unless negotiated international agreements are ratified.

In sum, striking balanced solutions that account for the first three E's (1, 2, and 3) through the process of the fourth E (engagement) is a critical aspect in creating long-range sustainability strategies that achieve disaster resiliency. However, there are major institutional and political obstacles to overcome to achieve balance.

## COPING WITH OBSTACLES TO LINKING SUSTAINABLE DEVELOPMENT TO DISASTERS

While the potential integration of sustainable development with disaster preparedness and mitigation is appealing, efforts to build consensus among organizations and citizen groups have often met with limited success. Aguirre (2002) observes that there is inadequate expertise to make the fundamental cultural and institutional changes required to implement the concept, and that ideologically driven norms associated with sustainability could lead to a discounting of real advances in disaster research and practice outside of these norms. In this chapter, the committee focuses on three major obstacles to achieving integration that make Aguirre's concerns apparent, including the low visibility of disasters in sustainable development policy making, the exclusion of sustainable development from the humanitarian aid delivery system, and a limited horizon of how we define disasters.

### Low Visibility of Disaster Issues

One obstacle is the low visibility of disaster issues in the sustainable development debate. Historically, there has been only limited attention toward integrating sustainable development with disaster reduction efforts. Assessments in the 1990s of mission statements and policies of national governments and multilateral development institutions (e.g., World Bank, Inter-American Development Bank, United Nations High Commission for Refugees) indicate that only a few incorporate disaster reduction as a component of sustainable development among the hundreds of organizations involved in applying sustainable development principles (Mitchell, 1992;

Berke and Beatley, 1997). More recently, many of these institutions have revised their mission statements and begun to integrate disaster loss reduction with their mainstream development activities (UNDP, 2004). Despite a shift in interest, disasters are still often used as indicators of nonsustainable development and as evidence that existing development practices are often not sustainable. There is a need for research on how and to what extent the recent shift toward disaster concerns has influenced long-range development practices and yielded measurable progress in human development.

Because of its historic low visibility, contemporary characterizations of the need for disaster reduction are often flawed. For example, while there is recognition that the connection between disasters and development is strong, this does not mean that disasters will disappear if sustainable development is translated into practice. Indeed, sustainable development does not necessarily translate to safe development. In many countries it is doubtful that improved development practices can prevent catastrophic events completely. Some built environments are too valuable or culturally significant to be abandoned or relocated. The capital cities Mexico City and Wellington, New Zealand are situated astride seismic fault zones, and New Orleans, as indicated by the Hurricane Katrina and other experiences, and Venice will remain susceptible to flooding.

Further, the built environments of megacities are too large and dynamic to be made completely safe (Mitchell, 1999). While in the 1950s there were only four cities with a population greater than 5 million, by 1985 there were 28 and in 2000 there were 39. New scales of vulnerability have emerged with the rapidly growing presence of megacities, including the new dimensions of large high-density concentrations of populations with immense sprawl and a serious increase in infrastructural, socioeconomic, and ecological overload. These cities may develop extreme dynamism in demographic, economic, social, and political processes. Both phenomena—the new scale and dynamism—make megacities highly vulnerable not only to natural hazards but also to technological hazards and terrorist attacks. Such agglomerations are highly complex and have major risks, which present significant challenges.

## Exclusion of Sustainable Development from Humanitarian Aid Delivery Systems

Another obstacle is the exclusion of sustainable development concerns by the international humanitarian aid delivery system, a vast network of emergency relief and development organizations. Harrell-Bond (1986:16) appropriately characterized these organizations as the "conscience of the world." Their primary task is to work in the poorest reaches of the world and to bring international attention to the plight of human suffering.

Until recently, these organizations had not acknowledged sustainable development in shaping their aid programs. Emergency relief organizations often consider disasters as isolated events that require unique, crisis-oriented, societal responses. Disaster-stricken people are often viewed as helpless victims, and aid is distributed free, as a form of charity. However, this perspective has recently been changing as international relief organizations have shifted more attention to building the capacity of local people to take control over the design of aid delivery programs that affect their lives. For example, *Strategy 2010*—the long-range plan for guiding aid delivery activities of the International Red Cross and Red Crescent Societies (IFRC)—reflects this change by emphasizing the linkage between emergency relief activities and local capacity building (IFRC, 2000). The IFRC has taken the place of the World Bank as the secretariat for the disaster reduction-oriented ProVention Consortium—a global consortium of governments and international organizations dedicated "to increasing the safety of vulnerable communities and to reducing the impact of disasters in developing countries" (www.proventionconsortium.org; accessed April 2, 2005).

The historical approach to emergency relief has been to meet short-term needs, but not the underlying problem of disaster vulnerability in poor countries. Studies have found that sometimes the impact of aid from emergency relief organizations can be counterproductive (Harrell-Bond, 1986; Oliver-Smith and Goldman, 1988; Berke and Beatley, 1997; Oliver-Smith, 2001). Aid recipients often adopt attitudes and behaviors that impede their progress toward self-sufficiency. The negative responses, sometimes called the "dependency syndrome," develop when aid recipients are considered helpless, needing outsiders to plan and take care of them. This assumption is the cornerstone of the "starving child" appeals for funds for relief organizations.

The primary objective of development organizations is economic growth and improving the ability of poor countries to cope with the challenges of poverty and underdevelopment. The underlying rationale was that project investment decisions should focus on immediate concerns associated with poverty and that investments would produce more resources to be available for disaster reduction. A report by the World Bank (1990) indicated that up to the 1990s this approach often ignored disasters. The report also indicated that internationally funded development projects during the 1980s were frequently designed for short-term exploitation of natural resources to generate exports to help repay massive foreign debts, but that the projects often exacerbated the severity of disasters by inducing substantial environmental degradation (e.g., increased flooding and landslides caused by excessive deforestation for timber production).

Another oft cited reason for failure to include disasters in development decisions is the common misperception that the devastating effects of disasters are a sign that only the poor suffer during from such events. This misperception is often used to justify denial of funding for disaster prevention activities and support the notion that development is the only solution to reducing vulnerability. Mounting evidence suggests that achieving disaster resiliency is far more complex than the poverty argument would imply since disasters do not only affect the poor (Rocha and Christopolis, 2001).

Since the 1970s, some disturbing trends have emerged. Countries experiencing rapid development suddenly lost momentum when disasters struck. Resources for development often became scarce when they were siphoned off for recovery and reconstruction. At first, it was assumed that more disaster relief from developed countries was needed. In response, annual worldwide development funding among donor countries grew dramatically during the 1980s and up to the peak year of 1992; however, economic losses expanded dramatically during the same period (UNDP, 2004).

Factors determining this outcome are highly complex and difficult to determine given our partial knowledge of the role of international aid delivery strategies and changing societal and environmental conditions. However, prior studies point to the failure of emergency relief and development organizations to link disasters to long-term development issues as an important contributor to the problem. As noted above, until recently emergency relief organizations have not addressed the underlying problem of disaster vulnerability in poor countries, nor have they dealt with resolving problems of underdevelopment. Up until the 1990s, development agencies have not been effective in accounting for disasters. The result has been inefficient uses of development funds, which reduce already scarce resources available for new development.

During the past decade there has been a change in funding plans and priorities of international humanitarian aid organizations (see, for example, UNDP, 2004). The change indicates that economic development should not contribute to the conditions that undermine human and environmental sustainability and increase disaster risk, and emergency relief should recognize the need to build local capacity. To move forward, many of these organizations recognize that there must be a clear understanding of the interaction of emergency relief and development plans with disaster risk. At issue is the need to systematically evaluate the results of these changes.

### Exclusion of Armed Conflicts from the Definition of Disasters

Another obstacle to linking sustainable development to disasters is the limited horizon in defining what is (or is not) a disaster. In contrast to the inclusive definition adopted by the committee in Chapter 1, disaster research has historically limited its definition of disasters to rapid-onset natural and technological events or to slow-onset stressors that continuously increase pressures on natural systems and increase the vulnerability of human populations. The field has given very little attention to slow-onset disasters brought on by armed conflict (Dynes, 2004). Slow-onset disasters created by violence remain understudied and are not connected with the sustainable development debate.

Slow-onset, conflict-driven disasters have been referred to as complex political emergencies or CPEs (Christopolis et al., 2001). CPEs frequently lead to displaced populations that are caught up in ongoing conflicts that often develop slowly. Recent examples during the past decade include the collapse of Yugoslovia, genocide in the Sudan, and places such as El Salvador that experience recurrent, rapid-onset disasters that take place in the midst of conflict. Of the 43 major armed conflicts throughout the world during the 1990s, 17 took place in Africa (Addison, 2000). Since 1990, approximately 70 million people have become international refugees, nearly 40 million people have struggled with starvation, and more than 20 percent of the population has been displaced in 15 developing countries (Addison, 2000).

Wisner (2001) refers to countries or regions that experience CPEs as in a "permanent" state of crisis. They experience a profound, intractable type of conflict, one of acute polarization. In these cases, ethnic and nationalistic claims eclipse social and economic equity claims at the local level. Governance is perceived by at least one ethnic community as either illegitimate or structurally incapable of producing fair outcomes for subordinate ethnic groups. In deeply fractured societies, CPEs are extremely difficult to overcome because fear of "the other" is not only felt at the level of individual behavior but becomes intertwined in specific development (and disaster) issues of every day life (Bollens, 2000).

Given the nature and location of CPEs, they have not generated much interest among the disaster research community in developed countries. CPEs are based on claims that conflict differs from earlier conceptual frameworks for understanding the links between disasters and development in that CPEs cannot be conceived in chronological and most certainly social time as temporary events in what are otherwise "normal" states. A central justification behind the need for new theories of human response to CPEs is that existing theories are not useful in understanding these events. Theories of human response have been borrowed unreflectively from natural disasters and applied to the very different phenomena that occur in the context of CPEs (Green and Ahmed, 1999). Theories of the disaster cycle, for example, are not relevant when it is impossible to differentiate between impact and recovery. The idea of a linear relief to a development continuum for natural disasters assumes that there are clearly defined roles for various organizations in the humanitarian aid delivery system. That is, there is some certainty as to who should do what when the disaster is over. In contrast, famine and drought intertwined with persistent conflict have not been salient topics to most in the Western disaster research and policy communities.

In sum, the preceding discussion identified three major obstacles to the integration of sustainable development with disaster preparedness and mitigation efforts: low visibility of disaster issues in the sustainable development debate; exclusion of sustainable development in the international humanitarian aid delivery system; and historical exclusion of CPEs from the definition of disasters by the research community. Improvement in our knowledge of the causes and consequences of these obstacles is critical to create long-range sustainability strategies that achieve disaster resiliency. Use of a more inclusive definition of disaster is only a first step. As suggested also in Chapter 1 (see Figure 1.2 and related discussion), the essential links between disaster risk and development must be expressed as relationships among the core topics of hazards and disaster research and their expression in social as well as chronological time.

## MODELS OF DEVELOPMENT AND HUMANITARIAN AID DELIVERY SYSTEMS

The greatest challenge to promoting disaster resiliency is to adapt strategies that map with the great variation of types of community vulnerabilities. Communities of refugees, indigenous people, women, children, minorities, and others within a society have different needs and opportunities for developing sustainable, disaster-resilient places. They vary in their capacity to deal with disasters as well as the strength of their ties with outside aid delivery systems.

As noted above, these communities are routinely labeled by external aid organizations as "vulnerable" populations or worse as in a "state of helplessness," rather than as active participants capable of taking self-directed development initiatives. Because all social systems have very different vulnerabilities and capabilities, they have different strategies to cope with vulnerability. Stereotypical generalizations are not only ineffective but are part of a discourse of disempowerment, wherein "they" are viewed as needing outsiders help to plan for them and take care of them (Oliver-Smith, 1990; Berke and Beatley, 1997; Bankoff, 2001). Oversimplified blanket representations of vastly different communities, through use of labels such as "the poor," have long been acknowledged in the discourse of humanitarian aid delivery organizations (Harrell-Bond, 1986). Christopolis et al. (2001:191) recently summed up the situation:

> These problems [of labeling] have continued to be used in the disaster discourse due to implicit assumptions and administrative structures that encourage outsiders to assume the "we" have a right to slot people into

categories as aid recipients. Simply equating vulnerability with poverty has led to a process of merely categorizing beneficiaries, rather than analyzing their situations. Without such analysis, risk tends to be over-shadowed by a pre-existing economic development agenda.

## Defining Success in Achieving Disaster Resiliency

Various conceptual models in development planning attempt to specify the key dimensions of effectiveness of aid delivery systems that are applicable to disaster contexts (e.g., Korten, 1980, 1984; Cuny, 1983; Uphoff, 1991). A useful and clear approach for focusing on disaster resiliency is Korten's (1980, 1984) experienced-based model for evaluating development aid strategies. As illustrated in Figure 6.1, the model consists of three broad dimensions: the need of aid recipients, the design of aid program, and the organizational capacity of both aid donors and aid recipients. Efforts are successful when the disaster mitigation and preparedness program is responsive to household needs and builds the strength of organizations so that they are capable of achieving program goals. That is, a high degree of "fit" among program design, local needs, and capacities of assisting organizations increases the chances of successful programs that link disaster to development.

Although derived more than two decades ago, Korten's (1980, 1984) model is still applicable. The concept of "fit" is of central importance to translating development initiatives into less vulnerable, more disaster-

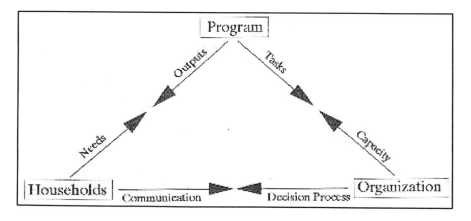

**FIGURE 6.1** Requirements for linking disasters to development programs. Source: Adapted from Korten (1980).

resilient, sustainable communities. Subsequent research has illuminated the important relationships among needs, program, and organizational capacity, concluding that the performance of an organization is a function of fit among these dimensions (e.g., Oliver-Smith and Goldman, 1988; Pelling, 2003; Wisner, 2001). For example, Pelling's (2003) study of disaster planning in the Dominican Republic found that local villagers maintained that "local needs" are what government authorities want to change through the imposition of national program requirements because such needs do not fit those requirements. This issue is exactly what development models like Korten's bring to light.

## A Model of Disaster Resiliency, Governance, and Social Capital

In formulating a model of how community capability and aid delivery strategies influence the achievement of disaster resiliency, the committee draws on several explanations of governance and social capital for understanding collective action to solve public issues. A useful approach for understanding how these factors affect disaster resiliency utilizes the concepts of local horizontal and vertical integration first introduced by urban sociologist Roland Warren (1963). This chapter also draws on Berke et al.'s (1993) conceptualization of how Warren's approach can be applied to disasters and development. Finally, emerging concepts of social capital are used to improve our understanding of the underlying dimensions of horizontal and vertical integration in the context of disasters and development.

Warren (1963) defines a community's horizontal integration as "the structural and functional relations among a community's various social units and subsystems." Such integration links local people and organizations in an egalitarian manner. The idea of social capital can be used to develop a more refined definition of these links and a deeper understanding of how they are formed. Social capital has recently been given prominence by the United Nations Development Programme, which set forth the concept as a central guidance framework for using aid to mobilize communities to deal with disasters and underdevelopment (UNDP, 2004).

Social capital is a general construct that links concepts that sociologists, political scientists, and community development planners have been defining and testing for nearly two decades, including citizen engagement, interpersonal trust, and collective action (Coleman, 1988; Putnam, 1995; Briggs, 1998, Dynes, 2002). Putnam (1995:67) offers a definition that draws on these concepts by stating that social capital involves "social organization such as networks, norms, and social trust that facilitate coordination and cooperation for mutual benefit." Putnam's definition is particularly useful for thinking about disasters and development.

Although Putnam does not present it in this way, social capital can be thought of as a multistaged model linking together civic engagement, interpersonal trust, and effective collective action. Figure 6.2 illustrates Rohe's (2004) conceptualization of the links among these concepts and examples of measures for each concept.[4] In Putnam's definition there is an implied set of relationships that begins with civic engagement. Engagement places people in a network of local social relationships, which affects interpersonal trust. Trust, in turn, affects collective action and ultimately both individual and social benefits.

Social capital is distinguished from other constructs, such as social networks and organizational capital (Rohe, 2004). Social networks represent patterns of interaction, but the social capital construct is more expansive. It embraces characteristics and consequences of interaction, including how interaction leads to trust and, ultimately, to collective action. Further, the interactions among organizations are sometimes thought of as social capital (or organizational capital), but organizational interaction and social capital are not equivalent. A nongovernmental organization charged with disaster mitigation responsibilities may have many community contacts, but if people are not participating and not attending meetings, the contacts do not benefit the community. Clearly, organizational interaction is not a sufficient indicator of social capital.

In keeping with the disaster context, a community with a high degree of horizontal integration (i.e., strong social capital) has an active civic engagement program that fosters more tightly knit social networks among citizens and local organizations. Stronger networks provide greater opportunity for creating interpersonal trust. The community is a viable, locally based problem-solving entity. Its organizations and individuals not only have an interest in solving public problems, but also tend to have frequent and sustained interaction, believe in one another, and work together to build consensus and act collectively. Thus, local populations have the opportunity to define and communicate their needs, mediate disagreements, and participate in local organizational decision making. Further, strong integration among local organizations can enhance the work of external organizations through use of field staff and their knowledge of local circumstances (Suparamaniam and Dekker, 2003). As a result, mitigation practices and disaster preparedness programs are more likely to fit the needs and capacities of the community.

A community with a low degree of horizontal integration has limited civic engagement and a weakly knit social network. Interaction is low among government agencies and social subgroups with an interest in collec-

---

[4]The measures on Figure 6.2 are not definitive. A more comprehensive approach to measuring social capital, which combines quantitative and qualitative research methods, has been developed by the World Bank. It is applicable to diverse social and cultural contexts (Krishna and Shrader, 1999).

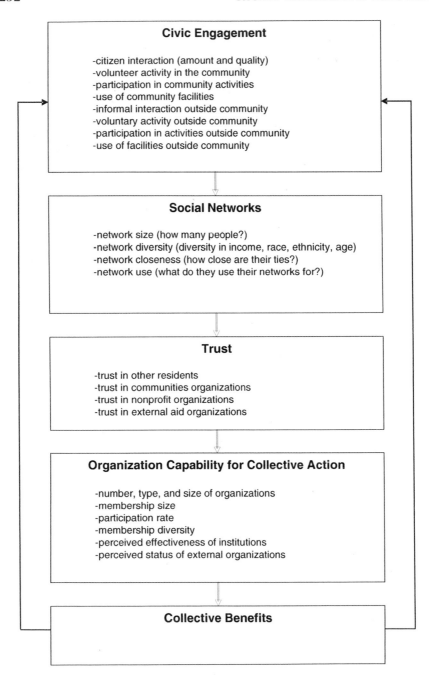

**FIGURE 6.2** Social capital model and examples of measures. Source: Adapted from Rohe (2004).

tive problem solving. Interpersonal trust is more likely to be low as people view ideas and actions of others with suspicion. The community thus lacks an ability to act with collective unity to solve local problems. Consequently, the fit between aid delivery programs and the needs and capacities of local people is likely to be weak.

Warren (1963) describes vertical integration as the "structural and functional relations of [a community's] various social units to extra-community systems." Under this form of integration, power differentials and inequality are evident. A community with a high degree of vertical integration has a relatively high number of ties through engagement with larger political, social, and economic institutions. Vertical integration helps expands networks with these institutions and creates trust between local people and larger institutions that are important in taking effective collective action. This form of integration, sometimes called "bridging social capital" (Briggs, 2004), helps to expand the resources (funds, expertise, influence, and so forth) potentially available to the community. Moreover, issues of local concern have a greater chance of being communicated to central authorities.

The extent to which vertical integration is beneficial relates strongly to the strength of horizontal relationships. When vertical integration is strong and horizontal integration is weak, outside aid organizations can work to build local networks and trust to enhance a community's ability to take collective action. However, when the community has strong horizontal integration in the face of weak vertical integration, there is likely to be tension as communities attempt to exert control over external interventions that are inconsistent with local needs. Weak vertical integration between communities and outside organizations can create severe problems when combined with a weak system of horizontal integration. In this situation, knowledge and degree of trust of the intentions, procedures, requirements, and benefits of outside programs are likely to be weak. Consequently, the likelihood of external programs fitting local needs and capacities to undertake collective action to advance disaster resilience initiatives is very low.

In societies with weak state administrative and judicial structures, notably in developing countries, weak vertical ties dominate and undermine formation of horizontal relations. The absence of laws and contracts that are enforced by the state is a precondition of the emergence of a patron-client system (Putnam, 1993; Krishna, 2002). Political patronage, bribes, and unpredictable use of sanctions generates uncertainty in agreements and mistrust. Lack of security and trust, ensured neither by the state nor by civic norms and networks, translates to powerful top-down patron systems. Vertical relations are defined by coercive authority and dependence, with little or no horizontal solidarity among equals. Organized criminality is frequently a result of the pattern of horizontal mistrust, vertical exploitation, and dependence that characterize societies with weak state structures. This poor state of vertical relations between patrons and clients (or local people) percolates throughout the social ladder and creates stagnation in economic development and a general reluctance to cooperate.

Based on the conceptualization of Berke et al. (1993), Figure 6.3 shows the potential relationships between horizontal and vertical integration as depicted by four types of communities. As noted, Berke et al.'s conceptualization is extended by using social capital concepts to explain these relationships. A Type I community is ideally suited for effective collective action. It possesses strong vertical and horizontal integration. It has well-developed bridging capital with external aid programs, while it has high levels of social capital that will allow it to exert influence in using aid in ways that meet local needs and capacities. A Type II community represents an autonomous, relatively isolated community with few vertical ties—an increasingly rare occurrence in the twenty-first century. While it has strong social capital, it suffers from a lack of bridging capital in terms of knowledge of and interaction with important external resources.

A Type III community is in a classic state of powerlessness and depen-

| Horizontal \ Vertical | Strong | Weak |
|---|---|---|
| **Strong** | Type I<br>• high potential for collective action<br>• high potential for aid to meet needs | Type II<br>• strong capability to define needs<br>• lacks ties with external aid opportunities |
| **Weak** | Type III<br>• powerless, dependent<br>• knowledge of aid, but weak ability for aid to meet needs | Type IV<br>• isolated and powerless<br>• limited access to external aid |

FIGURE 6.3 Community types by degree of horizontal and vertical integration. Source: Adapted from Berke et al. (1993).

dence. Lacking a viable level of social capital, it has less chance to be able to influence the direction of development efforts and define how they are tied to disaster resiliency. Thus, it is more likely that such efforts will not be consistent with local needs and capacities. A Type III community has the advantage of at least having bridging capital with external aid programs. A Type IV community is confronted by significant obstacles to undertaking advancement of disaster resiliency initiatives as it is devoid of access to external resources. However, if vertical channels are activated, it still lacks a viable level of social capital for effectively making collective decisions on how to use external aid or influencing the goals and policy directions of development programs. Moreover, Type III communities and especially Type IV communities are likely to experience many of the conditions of CPEs that are in a constant state of conflict and extreme polarization.

To demonstrate the conceptual and practical significance of this parsimonious model of horizontal and vertical integration, three case studies of local experiences in linking disasters to development issues aimed at supporting disaster resiliency (see Sidebar on linking development to disaster resiliency supported by the National Earthquake Hazards Reduction Program [NEHRP]). The case studies demonstrate how horizontal and vertical integration (or lack of it) has influenced disaster resiliency outcomes. These cases cut across domestic and international settings as well as developed and developing societies.

### What Causes the Formation of Social Capital?

There are several unanswered questions about the causes of social capital and its transformation across different types of communities. Sometimes it is incorrectly assumed that strong social interactions fostered by active civic engagement programs will enhance interpersonal trust. In this case, emphasis is placed on the structural dimension of social capital (i.e., networks), without giving attention to the substantive content of interactions and power relationships among participants in the network. It is not just the frequency of interactions, but the sentiments, actions, and reactions of participants to the content of the interaction. There may be high levels of interaction but minimal trust or even mistrust if the content creates suspicion and ill will. A patron-client system that is fraught with corruption is an obvious case. Suspicion and mistrust also often occur when projects are initiated by outside organizations and local people have doubts about the underlying motives of these organizations. Resident distrust and cynicism may increase if residents are simply informed of a particular mitigation policy but not involved in the policy decision. Even in communities with high levels of engagement and interaction, government officials may not be listening and acting in response to what stakeholders are saying

## SIDEBAR:
## Case Studies Linking Development to Disaster Resilience

**Montserrat, Caribbean.*** This case illustrates a change from a Type III to a Type I community (Berke and Beatley, 1997). Before Hurricane Hugo struck in 1989, this poor village on the island state of Montserrat was a Type III community. Vertical integration was moderate, but horizontal integration was very weak. After hurricane landfall, a collaborative recovery effort evolved between an international nongovernmental organization from Canada, an intermediary NGO from the region with long-standing external ties to foreign donor organizations, and a local community action group. The Canadian NGO sought to provide housing recovery assistance after Hugo by establishing a cooperative arrangement with the intermediary NGO, which had been involved in community development work in Streatham Village for several years before the disaster. The arrangement involved the Canadian NGO providing funds to the intermediary for undertaking reconstruction activities in Streatham. The intermediary, in turn, worked with the community action group to initiate a new housing assistance program. The intermediary NGO trained local people and provided funds to temporarily employ local people to undertake reconstruction activities. The Canadian NGO also supplied the program with building materials and logistics for transporting the materials. The accomplishments of this program were substantial, with numerous training workshops on carpentry and structural strengthening techniques, 20 homes rebuilt, and many others were repaired. Of greatest significance were the long-term development accomplishments. The local visibility and sense of importance of the community action group were raised considerably due to its reconstruction work. The voluntary participation of local people in group activities also increased. This strengthened the community action group's capacity to undertake several development projects not directly related to disaster recovery (e.g., new farming practices, building a community center, improving potable water distribution systems).

**Santa Cruz County, California.** The case represents a successful change from a Type II to a Type I community (Berke et al., 1993). Before the 1989 Loma Prieta earthquake, Santa Cruz County could be classified as a Type II community. Horizontal integration was high, and the county had a high degree of citizen and group political activity and experience in seeking responses from government. Much of this activity can be traced to the occurrence of three major disasters in the county in the 1980s. These disasters induced the county government to develop new partnerships and capabilities with its citizens. Specifically, a cooperative association of households, known as the Neighborhood Survival Network (NSN), was established to organize citizen self-help in future disasters. After the Loma Prieta disaster, this high degree of horizontal integration was vital in aiding the overlooked minority and low-income population in rural areas of the county and providing a basis for increasing vertical integration. When the Federal Emergency

Management Agency (FEMA) initially opened a disaster assistance center in the City of Santa Cruz, citizen leaders maintained that they could make household recovery aid more accessible to the county's rural population by opening a satellite center in conjunction with the NSN. FEMA officials realized that NSN could use its well-established ties with local people to assess needs and distribute assistance. FEMA accepted the offer after NSN members pointed out that numerous rural households that sustained damages had been overlooked because of FEMA's initial assumptions about local conditions.

**Invercargill, New Zealand.** This case represents an unsuccessful shift from a Type I to a Type IV community (May et al., 1996). The City of Invercargill with a population of about 50,000 experienced a devastating flood disaster in 1984. After the disaster, horizontal integration was strengthened as a collaborative recovery effort evolved among stakeholders and public officials in the city and was further reinforced with stronger vertical integration between the city and the national government. During the disaster aftermath, there was consensus among city leaders and stakeholder groups to build long-term risk management considerations into the reconstruction of devastated areas of the city. The National Water and Soil Conservation Authority (NWSCA) took a fresh approach to flood mitigation by developing a cooperative arrangement with the city focused on long-range planning, rather than the traditional approach of structural mitigation that supports floodplain redevelopment. The arrangement involved NWSCA providing recovery subsidies to the city in return for city adoption of a long-term-risk planning approach. In 1985, a comprehensive approach was adopted by the city, making Invercargill a national leader. Planning measures included rezoning of hazardous land, relocation of damaged properties, hazard disclosure requirements in future real estate transactions, and minimum building elevation levels. However, support from the national government for planning collapsed at a critical point in the implementation of Invercargill's comprehensive program. Under the 1991 Resource Management Act the new lead national planning agency, the Ministry for the Environment, opposed the city's program and instead took an antiregulatory, free market approach to land development. As the memory of the disaster faded, local commitment waned, and without national support, local consensus for long-range risk management planning disintegrated. The city reverted to a strategy of allowing floodplain development with levee and dam protection. Thus, vertical integration declined, which stimulated the decline of horizontal integration.

---

*Since field data and analysis for this case study were completed in 1993 (Berke and Beatley, 1997), most people in Montserrat were evacuated due to an ongoing series of volcanic eruptions.

about disaster mitigation needs in their communities, as suggested by a study of indigenous people (the Maori) in New Zealand (Berke et al., 2002).

There are other unanswered questions involving the development and continuation of community social capital. First, do changes in phases of the disaster policy cycle (mitigation $\rightarrow$ preparedness $\rightarrow$ response $\rightarrow$ recovery) influence social capital? How can high levels of social capital in one phase be sustained across phases?[5] Dynes (2002) observes that social capital during the emergency response phase is high, but dissipates during the recovery, mitigation, and preparedness phases. Given the rising losses from disasters, it is important to improve our understanding of how to sustain the peaks of social capital and limit its dips across phases. Dynes also indicates that research on social capital theory has not been applied to any of the phases of the disaster policy cycle, which offer classic situations involving collective action for mutual benefit.

Second, how does engagement among different groups impact trust? That is, does it matter who is engaged with whom? Racial, ethnic, gender, age, and income differences may be an important factor. Moreover, unequal power among participants, such as traditionally powerful real estate interests or corrupt patrons in a weak state structure versus low income residents, may create mistrust, because the less powerful see no benefit in their participation. More research is needed on the types of civic engagement techniques, and on the nature and content of the engagement to understand how they affect trust among groups and institutions.

Third, the question of vertical integration is also relevant. Is horizontal integration sufficient to create effective social capital, or do members also need to be engaged with external organizations? If external organizations are important, what role should they play? Peter May and his colleagues' (1996) study of local implementation of national (and state) hazard mitigation policies in Australia, New Zealand, and Florida offers useful insights that begin to answer these questions. Although May's study does not address disasters in the context of underdevelopment, its findings suggest how external organizations can strengthen vertical integration through techniques that foster negotiation and consensus building, plus technical capacity

---

[5]One can contend that different dimensions of social capital influence community capability to cope differently, especially across phases of the disaster policy cycle. Given the multiple dimensions of social capital, prior research suggests that some dimensions may be more powerful than others by type of social domain. For example, Messner et al. (2004) found that one dimension of social capital—social trust—exhibited a significant direct effect on reducing community homicide rates. However, 11 other dimensions (e.g., political activism aimed at affecting change and community involvement) had no effect on homicide.

building and selective use of penalties to deter noncompliance.[6] Given the large number of factors that can influence vertical integration, more cross-cultural comparative case analyses similar to May's study are needed to detect the independent effects of factors identified by this research.

Fourth, what does the role of the historical social and political context play in framing how people think about engagement or trust? How do changes in human life support functions of natural systems influence a population's ability to act collectively? Social capital can be influenced strongly by these local contexts. Can social capital be changed, or is it more strongly influenced by these historical contexts? In particular, context might play a strong role in deeply polarized, conflict-ridden societies experiencing CPEs. What roles should community development planners and emergency managers play to build social capital in these situations? Bollens' (2002) penetrating analysis of conflict in Belfast, Jerusalem, and Johannesburg suggests several roles for urban planners (neutral, partisan, equity, and resolver), but research is limited in the CPE arena.

## COLLABORATIVE INTERNATIONAL RESEARCH

Collaborative international research is important for at least two reasons. First, as discussed earlier in this chapter and in Chapter 2, developing countries account for the preponderance of human losses from disasters on a global scale, and these losses are expected to increase in future decades. The National Science Board (2001), in discussing the need for collaborative research links between developing and developed countries, specifically mentions the potential for science and technology to address the problem of natural and human-induced disasters.

Second, there is great potential for collaborative international research to advance knowledge on the social science dimensions of hazards and disasters, particularly through cross-cultural comparisons. This potential remains largely unrealized. Bates and Peacock (1993:120) argue that:

> Disasters are relatively rare events in any given geographical or cultural setting. Therefore, the accumulation of knowledge on disasters as social

---

[6]The major lesson from the May et al. (1996) study is that the approaches to mitigation can learn from one another. For instance, Florida's approach that emphasizes technical capacity building, coercion, and funding leads to a strong fact basis and regulatory policies in local plans, but could benefit from New Zealand's (and Australia's to a lesser degree) strong point of cooperation and consensus building that strengthen the local commitment needed to advance mitigation in the long run. New Zealand's approach leads to strong local plan goals, but does not emphasize building local capacity to generate strong fact bases of plans and threatening coercion to ensure adoption of strong local regulatory policies. Thus, effective external programs would represent a mix of the two approaches.

as well as physical processes requires the accumulation of knowledge by the comparison of cases occurring in many different sociocultural and geographic contexts. In addition, culture and social organization, as well as the affected community's level of social and economic development, are known to play significant roles in the disaster process, and we need to understand these roles through comparative cross-cultural research.

While single case studies can help formulate hypotheses on what factors promote community disaster resilience, cross-cultural case research designs can allow for the testing of these hypotheses to uncover unique and interactive effects of these factors. Comparative research must involve international collaboration because this type of research can be very expensive and requires in-country investigators who are sensitive to local cultural contexts. Although it was traumatic and caused untold human suffering, the devastating 2004 Indian Ocean tsunami disaster offers a unique opportunity to apply multiple case designs that capture relationships among horizontal integration, vertical integration, and dimensions of social capital

## NEHRP Exemplars

The National Earthquake Hazards Reduction Program (NEHRP) has supported several studies that exemplify the benefits and importance of collaborative international research. One strong collaborative effort was the long-term partnership between the Texas A&M University Hazard Reduction and Recovery Center (HRRC) and the National Center for Hazards Mitigation (NCHM) at the National Taiwan University. The two centers conducted research early-on by supporting each other's staff in field studies after the 1999 Chi Chi earthquake in Taiwan. This led to annual exchanges between faculty and graduate students, as well as creating opportunities to conduct additional collaborative studies in the disaster field. Box 6.1 on the U.S.-Taiwan collaboration further explores this successful partnership.

Other studies have also shed light on successful collaboration. Bates and Peacock (1993) developed and validated a standardized index of living conditions (the Domestic Assets Scale) for allowing comparisons of disaster impacts across cultures and over time. This study, an outgrowth of earthquake investigations in Guatemala, involved collaboration between researchers in the United States and Peru, Mexico, Turkey, Yugoslavia, and Italy. May and his colleagues (1996) developed insights on the role of intergovernmental structures in hazards management and environmental sustainability by comparing more "coercive" approaches in the United States with more "cooperative" approaches in Australia and New Zealand. Collaboration between researchers made it possible to conduct cross-national comparisons of higher-level government policies and local government hazard mitigation plans that fostered tight control for differences in

---

### BOX 6.1
### U.S.-Taiwan Collaboration

The Texas A&M University Hazard Reduction and Recovery Center (HRRC) has had a five-year collaborative relationship with the National Center for Hazards Mitigation (NCHM) at the National Taiwan University. The Memorandum of agreement calls for collaboration between the two centers in the conduct of research as well as the exchange of faculty and graduate students. Close collaboration achieved an early success when NCHM staff supported HRRC staff during their NSF-sponsored research on the 1999 Chi Chi earthquake. Since that time the staff of the two centers have conducted faculty exchanges, approximately annually, in which researchers discuss their current work and identify opportunities for knowledge transfer. Recent survey research on landslide evacuation in Taiwan has adapted a questionnaire used at HRRC to study hurricane evacuations. This makes it possible to perform cross-national comparisons in household response to disasters that more tightly control for differences in governmental systems and local circumstances (e.g., land-use, building construction, and emergency preparedness practices). In addition to knowledge transfer at the faculty level, this collaboration has also been successful in producing new scholars because two graduate students who studied at HRRC have returned to faculty positions in Taiwan.

The success of this program to date can be attributed to a number of factors. First, the initial attraction of the two centers was based on the similarity of hazard interests. Taiwan is vulnerable to earthquakes and typhoons (whose secondary hazards are floods and landslides) whereas HRRC was funded by the Texas Division of Emergency Management for work on hurricanes and by NEHRP (through NSF) to study earthquakes. Second, the two centers have had continuing sources of independent funding (the National Science Foundation for HRRC and the National Science Council for NCHM). This funding continuity has made it possible to fund travel expenses and, in the case of HRRC, financial support for most of the Taiwanese graduate students who attend Texas A&M. Third, there is sponsor support for this collaboration: NSF supports international collaboration for U.S. investigators and National Security Council (NSC) funding strongly emphasizes it for Taiwanese researchers.

---

cultural and local circumstances (e.g., pressures of urbanization, plans, plan making processes, permitting procedures). A significant outcome of this research was the development of a mitigation plan quality evaluation index that has been applied in subsequent studies by various investigators in New Zealand and the United States (Ericksen et al., 2004; Godschalk et al., 1999).

The success of these initiatives results from several factors. First, individual investigators and their hazard research centers had prior experience in cross-national research, thus recognizing the mutual benefit of such partnerships. Second, sustained sources of independent funding created continuity that strengthened the partnerships. This funding led to more opportunities for exchanges of faculty and graduate students among the investigators' universities, as well as application of standardized data collection instruments. Third, the sponsors of these studies recognized the importance of international collaboration and the need for sustained multi-year research projects, as well as post-study exchanges of findings.

### Challenges of Collaborative International Research

Productive international collaborations such as these exemplars are relatively rare due to a number of challenges. One challenge is the difficulty of making cross-cultural comparisons. For example, different countries have different reporting practices for even seemingly straightforward data on disaster deaths. Looking at households displaced by disaster requires understanding such issues as household structure and housing norms that differ between cultures. The need to develop standardized methods for data collection that are applicable across cultures is therefore a central and complex problem in and of itself (see also Chapter 7). One successful effort in developing standardized disaster loss data is the DesInventar project of LA RED (Network for Social Studies on Disaster Prevention in North America). DesInventar is software for the collection and classification of spatially referenced data on disaster occurrence and loss that can be used without specialized programming skills.

A second challenge is the need to identify appropriate research counterparts. As discussed in Chapter 9, the social science hazards and disaster research community is small, even in the United States, which can be considered a world leader in this area. Research on hazards and disasters in the United States is dominated by natural scientists and engineers in terms of both personnel and resources. This circumstance is even more pronounced in other countries, where addressing the hazards and disaster "problem" is often considered synonymous with developing mainly technical "solutions." Moreover, in other countries, organizations focusing on disaster studies almost always have a physical science and engineering mission. Social scientists who study disasters in these countries are often not well connected to these other organizations. Experience has shown that identifying appropriate research counterparts and developing relationships with them are essential to successful international collaboration. Box 6.2 provides an example of a relatively successful partnership across 14 countries involving multidisciplinary teams. The project, which went by the name of EQTAP, was funded by the Japanese government.

---

**BOX 6.2**
**Japan's EQTAP**

In 1999, with funds from the Japanese government, a multiple-year collaborative research project was initiated with the distinguishing title of Earthquake and Tsunami Disaster Mitigation Technologies and Their Integration for the Asia-Pacific Region (EQTAP). The project grew out of the Japanese experience a few years earlier with the 1995 Kobe earthquake, which killed more than 6,000 residents and caused billions of dollars in losses. EQTAP's stated goal was to achieve safety and sustainability by reducing preventable deaths and injuries from earthquakes and tsunamis as well as social and economic disruption, psychological impacts, and environmental damage. The project was coordinated by Japan's Earthquake Disaster Mitigation Research Center in Miki, Japan, which is just outside of Kobe. EQTAP, which completed its work in March 2004, was innovative in a number of important respects, proceeding in a manner seldom seen in the research community. It was extensively multilateral, involving research collaborators in 14 different countries, including China, the Philippines, Peru, Chile, Papua New Guinea, Indonesia, the United States, and Mexico in addition to Japan. It was also multidisciplinary, involving investigators from earthquake engineering, the earth sciences, and the social sciences. A common risk management model originally developed in New Zealand and Australia was used as a mechanism to further the integration of project activities, which included research on such topics as hazard and vulnerability assessment, structural mitigation, and urban disaster planning. Joint case study activities centered in Manila and periodic workshops were also used for this purpose. EQTAP was also unique in that the research team was expected not only to develop new knowledge, technologies and procedures for risk reduction but also to identify and work with stakeholders in the Asia-Pacific region, such as practitioners and local decision makers, who were potential users of project outputs. A monitoring and assessment panel that included international experts was appointed to begin overseeing the evaluation of the project two years before its completion. In its final assessment, the panel described EQTAP as highly innovative and reasonably successful given the challenges it faced as a multilateral and multidisciplinary effort, especially since it involved integrating engineering, physical science, and social science disciplines while addressing the needs of stakeholders in many different countries.

SOURCE: Hiroyuki, Kameda, 2003.

---

## RECOMMENDATIONS

This chapter has related findings from empirical studies to a conceptual model of disaster risk and development in an effort to identify the contributions and weaknesses of international research on disasters. Scholarship on the links between disaster risk and development has made distinctive contributions to improving the understanding of disaster-resilient communities by focusing on basic premises of development, governance, and social capital. The conceptual writings and empirical research highlighted above have provided a basis for rethinking the role of disaster risk in development activities.

Research shows that robust institutional performance, manifested by strong horizontal ties within communities and vertical ties between communities and external aid institutions, is related to policy choices that encourage social capital wherein the elements of civic engagement, social networks, and trust are high. These conditions are likely to foster strong levels of collective action where there is a high degree of fit among aid program design, local needs, and capacities of assisting organizations. Further research is needed to test these empirically grounded hypotheses. Moreover, as discussed above, many unanswered questions remain about the efficacy and formation of social capital. In disaster contexts dealing with armed conflict, for example, other complementary conceptual models may be needed.

This assessment of current knowledge on disasters and development has therefore been aimed at evaluating the theoretical underpinnings and methodological assumptions of research in this field. After reviewing key studies and important writings, the committee concludes that (1) with NEHRP and other support, disaster research has made some advances in this area and (2) that it is important to undertake further comparative research based on careful research designs and sample selection. Too often in the past, research on disasters and development has involved one or a few case studies of pre- and post-disaster activities. When the number of casual factors to be examined is greater than three or four, the small-scale case approach is inadequate.

> **Recommendation 6.1:** *Priority should be given to international disaster research that emphasizes multiple case research designs, with each case using the same methods and variables to ensure comparability.*

Studies that explicitly hypothesize casual links between disaster risk and development can be completed only through highly structured multicase comparisons. A relatively large number of cases must be selected, based on variation in dimensions of horizontal integration, vertical integration, social capital, aid delivery, and local and national contexts. As illustrated in Figure 6.3, knowledge about interdependent relationships among social capital and humanitarian aid delivery strategies has advanced to the point where hazards and disaster researchers have made significant gains in identifying factors that influence the development of disaster-resilient communities. Multicase research designs are needed to test the independent effects of these relationships on achievement of local disaster resiliency. The primary limitations raised in this chapter relate to the broad range of factors that are necessary for testing relationships between disaster risk and development. As indicated in Figure 6.2, some factors that effect collective action depend on the influence of others, and both sets of factors must be included in the research designs.

The potential for devising research designs that include larger samples of disaster sites has never been greater. As noted above, the 2004 Indian Ocean tsunami disaster offers a significant opportunity to apply multiple case designs. Furthermore, in the past decade, national governments throughout the world have been experimenting with decentralized local planning approaches to disaster mitigation and preparedness (United Nations Commission on Sustainable Development, 2001; UNDP, 2004). A National Research Council report *Drama of the Commons* indicates that governments in both developed and developing countries are increasingly adopting decentralized approaches to common pool resource management (Ostrom et al., 2001). These emerging institutional arrangements offer opportunities to conduct comparative analyses that systematically explore the links between underdevelopment, disasters, and ecosystem degradation.

**Recommendation 6.2:** *Common indicators of disaster risk and development should be constructed.*

Methodological advances in measurement and data collection clearly indicate that the current state of research on disasters and development has come of age. Investigators are now better equipped to formulate research designs that rely on larger number of cases selected on the basis of variation in casual variables. Improved multitiered data systems are needed to advance our understanding of the links between disaster risk and development. In these systems, disaggregated disaster data collected at the local level should be progressively aggregated into regional, national, and global disaster data sets. Aggregated data sets such as the mortality data used by UNDP (2004) deal with entire nations, but not with specific individuals, subgroups, or regions. The aggregations conceal internal variations that are important for interpreting differential vulnerability and exposure, as well as the effects of aid delivery strategies on communities. Thus, a full analysis of the dynamics of disaster risk and development requires data sets that provide richer and more fine-grained assessments of trends and outcomes. Aggregated data sets remain important for detecting regional, national, and global trends of hazard vulnerabilities and their associated risks. They can also be used to evaluate the effects of national policy and aid delivery strategies of multilateral humanitarian aid organizations.

A potentially useful methodological advancement is the development of a comprehensive approach to measuring social capital, one that combines quantitative and qualitative research methods and is applicable to diverse social and cultural contexts. Krishna and Shrader (1999), for example, have developed a set of cross-cultural indicators and systematic data collection techniques for the study of the formation of social capital at the group level within communities in developing societies. Similarly, Bates and Peacock (1993) have derived cross cultural measures at the household level of disaster impacts and measures to track progress toward recovery once disaster losses have been calibrated. Together, these complementary methodological approaches can be used to examine the effects of recovery aid delivery strategies on physical recovery of communities and on the ability of households within communities to act collectively in dealing with recovery and development issues.

**Recommendation 6.3:** *Collaborative international research projects should be the modal form of cross-national research on disasters and development.*

The next generation of studies should exploit the advantages of collaborative partnerships with investigators from other countries. A major advantage is the leveraging of pooled resources to reduce costs to any particular stakeholder. As noted earlier, conducting multicase research in different countries is expensive and requires in-country investigators who are sensitive to the cultural context. Supporting collaboration among researchers in different societies can address these limitations. The National Science Foundation (NSF) has had a long tradition of encouraging collaborative international research in the social sciences, but there has unfortunately been a drop-off in support during recent years. Nevertheless, relatively small NSF post-disaster reconnaissance grants have been important in furthering some international collaboration (see also Chapter 7). Bilateral programs have also provided a vehicle for encouraging the development of collaborative relationships. For example, NSF support for U.S.-Japan collaboration over at least two decades has fostered long-term relationships between the hazards research communities in the two countries. As social scientists have become more involved in the U.S.-Japan activities (e.g., through bilateral workshops), acceptance and recognition of the social sciences has grown. This involvement may have contributed to the increased salience of social scientists in the Japanese hazards and disaster research community. NSF should continue to support international collaborative research on hazards and disasters and ensure that social scientists play a major role in such efforts.

The above recommendations apply to the study of the full range of hazards and disasters—natural, technological, and willful, including those circumstances brought on by complex political emergencies. In tandem with NSF, DHS should also assume a major role in supporting the work of

U.S. investigators in international hazards and disaster research. Given its mission, it seems reasonable to expect DHS to play a growing role in enabling U.S. investigators to conduct collaborative international research on willful hazards and disasters. Indeed, there are already signs that this is being considered for its agenda.

# 7

# The Role of State-of-the-Art Technologies and Methods for Enhancing Studies of Hazards and Disasters

Technical and methodological enhancement of hazards and disaster research is identified as a key issue in Chapter 1, and computer systems and sensors are discussed in Chapter 2 as technological components of societal change having important implications for research on societal response to hazards and disasters. As summarized in Chapters 3 and 4, pre-impact investigations of hazard vulnerability, the characteristics and potential impacts of alternative hazards, and related structural and nonstructural hazard mitigation measures have been the sine qua non of hazards research. Post-impact investigations of disaster response, recovery, and related disaster preparedness measures have been the hallmark of disaster research. Indeed, post-impact investigations have been so prominent historically that special attention was given in the committee's statement of task to offer strategies for increasing their value. Yet as highlighted in both Figure 1.1 and Figure 1.2, the committee believes that hazards and disaster research must continue to evolve in an integrated fashion. Thus, any discussion of state-of-the-art technologies and methods must ultimately be cast in terms of how they relate to this field as a whole.

Post-impact investigations inherently have an ad hoc quality because the occurrence and locations of specific events are uncertain. That is why special institutional and often funding arrangements have been made for rapid-response field studies and the collection of perishable data. However, the ad hoc quality of post-impact investigations does not mean that their research designs must be unstructured or that the data ultimately produced from these investigations cannot become more standardized, machine readable, and stored in accessible data archives. Having learned what to look for after decades of post-disaster investigations by social scientists, the potential for highly structured research designs and replicable datasets across multiple disaster types and events can now be realized. As noted in Chapter 1, post-impact studies also provide a window of opportunity for documenting the influence of vulnerability analysis, hazard mitigation, and disaster preparedness on what takes place during and after specific events. However, pre-impact investigations of hazards and their associated risks are critically important on their own terms, less subject to the uncertainties of specific events, arguably more amenable to highly structured and replicable data sets, and no less in need of machine-readable data archives that are accessible to both researchers and practitioners.

So what has been referred to in Chapter 1 as "hazards and disasters informatics" (i.e., the management of data collection, analysis, maintenance, and dissemination) is a major challenge and opportunity for future social science research. This chapter begins with an overview of how social science research on disasters and hazards has been conducted in the past, and consistent with Figure 1.2, a case is made for the essential relatedness in chronological and social time of post-disaster and pre-disaster investigations. This section also illustrates the influence of changes in technologies and methods in hazards and disaster studies. Survey research is highlighted specifically in this regard because of its historical prominence within hazards and disaster research as well as mainstream social science. Consistent with the committee's statement of task, the second section provides a specific discussion on the challenges of post-disaster investigations and ways to increase their value. The third section discusses "hazards and disaster informatics" issues such as dealing with institutional review boards (IRBs), standardizing data across multiple hazards and events, archiving resulting data so that they accumulate over time, and facilitating access of accumulating data from original researchers to those engaged in secondary data analysis.

The fourth section provides examples of how state-of-the-art technologies and methods enhance hazards and disaster research and, in so doing, relate directly or indirectly to these informatics issues. Although this chapter cannot cover everything in what amounts to the very broad terrain of "nuts and bolts" research matters, special attention is given to increased use of computing and communications technologies, geospatial and temporal methods, statistical modeling and simulation, and laboratory gaming experiments. Sensitivity to the roles of these technologies and methods will contribute to more focused attention and advancing solutions to hazards and disaster informatics issues. The chapter closes with specific recommendations for facilitating future hazards and disaster studies.

## DOING HAZARDS AND DISASTER RESEARCH

In examining hazards and disasters through disciplinary, multidisciplinary, and interdisciplinary lenses and perspectives (see Chapters 3 to 6), social science researchers have used a variety of technologies and methods. They have employed both quantitative and qualitative data collection and data analyses strategies. They have conducted pre-, trans-, and post-disaster field studies of individuals, groups, and organizations that have relied on open-ended to more highly structured questionnaires and face-to-face interviews. They have used public access data such as census materials and other historical records from public and private sources to document both the vulnerabilities of social systems to hazards of various types and the range of adaptations of social systems to specific events. They have employed state-of-the-art spatial-temporal, statistical, and modeling techniques. They have engaged in secondary analyses of data collected during previous hazards and disaster studies when such data have been archived for this purpose or otherwise made accessible. They have run disaster simulations and gaming experiments in laboratory and field settings and assessed them as more or less realistic. As research specialists, hazards and disaster researchers have creatively applied mainstream theoretical and methodological tools, thereby contributing to their continuing development and use.

### The Commonality of Hazards and Disaster Research

The technologies and methods of hazards and disaster research are indistinguishable from those used by social scientists studying a host of other phenomena (Mileti, 1987; Stallings, 2002). That is as it should be. However, the simultaneity of hazards and disasters core topics within chronological and social time is a source of theoretical complexity, the consideration of which calls for creative applications of the most robust technologies and methods that are available. As noted in Chapter 1 (see Figure 1.2 and its related discussion), chronological time allows partitioning of collective actions by time phases of disaster events (pre-, trans-, and post-impact). The primary explanatory demands of hazards research in chrono-logical time are to document interactions among conditions of vulnerability, disaster event characteristics, and pre-impact interventions in the determi-nation of disaster impacts (see Chapter 3). The primary explanatory demands of disaster research in chronological time are to document interactions among disaster event characteristics, post-impact responses, and pre-impact interventions in the determination of disaster impacts (see Chapter 4). How-ever, such straightforward partitioning in chronological time is not feasible with social time because, as discussed in Chapter 1, pre-, trans-, and post-disaster time phases become interchangeable analytical features of hazards on the one hand and disasters on the other. In social time, in effect, the respective explanatory demands of hazards and disaster researchers become one and the same.

So in considering how state-of-the-art technologies and methods can enhance studies of hazards and disasters, there must always be sensitivity to the way specific applications and findings within disaster research inform applications and findings within hazards research and vice versa. For example, post-impact field interviews and population surveys seek data on "present" behaviors during the disaster, relationships between these behaviors and "past" experiences with hazards and disasters, and links between present behaviors and past experiences with "future" expectations of vulnerability. Pre-impact field interviews and population surveys seek data on relationships between past experiences with hazards and disasters, future expectations of hazard vulnerability, and links between these experiences and expectations with decisions to locate in harm's way, adopt hazard mitigation measures, or engage in disaster preparedness. Pre- and post-disaster uses of public access data and other historical materials, as well as searches for unobtrusive data (e.g., meeting minutes, formal action statements, communications logs, memoranda of understanding, telephone messages, e-mail exchanges), are undertaken with these same objectives in mind. Computer simulations and gaming experiments are always subject to reality checks, and with respect to hazards and disasters, these checks are subject to present behaviors, past experiences, and future expectations.

Thus, taking an integrated approach to research on disasters and hazards requires that any assumed impediments of data production during post-impact investigations—such as the ad hoc selection of events, special pressures of the emergency period, lack of experimental controls, difficulties in sampling population elements, and perishable data (see Stallings, 2002)—should be considered also in terms of their consequences for hazards research. In the final analysis, it is because the explanatory demands of disaster and hazards studies are essentially inseparable that these impediments, whatever they may be, are of concern within this entire research community. Also, the impediments are not simply confined to doing either post-disaster or pre-disaster field research. They encompass the way data are collected, maintained, retrieved, and used for purposes above and beyond those of the original studies. The resulting informatics demands on state-of-the-art technologies and methods are major.

## Influence of Technology on How
## Hazards and Disaster Research Is Conducted

Mainstream social science technologies and methods used to study hazards and disasters have changed over the years, and the role of technology has been singularly important. A useful illustration because of its importance to hazards and disaster research is technological change in the administration of social surveys. As summarized in Chapters 3 and 4, survey research has provided an excellent source of data for post-impact investigations of the physical and social impacts of disasters, as well as individual and structural responses to these impacts (i.e., disaster research). No less important, survey research has provided an excellent source of data for pre-impact investigations of vulnerability expectations, as well as individual and structural responses to these expectations (i.e., hazards research). Over time, therefore, in hazards and disaster research the survey has been increasingly recognized as a valid form of quantitative data collection (Bourque et al., 2002). Yet like all other methodological tools, the use of surveys is subject to technical, methodological, and societal changes that can affect, both positively and negatively, the ability to collect high-quality data.

Surveys of human populations threatened by hazards or actually experiencing disasters may be conducted using a number of different administration forms. They can be administered in face-to-face interviews, through telephone interviewing, or through self-administration of questionnaires. Each of these forms has its own merits and drawbacks, and new technologies are influencing the way they are implemented. Survey research has changed over the past three decades. In the 1970s, most surveys were administered using traditional face-to-face interviews or through mailed questionnaires. However, the near universal access to telephones by the 1990s made telephone interviewing a more attractive administration format. By 1998, 95 percent of U.S. households had telephones, with most of the remaining households having access to a phone. Telephone coverage is lowest in the South, with approximately 93 percent of households having a phone (Bourque et al., 2002). Moreover, the availability of computers and access to the Internet by the late 1980s and early 1990s for both the general population and, more notably, hazards and disaster management practitioners, has led to increased use of self-administered e-mail and web-based surveys.

Survey research has become increasingly difficult during the more recent past. Response rates for all forms of administration are dropping, and the costs of conducting survey research are increasing. More people live in gated communities, have guard dogs, have answering machines or caller ID, or live in a "cell phone-only" home. All of these trends, along with increases in the elderly and non-English-speaking immigrants in the general population of the United States (see Chapter 2) are affecting interview completion and response rates. While the rates of nonresponse of all types are increasing, this does not appear to increase bias in the studies (Tourangeau, 2004).

Certainly surveys have become more difficult to implement; however, there have been significant changes in technology that have increased the choices of administration methods. In the 1970s, computer-assisted telephone interviewing became available. This methodology allowed researchers to load a questionnaire onto a computer from which interviewers could read and enter data directly into a database during the interview process. By the 1980s, similar systems for in-person interviewing became available. This methodology allows complex skip patterns to be programmed into the questionnaire and reduces the need for interviewers to find the correct question. It also eliminates the data entry step, creating a complete data set at the close of interviewing. However, this also means that paper interviews are not available for double entering of data. If errors are made in data entry during the interview process, there is no way to verify accuracy.

By the time of the Second Assessment in 1994 (Mileti, 1999b), computers had gained widespread uses, and access to the Internet had just taken off. With the rise of the Internet, e-mail surveys quickly became available. The earliest form of e-mail surveys were questions typed into the body of an e-mail. When replying to the e-mail, the participant simply typed in his or her responses. Then in the 1990s, Web-based survey technology became available. In Web surveys, questions are programmed with response options. Although the methodology shows promise as a low-cost survey method, there are questions about its applications in academically sound research. While Internet access is increasing, the coverage is not currently sufficient to be able to adequately sample the general population without significant bias. Furthermore, unlike telephone samples, a sampling frame for all people who have access to the Internet does not exist currently. As a result, at present a probability sample of all Internet users cannot be determined.

Web surveys may be useful for specific populations in which Internet use is high and there is a list of users in a closed system, such as a university. It is also possible to utilize Web surveys in a mixed-mode fashion. For example, in a survey of health care providers in California regarding their training needs for bioterrorism response, a list of all licensed providers was obtained from the licensing agency in the state. A sample was selected from the list and mailed an invitation to log into a Web site to participate in the survey. Each invitation letter included a unique password so that responses could be tracked.

Notwithstanding problems of administration, technically enhanced and highly structured survey research has been used increasingly to produce quantitative data about hazards and disasters. When combined with more traditional qualitative field research methods, geospatial and temporal methods, considerable use of public access data and historical records, and some simulation and experimental work, the picture that emerges over the past half century is one of an ever-expanding volume of data on hazards and disasters. The production of these data has been and will continue to be facilitated by state-of-the-art technologies and methods within mainstream social science. However, the data being produced are largely not standardized across multiple hazards and disasters, not archived for continuing access, and underutilized once the original research objectives have been met. Therein lies the "hazards and disasters informatics" problem discussed in the third section of this chapter.

In 1954, a National Research Council (NRC) committee charged with writing a volume similar to this one gave highest priority to exploratory research to define major variables and discover trends (Williams, 1954). It is safe to say that in the ensuing 50 years that goal has been achieved through a host of descriptive and often comparative case studies. With that foundation, and through the National Earthquake Hazards Reduction Program (NEHRP) support during the past 25 years, the transition from descriptive work to the more integrated explanatory work demanded by Figure 1.2 is certainly well under way.

## THE CHALLENGES OF POST-DISASTER INVESTIGATIONS AND INCREASING THEIR VALUE

Post-disaster investigations, especially the field work required for the collection of data on disaster impacts as well as activities related to emergency response and disaster recovery, are undertaken in widely varied contexts and often under difficult conditions. As suggested earlier, the selection of events to be studied is necessarily ad hoc. The timing and location of field observations are heavily constrained by the circumstances of the events themselves as is the possibility of making audio and video recordings of response activities. There are special constraints and difficulties in sampling and collecting data on individuals, groups, organizations, and social networks. Unobtrusive data such as meeting minutes, formal action statements, communications logs, Memoranda of Understanding, telephone messages, and e-mail exchanges are sometimes impossible to obtain, and so on.

Post-disaster investigations rely heavily on case studies (the "events"). These case studies have accumulated over time, providing incomplete albeit often sufficient data upon which to base theoretical generalizations about community and societal responses to disasters. In so doing, they have often confirmed and reinforced existing knowledge about response to disasters and hazards (including the continued existence of hazard exposure and specific vulnerabilities). In documenting planned as well as improvised post-disaster responses, they have shed light on hazard mitigation and disaster preparedness practices. In addition, they have served as experience-gaining and training mediums for hazards and disaster researchers.

While the analysis of hazards and disasters in social time requires a historical perspective and research, post-disaster investigations can be characterized loosely by five principal and frequently overlapping chronological stages: (1) early reconnaissance (days to about two weeks), (2) emergency response and early recovery (days to about three months), (3) short-term recovery (three months to about two years), (4) long-term recovery and reconstruction (two to about ten years), and (5) revisiting the disaster-impacted community and society to document any other longer-term changes (five to at least ten years). Not all of these chronological stages necessarily require field research, and post-disaster investigations may not even take place in the stricken community. For example, studies of post-disaster national response, recovery, and public policy actions may best be completed in capital cities where decision agendas are established and resources are allocated. The level of funding, research foci, methods, availability of data, their quality, and the duration of the study vary greatly across these chronological stages, but resources permitting, the net long-term results can provide important advances in knowledge. Each chronological stage is described briefly below:

1. *Early Reconnaissance*: Although of primary interest to physical scientists and engineers because of their need to examine and collect data about the direct physical impacts of a disaster, this stage presents social scientists with opportunities to identify the physical causes of social impacts, learn from scientists and engineers about why and how such physical impacts occurred, observe and document emergency response and immediate relief operations on an almost real-time basis, and define potential responding individuals, groups, organizations, and social networks for more structured follow-on research.

2. *Emergency Response and Early Recovery*: Observing planned and improvised actions at the height of the emergency response stage provides knowledge about the analysis and management of disaster agent- and response-generated problems, the availability and allocation of local and externally provided resources, the types and effectiveness of individual and structural responses, and the transition from emergency responses (e.g., search and rescue) to early recovery (e.g., temporary shelter) activities.

3. *Short-Term Recovery*: Studying the evolution from the emergency response and early recovery stages to the short-term recovery stage is particularly interesting because researchers can identify more clearly the characteristics of key responding groups and organizations, how these social units influence decisions, and how short-term decisions (e.g., location of temporary housing) influence the allocation of resources for long-term recovery and reconstruction.

4. **Long-Term Recovery and Reconstruction**: During this period the sometimes permanent consequences of earlier decisions (or non-decisions) and the application of resources to implement them become visible to researchers, as disaster-related workings of the marketplace. It is then possible to reconstruct how the host of earlier commitments (or noncommitments) combine to shape the previously stricken area spatially, demographically, economically, politically, and socially. This stage also provides the opportunity to document how influential leaders, groups, and organizations have affected the outcomes and why.

5. **Revisiting the Stricken Area**: After significant time has passed (probably five years to more than a decade) and the disaster-related issues have receded largely from the public's and decision makers' agendas, post-disaster investigations of how the "new equilibrium" came to be and how and why the impacted social system is functioning the way it is can help researchers and users understand the anticipated, real, and unintended consequences of the full range of earlier decisions and their implementation. Research at this interval can include, for example, examining the effectiveness of mitigation/loss prevention measures instituted after the previous disaster and understanding who benefited and who did not from the entire process.

Sometimes operating alone or in partnership with engineers, earth scientists, and representatives from other disciplines, social scientists have been part of the continuing history of post-impact investigations. Within the context of post-earthquake studies, it was the National Academies' comprehensive study of the March 1964 Alaska earthquake that saw a fully integrated social science component (NRC, 1970). To varying degrees, this model was repeated for subsequent events, such as the National Oceanic and Atmospheric Administration (NOAA) study of the 1971 San Fernando, California, earthquake, and it continues to serves a model for post-disaster investigations of earthquakes as well as other natural and technological disasters.

As noted in Chapter 1, post-disaster investigations have been seen historically as so important to advancing knowledge that special institutional arrangements have been made and special funding has sometimes been made available (particularly for earthquake research) to enable social scientists and other researchers to enter the field and collect perishable data or conduct more systematic research. As suggested in Box 7.1, support for post-impact investigations of willful disasters is now part of the funding mix at the National Science Foundation (NSF).

A possible model for enhancing the value of post-disaster investigations

---

**BOX 7.1**
**National Science Foundation Support for**
**Post-September 11 Research**

The National Science Foundation's (NSF) Division of Civil and Mechanical Systems in the Directorate for Engineering has a long history of supporting post-disaster investigations, particularly those induced by natural and technological hazards. For example, funding from NSF has enabled social science and engineering researchers to carry out post-disaster investigations to gather information (perishable data) that might be lost once the emergency period is over. Such research is funded through NSF's Small Grants for Exploratory Research Program and with funds made available for rapid response research programs administered by the Earthquake Engineering Research Institute (EERI) and the Natural Hazards Research and Applications Information Center (NHRAIC). The largest of the latter efforts is EERI's Learning from Earthquakes Program, whose funds are used to support multidisciplinary reconnaissance teams after significant earthquakes in the United States and overseas. NHRAIC's activity, called the Quick Response Program, supports primarily social science investigations. All three of these NSF funding mechanisms were put in play after the September 11, 2001 attacks on the World Trade Center and the Pentagon, and the plane crash in Pennsylvania, resulting in important social science and engineering studies. Upon completion, the results of these studies were published in a book (NHRAIC, 2003). This book includes social science analyses of the disaster responses following the September 11 attacks, such as individual and collective actions, public policy and private sector roles, and engineering analyses on physical impacts on physical structures and infrastructures. No less important, the book documents similarities and differences between the September 11, 2001 event and past disasters, offers policy and practice recommendations for willful and other kinds of disasters, and provides guidance for future research. An appendix includes a list of awards for social science and other studies funded by NSF that were published in the book as well as other awards related to homeland security made directly by NSF or through NHRAIC in fiscal year 2002.

---

SOURCE: NHRAIC (2003).

---

of natural, technological, and willful disasters is the Earthquake Engineering Research Institute's (EERI's) Learning from Earthquakes Program (LFE). When federal funding through NSF became available to support field investigations of (primarily) earthquakes, such studies were small in scale, of very limited duration, and composed virtually exclusively of engineers and earth scientists, and the dissemination of the knowledge gained was limited, for all practical purposes, to the earthquake engineering community. The paradigm shifted in 1973, resulting in more sustained federal support for post-disaster investigations and the inclusion of social scientists. The effectiveness of today's LFE program can be traced directly to that paradigm shift. Within the normal constraints of NSF funding, combined with the support capabilities of other organizations (such as the Natural Hazards Research and Applications Information Center [NHRAIC], the three earthquake research centers, and independent researchers from universities, nonprofit and consulting organizations), the availability of principal investigators (who are expected to and do contribute their time) continues to advance research and knowledge about earthquakes and other types of disasters.

Drawn from its researcher and practitioner members, EERI (2005) has produced a thoughtful retrospective *The EERI Learning from Earthquakes Program*. This retrospective captures succinctly LFE's significant accomplishments during the past 30 years. Among those accomplishments are 11 subjects identified and documents that have benefited directly from investments in social science post-disaster investigations. These 11 subjects have nearly universal application, transcending earthquakes as well as other natural, technological, and willful hazards and disasters. The initial four subjects include (1) strengthening research methods and broadening the mix of social science disciplines involved; (2) applying lessons learned to improve the development of loss estimates and their implications for planning scenarios, emergency operations plans, and training; (3) increasing the understanding of cross-cultural disaster impacts that have demonstrated both commonalities and differences related to key societal variables and levels of development; and (4) providing lessons learned that have or are being applied to improve emergency response capabilities, recovery and reconstruction plans, search and rescue actions, understanding the epidemiology of casualties, measures to reduce life loss and injuries, managing large-scale shelter and temporary housing services, and organizing mutual aid programs. The remaining subjects include (5) applying organizational response lessons learned to improve and standardize emergency response procedures; (6) developing clearer and more effective warning procedures and messages, a necessary component of improving warning system technologies; (7) applying lessons learned about fault rupture and other geologic hazards to land-use planning and zoning; (8) carefully examining the adaptive organizational and decision making processes involved in recovery; (9) understanding the need for and measures to organize and manage large-scale temporary shelter programs; (10) improving management related to the flow and on-site handling of inappropriate donations to impacted areas; and (11) adapting scientific data from instrumental networks to support real-time decision making and emergency operations.

It is notable that all of the above subjects relate directly to the social science research summarized in Chapters 3 to 6 of this report. One implication is very clear: The future development and application of social science knowledge on hazards and disasters depends heavily on implementing recommendations included in these chapters. As noted in Chapter 5, some of these recommendations are disciplinary based, some involve interdisciplinary research among the social sciences, and some require interdisciplinary research that connects the social sciences with natural science and engineering fields. However, the planning and funding needed to implement these research recommendations must be sensitive (1) to the essential relatedness of post-disaster and pre-disaster investigations, (2) to the need for a cross-hazards and cross-societal approach, and (3) to addressing the hazards and disaster informatics issues discussed below.

## THE HAZARDS AND DISASTERS INFORMATICS PROBLEM

Informatics refers generally to the management of data—from its original collection and analysis, to its longer-term maintenance, to ensuring its accessibility over time to multiple users. As noted in Chapter 1, the 2003 NEHRP plan (Department of the Interior, 2003) makes it clear that hazards and disaster informatics is an essential planning consideration. The plan speaks, for example, of the need for searchable Web-based data systems, but it is not precise about how these systems should be constructed, the kinds of data that should be included in them, when these data should be collected (pre- or post-disaster), where they should be stored, or how the demands for information from multiple audiences will be met. Hazards and disaster informatics, therefore, is an enormously significant problem. The problem is summarized below in terms of a series of specific trends and related issues: the changing conditions within which hazards and disaster research is conducted, with IRBs, standardizing data across multiple hazards and events, data accumulation and storage, and providing data access to researchers and practitioners that is user friendly.

### The Changing Environment of Research on Hazards and Disasters

Fieldwork remains fundamental to hazards and disaster research as this field enters the twenty-first century. Skillful field researchers continue to gain access to individuals, households, and representatives of organizations in the public and private sectors, and respondents more often than not want to be cooperative and helpful. The result is often an effective blending of field interviews, broader population surveys, spatial and temporal data, census materials and other public access information, and unobtrusive data. Such blending is essential to the development of knowledge about the five core topics of hazards and disaster research identified by this committee. Tierney (2002) notes, however, that six important societal trends—mostly challenging, but sometimes facilitating—are affecting the practice of fieldwork. The first of these, (1) human subjects regulations, is of such importance that it is discussed separately. The others include (2) legal complexities affecting social science research, (3) organizational perceptions of and attitudes toward social science research, (4) the significant expansion of post-disaster research activities, (5) increasing ethnic and gender diversity within the research community and among those being studied, and (6) the increasing professionalism of hazards and emergency management.

The increasingly litigious environment within the United States will continue to affect studies of hazards and disasters. Because researchers are potential sources of information about legal issues, they may become part of a larger pool of people named in complex, controversial, expensive, and lengthy court proceedings. Tierney provides several recent sobering examples from hazards and disaster research as well as mainstream social science and notes that "courts are increasingly faced with balancing the privilege offered to researchers and research participants with the needs of litigants, often to the detriment of the former" (Tierney, 2002:355).

The approach adopted by the National Institute of Standards and Technology (NIST) on use of research findings and reports in civil actions may have broad applicability to hazards and disaster research. Specifically, NIST's recent draft report on structural and life safety systems at the World Trade Center (Lew et al., 2005) contains this disclaimer: "No part of any report resulting from NIST investigation into structural failure or from an investigation under the National Construction Safety Team Act may be used in any suit or action for damages arising out of any matter mentioned in such report (15 USC 281a; as amended by P.L. 107-231)." The National Construction Safety Team Act (P.L. 107-231, 15 U.S.C. 7301 et seq.) was enacted by Congress in 2003 as a direct result of the collapse of the World Trade Center. With respect to federally funded social science research on natural, technological, or terrorist-induced hazards and disasters, the NIST disclaimer merits careful consideration and for the same reason: to allow social scientists to conduct the best possible science.

Impression management is a related issue that affects organizational studies, in particular, because of the heightened mass media scrutiny that attends management of and accountability for hazards and disasters, and the possible importance of research findings for assessment of organizational performance (Tierney, 2002:359-362). An increasingly litigious environment and related concerns about impression management are exacerbated by the convergence of field researchers, particularly following disasters of significant magnitude and scope of impact (Tierney, 2002:362-365). The need for coordination becomes increasingly important to reduce the burden on disaster impacted communities and regions, as is the need to communicate clearly the purposes and rationale for social science research.

On the more positive side, Tierney (2002:365-370) identifies gender and ethnic diversity as having significant implications for knowledge development and the capabilities of the disaster research community. As summarized in Chapters 3 to 6 of this report, the focusing on gender, ethnic, and cross-cultural diversity has several positive results. These include improved access to and reliable information from and about groups that were outside the mainstream of earlier hazards and disaster research, improved understanding of how hazards and disasters affect a broader spectrum of people, increased attention to the impacts on and roles of more informal and community-based groups compared to highly structured formal (particularly government) organizations, and (as discussed in Chapter 9) a more representative and capable research community.

Finally, one of the most interesting, albeit uneven, trends has been the increasing professionalism of emergency management during the latter decades of the twentieth century (Tierney, 2002:370-372). The largely ex-military background of emergency managers following World War II is explained largely by the civil defense and Cold War orientations of the nation's civil emergency management programs. In more recent decades, however, academic instruction and professional development activities have raised the level of knowledge of emergency management practitioners, provided opportunities for continuous education, created closer connections between hazards and disaster researchers and emergency management practitioners, and otherwise contributed to greater prestige and professionalism in the emergency management field. These kinds of developments facilitate access in pre-disaster as well as trans- and post-disaster contexts and increase communications and understandings about the purposes and rationale of social science research.

## Dealing with Institutional Review Boards

The current requirements governing research on human subjects extend from experimental research and studies of "at-risk" populations under normal and controlled conditions to the messier, less structured, and often more fluid conditions encountered by hazards and disaster researchers. For some time the trend has been moving in the direction of defining most contacts in the field as being within the domain of human subjects regulatory procedures. This inclusion complicates the process of doing fieldwork and ensuring confidentiality, particularly during post-disaster reconnaissance studies, where highly formalized approaches to informed consent and confidentiality are inconsistent with the fluid, and often unstructured, data collection strategies and techniques that are required in these contexts (Tierney, 2002:353).

Protecting the rights of research participants and the formal necessity of informed consent have been the major historical issues in studies of human subjects since World War II. The experiments performed by the

Nazis during the war focused attention on the need to protect participants in research. Although initially concerned primarily with biomedical research, by the 1960s federal agencies had begun to consider the potential risks of sociobehavioral research. In May 1974, the Department of Health Education and Welfare issued regulations requiring the review by an IRB of all funded research on human subjects. The IRBs were mandated to determine if research participants were at risk for harm; whether risks were outweighed by benefits (to the individual or society); whether the rights and welfare of research participants were adequately protected; and whether "legally effective informed consent" would be obtained (NRC, 2003c).

In January 1981, following concerns about the impact of the existing regulations on sociobehavioral research, a revised set of regulations was issued by the Department of Health and Human Services. These regulations narrowed the definition of human subjects and allowed certain broad categories of research to be exempted from IRB review or to be subject to an expedited review process. In 1991, the Common Rule was published, which again changed the requirements for exemption and expedited review, allowing IRBs to decide more easily not to exempt certain research from review. Since that time, a number of highly publicized tragic events associated with biomedical studies have occurred. These events underlie what many researchers believe is a tightening of restrictions on exemption and expedited reviews. An NRC panel convened to review the participation of human subjects in social and behavioral research identified three broad areas for improvement of IRB procedures: (1) enhancing informed consent; (2) enhancing confidentiality protection; and (3) improving the effective review of minimal-risk research (NRC, 2003c).

Hazards and disaster researchers are particularly affected by the definition of minimal-risk research. Anecdotal reports of researchers in the field include instances of research not being conducted because of the restrictions placed on researchers. Box 7.2 provides an illustration of the challenge.

IRBs may view research on hazards and disasters, and especially on terrorism, as inherently being of significant risk, thereby requiring full review of studies that otherwise would meet the requirements of an exempted study. Full review takes time and may limit the ability of field researchers to successfully gather potentially perishable data in the immediate post-disaster period. The issue of dealing with IRBs is of such significance that the committee has developed an explicit recommendation at the end of this chapter. Following the committee's recommendation will not necessarily solve the problem, but it could lead to the development of workable guidelines that will be of educational value to hazards and disaster researchers and the IRBs that oversee their studies.

---

**BOX 7.2**
**Impact of IRB Requirements**

One example was a proposed study of the perceived effects of convergence behavior in hospital emergency departments following a well-publicized mass casualty event. Researchers proposed to conduct a study using anonymous self-administered questionnaires. The questionnaires were to elicit the respondents' (professional staff in the emergency department) perceptions of the impact of the convergence of staff, the media, and patients' families on the ability to respond to the event. While such research is generally considered exempt under the Common Rule, the local university IRB not only would not exempt the study, but required full review of the project. After three rounds of trying to meet the changing requests of the IRB to the researchers, the researchers decided that too much time had passed since the event to effectively retrieve the perishable information from the respondents (NRC, 2003c).

---

### Standardizing Data Across Multiple Hazards and Events

Over the years, there have been calls to standardize data collection across multiple hazards and disasters. This call was formalized in the previously discussed NEHRP plan (Department of the Interior, 2003). Specifically, the plan recommends that data collection strategies and instruments become standardized so that comparisons can be made over time and across earthquake events. The committee concludes that this formal call for standardization for earthquakes applies equally to hazards and disasters of all types, and to social science as well as natural science and engineering studies of them.

In interpreting the NEHRP plan from a cross-hazards perspective, disasters having major significance can provide findings of relevance to the United States as a whole. Significance can be defined as events having relatively high magnitude and scope of impact (see Chapter 4, Recommendation 4.4). Such high-impact events ensure a presidential declaration of disaster and provide the opportunity to examine much more comprehensibly the interrelationships among all dimensions of Figure 1.2 (i.e., interrelationships among conditions of vulnerability, event characteristics, predisaster interventions, and post-disaster responses as determinants of physical and social impacts). Yet to optimize the value of research on these relatively rare events, a more coordinated and integrated approach to research design, data collection and analysis, data archiving, and dissemination of findings is needed. Smaller-scale research of less severe but still locally damaging events certainly should continue because findings related to them remain valuable to researchers and practitioners.

The NEHRP plan recognizes that modern post-disaster investigations are far more complex and sophisticated than they were just a few decades ago when teams were small, often funded voluntarily by their members, and of short duration. It also recognizes the need to avoid overwhelming local contacts and organizations in the interest of learning about hazards and disasters, particularly in foreign settings when local officials and residents—many of whom might have experienced losses in the disaster—could be operating under very stressful conditions. Consistent with the above discussion of the changing environment of hazards and disaster research, the committee endorses these assessments. Indeed, the changing environment of research compels the coordination envisioned by the 2003 NEHRP plan.

A few other features of the plan deserve mention because they represent a better understanding of post-disaster contexts, data collection and archiving, and the importance of the timely dissemination of findings in multiple formats and media. Noteworthy, the plan anticipates studies of significant disasters as lasting about five years. Social science researchers have been aware of this need, and at least for some recent earthquakes, such as Loma Prieta (1989), Northridge (1994), and Kobe, Japan (1995), support has been provided for longer-term social science research. Recognizing the need to expand traditional contexts and chronological time frames is consistent with Figure 1.2 and the research summarized in the preceding chapters of this report. The related challenges, of course, are ever-expanding data and the need for more standardized and predictable data collection, archiving, and dissemination. The committee encourages recognition of this informatics problem within the social sciences.

In this regard, the NEHRP plan lists several newly available technologies that can assist with the early collection, rapid transmission, and archiving of field data related primarily to building and lifeline performance. Less is said about the value of technologies and methods to support social science research. Nevertheless, the NEHRP plan recognizes the need to improve the quantity and quality of social science data through new, more standardized protocols on the socioeconomic and health impacts of hazards and disasters. The committee concurs that these improvements are essential for high-quality comparative research.

In sum, the 2003 NEHRP plan provides a focused statement on the need to address hazards and disaster informatics issues that are of central importance to the social sciences as well the natural science and engineering fields. Following the guidance provided by the plan can help to optimize resources, achieve greater efficiencies, avoid duplication, minimize burdens on those being studied, and yield cumulatively greater comparative knowledge about hazards and disasters. To every extent possible, standardization of social science data collection is essential for achieving these objectives.

Attempts to standardize data collection efforts and instruments have occurred intermittently within the social science research community and with variable success. Most of the efforts have been made by individual researchers attempting to cross-validate over time their own studies. One example is the NEHRP-supported archival work of Kreps, Bosworth, and Webb over two decades on organizing and role enactment during the emergency periods of multiple types of events (Kreps, 1985; Bosworth and Kreps, 1986; Webb, 2002). Another example is NEHRP-supported work that responds to the long-standing call by disaster epidemiologists and medical personnel for standardization in collecting casualty data. Here work by Shoaf et al. (2000) involves efforts to standardize the collection and reporting of casualty data on earthquakes. The work (available at http://www.ph.ucla.edu/cphdr/scheme.pdf) recommends standards for data on the hazard, the building and the person and, where possible, makes use of existing standards, such as the International Classification of Diseases manual for coding injuries (Shoaf et al., 2000). It also makes recommendations for expanding and otherwise improving protocols where existing coding schemes are not sufficient.

While previous attempts to standardize social science data on hazards and disasters have generally been intermittent and not coordinated among respective individual researchers and teams working on the same or related topics, the potential for increased standardization in future research is enormous. The above two examples of standardization efforts highlight again a fundamental point: Knowing what to look for in studies of hazards and disasters enhances the possibility of developing modular protocols and data collection instruments.

As highlighted in Chapters 3 to 6, with NEHRP support a fairly solid knowledge base has developed on physical and social vulnerabilities and their associated risks (both objective and subjective) as well as the standard data requirements to produce critically needed loss estimation models. A great deal has been learned at the individual and household levels about risk communication, warning dissemination and response, evacuation, and other forms of protective action. The preparedness and response activities of disaster-relevant organizations have been the foci of post-impact investigations for decades, to the point that over time codification of knowledge has become increasingly possible. Findings at the multiorganizational response network level of analysis have expanded rapidly during the past two decades and they are based on highly structured methods and protocols. And while less is known about the behavior of firms, other community-based organizations, and intergovernmental relationships before, during, and after major disasters, existing conceptual and methodological tools that have been used to study individuals, households, disaster-relevant organizations, and multi-organizational response networks can be readily applied to these related topics.

So the groundwork has been established through past social science research under NEHRP and other funding sources for standardizing data across multiple hazards and events. Figure 1.2 provides a useful conceptual framework for building on that foundation. However, individual researchers and teams engaged in studies of the same or related topics need to go beyond the traditional reviewing and sometimes discussion of their respective papers and publications. For this to happen, however, structural mechanisms will be needed to focus, motivate, and support collaborative efforts to produce modular research designs and data collection instruments. As made clear in Chapter 5, such collaboration is essential at both intra- and interdisciplinary levels.

### Data Archiving

Creating and maintaining data archives have not heretofore been preoccupations of social science research on hazards and disasters. A notable exception with respect to post-disaster investigations is Disaster Research Center archives. The center was founded during the mid-1960s at the Ohio State University and, since 1985, has based its research activities at the University of Delaware. At its founding, the leadership of the center made the decision to create archives of transcribed field interviews (from audiotapes) and documents from its post-disaster field research and then developed a rudimentary system of cataloguing and retrieving research materials on specific events. The transcribing of field interviews continued until the late 1970s, when it became too expensive; however, since then, audio tapes and documents have continued to be catalogued, stored, and made available to other researchers. The wisdom of that early decision at the Disaster Research Center is documented in Box 7.3 on NEHRP-sponsored secondary research using these archival materials. The archival materials discussed in the box were composed almost exclusively of unstructured field interviews and unobtrusive data until the mid 1980s when survey research became a more prominent tool at the center.

Archiving highly structured population surveys is much easier to accomplish and the resulting data are much easier to work with. The availability of computer-assisted telephone interviewing (CATI) systems and computerized data entry programs allows for the rapid development of clean data sets that can be made available to both the original researchers and other researchers for secondary data analysis. For example, surveys from the Whittier Narrows and Loma Prieta earthquakes have been housed

## BOX 7.3
## Archival Research

Funded under NEHRP for nearly two decades (1982–2001), a series of secondary analyses using the Disaster Research Center (DRC) data archives have been completed by a research program at the College of William and Mary. The archives contain descriptions of planned and improvised post-impact responses to multiple types of disasters. The goal of the archival research program has been to extract qualitative descriptions from the DRC archives that allow for quantitative comparisons of organized responses, social networks that connect them, and the performance of post-disaster roles within organized responses and social networks (see Kreps, 1985, 1991b, 1994; Bosworth and Kreps, 1986; Saunders and Kreps, 1987; Kreps and Bosworth, 1993; Noon, 2001; Webb, 2002). The starting point for the William and Mary research program was the DRC typology of organized disaster responses (Dynes, 1970). That typology distinguishes organizations that are expected to be involved post-impact (established and expanding) from other existing organizations whose involvement is not expected (extending), and from completely new organizations (emergent) whose involvement is totally ad hoc.

The William and Mary research program has employed a structural code and logical metric to measure the origins of emergent organizations as falling along a continuum of formal organizing to collective behavior. The structural code and related metric have also been used to describe the restructuring of existing organizations (established, expanding, and extending) as well as social networks among all four types of organized responses in the DRC typology. Additionally, the research program has developed a methodology to isolate individual role behaviors in organizations and social networks as either consistent or inconsistent with pre-disaster positions, as either continuous or discontinuous with pre-disaster relationships among positions, and as performed either conventionally or improvised.

Both the findings and the methodology of William and Mary archival research have drawn the interest of researchers from the Rensselaer Polytechnic Institute and the New Jersey Institute of Technology who have expertise in disaster research, information science, and decision science. These researchers have three primary interests: first, studying the dynamics of conventional and improvised role enactments during the emergency periods of disasters; second, applying state-of-the-art communications technologies to advance archival methods for analyzing post-disaster roles within organizational and social network contexts; and third, using these advanced archival methods to develop simulations and other decision support tools for emergency management practitioners. These tools can both increase practitioner understanding of post-impact improvisations and improve their ability to plan for improvisation prior to impact (Mendonca and Wallace, 2002, 2004). Maximizing the utility of decision support tools in the future will require standardized data collection protocols and data archiving on, in particular, the responses of established (e.g., law enforcement agencies, fire departments, hospitals and public health agencies, public utilities, departments of public works, military units, mass media) and expanding (e.g., emergency management agencies, Red Cross, Salvation Army) organizations from the original DRC topology, whose involvement is expected in natural, technological, and willful disasters.

at the Earthquake Engineering Research Center Library at the University of California, Berkeley and the Social Science Research Archive at the Institute for Social Science Research, University of California, Los Angeles. Such archiving of general population surveys is consistent with mainstream social science practices generally. Standardized population-based surveys are particularly useful for examining individual perceptions and behaviors (e.g., perceptions of community vulnerability and personal risk, individual and household preparedness and mitigation measures, sources and uses of warning information, evacuation and other types of protective action, estimates of damages, uses of disaster services, support from relatives and friends). Of essential importance, these survey data are quantified and therefore amenable to comparisons with other quantitative data using geographic information system (GIS) and other state-of-the-art technologies on the spatial and physical features of impact (e.g., proximity of households, neighborhoods, census tracts) to areas of varying physical impact (Bourque et al., 2002).

Issues of standardization and data archiving, when combined, pose perhaps the most significant informatics challenge facing social science hazards and disaster research. Simply put, there historically has been a lack of attention to standardizing quantitative or qualitative data and a lack of support for archiving these data in an orderly way for short- and longer-terms uses. The result is unavailability of and access to useful information (Thomas, 2001; Goodchild, 2003). For some important problem areas, data are not available in a form that is of use for the research community. Perhaps the most significant case in point is the lack of consistent and standardized data on economic losses attributed to natural and technological hazards and disasters in the United States. We simply do not know with any certainty what hazards and disasters cost this nation on an annual basis. Further, we do not have a standardized reporting method for losses (nor a clear or consistent definition of what "loss" means), despite repeated attempts to do so (NRC, 1999). Missing as well are archives of general population surveys and field research data on what the committee has termed the hazards and disaster management system.

### From Data Standardization and Data Archiving to Data Sharing

Plans and strategies related to the output functions of hazards and disaster informatics are no less important than those related to its inputs. It is reasonable to assume, however, that future advances in data standardization in hazards and disaster research will compel the application of technical tools to support management of archives and mining data from them. Much can be learned about these functions from ongoing research and development activities in the physical and life sciences, in engineering, and in interdisciplinary work in computational science (e.g., software solutions and professional services that support extraction of data, visual imaging, and Web browsing). Bioinformatics issues, broadly defined, have become sufficiently important within the life sciences that the National Academies has focused attention on them within the context of future research and development initiatives (see, for example, National Research Council [NRC] 2003a, 2002c). The growing technical capabilities and required bandwidth for data transmission through the Internet certainly will facilitate data sharing efforts within all natural science, social science, and engineering fields if related administrative and policy issues can be resolved.

Simply put, the technical means of data archiving, data mining, and data transmission by researchers must be augmented by formal management of data sharing. The "rules of the game" on data sharing are not nearly as clear and agreed upon as those related to the control of data by original investigators. Formal data control values and norms such as those related to standards of validity and reliability, proprietary access and intellectual property, human subject protection, confidentiality of information, and anonymity of sources must translate as formal "terms of use" in the sharing of data between original researchers and secondary data analysts. While, once again, lessons can be learned from experiences in the life sciences and other fields, it is important for the hazards and disaster research community within the social sciences to consider the management of data sharing and promulgate formal standards before rather than after data standardization and data archiving gain the momentum this committee hopes will occur during the early twenty-first century.

The informatics issues of data standardization, archiving, and sharing are generic as are potential solutions to them. The solutions developed collaboratively by researchers lead inevitably to questions of how best to disseminate findings from primary researchers and secondary data analysts to management professionals at national, state, and local levels. The dissemination issue is of sufficient importance to the committee's charge that Chapter 8 is devoted to its consideration. The technical capabilities to disseminate findings in more "user-friendly" ways and through multiple media will continue to increase.

## RELATIONSHIP OF STATE-OF-THE-ART TECHNOLOGIES AND METHODS TO HAZARDS AND DISASTERS INFORMATICS ISSUES

This section considers four state-of-the-art technologies and methods that relate directly or indirectly to the above hazards and disasters informatics issues: computing and communications technologies; geospatial and temporal methods; modeling and simulation; and laboratory or field gaming experiments.

## Computing and Communications Technologies

Much of the change in qualitative data collection in hazards and disaster research has resulted from improvements in audio and video recordings of data collected in the field. High-fidelity microphones and the ability to digitally record images and sounds have become accessible to all researchers in the last few years. Video and audio data can provide all of the details collected from key informant interviews and, when they are feasible or required by circumstances, focus groups of respondents. They also allow for matching verbal statements with nonverbal cues of research partici-pants. In addition, there is the possibility of gathering data without the presence of a researcher, who potentially can bias the responses of interviewees of focus group members.

As described below in the section on gaming experiments, both audio and video recordings can also be made of participant responses to labora-tory or field experiments. New qualitative analysis software such as ATLAS.ti (ATLAS.ti Scientific Software Development GmbH, 2002) and Qualrus (Idea Works, Inc., 2003) allow for effective use of these enhanced forms of data processing. With these existing and pending new versions of qualitative analysis software, researchers can build highly structured proto-cols for text data, such as transcribed interviews and documents in the Disas-ter Research Center archives discussed in Box 7.3 (Mendonca and Wallace, 2004). Such protocols can also be applied to video and audio data that are exclusively in these forms. Existing software also allows for exporting coded data into state-of-the-art statistical software packages. Using these same statistical packages, the potential then exists to integrate highly structured visual and audio data with highly structured data produced through general population or subpopulation surveys as discussed earlier in the chapter. In effect, the technical and methodological means to merge qualitative and quan-titative information in standardized data sets is substantial, and this poten-tial exists for both pre-disaster and post-disaster investigations. By employ-ing computing and communications technologies, such standardization also facilitates solutions to data archiving, mining, and transmission issues.

Possibly the greatest influence both on researchers and the population as a whole during the past three decades has been access to computer technology and the Internet. Indeed, changes in computation and commu-nications are arguably among the most rapidly diffusing technologies in America. In the year 2000, for example, 51 percent of households in the United States had access to a computer in the home, which compares to only 8 percent in 1984, the first year the question was asked in a U.S. Census Bureau (2001) current population survey. Today, Internet access is practically synonymous with computer access, with nearly 42 percent of households reporting Internet access at home. Households with children are the most "plugged in," with two-thirds having computers and 53 percent having Internet access. With telephone and cable companies offering access to DSL and broadband Internet, more and more homes have high-speed access to the Internet, allowing more complex forms of information to be accessed (e.g., streaming video).

Changes in computing and communications technologies during the past decade are not simply a matter of increased access. There is also greater computing capacity in smaller and smaller computers. The handheld computers of today, whether powered by a Palm operating system or Windows CE, are as powerful as desktop computers were 10 years ago. The advent of compact memory cards and USB memory drives allow for storage of large amounts of data that are easily transferred from computer to computer. Likewise, the advances in microprocessor technology have resulted in improved digital imaging as well as audio and video recording. As discussed earlier, all of these advances have improved the ability to conduct research and have greatly facilitated more highly structured data collection in the field. Such technologies also increase enormously the ability to archive, mine, and transmit data among researchers.

Access to the Internet has increased the speed with which field reports become available to other researchers and the general public. The Quindio, Colombia, earthquake of January 1999 was one of the first times that an EERI Learning from Earthquakes (LFE) reconnaissance team filed its initial report from the field. It is now standard practice for field reports to be sent back to research centers from the field via the Internet.

In addition, there have been advances in wireless communication technologies. In 2001, more than 62 percent of Americans owned a cellular phone. Additionally, cellular telephone coverage is becoming ubiquitous in even the least developed countries. Indeed, in less developed countries cellular telephones are popular because people do not have to wait for the installation of standard national telephone services (McFarland, 2002).

A definite asset in conducting research is the almost universal coverage of cellular service and the capability of the newest phones to be used on multiple network formats. This capability allows U.S. researchers to have phone service, using a single phone number that provides access across the country and internationally. Wireless communication technology has also impacted the computing world. It can now be included in notebook computers to take advantage of the more than 25,000 publicly available wireless access points. This wireless access allows the transfer of information from a remote location to other researchers and to centralized data storage points. Wireless computers can take advantage of publicly available wireless access points, connections through cellular telephones, or similar technology built into a wireless modem. Although wireless technology is still limited by the number of access points or the location of cell sites, as cell sites increase so will the usefulness of wireless computing.

## Geospatial and Temporal Methods

As noted throughout this report, hazards exist and disasters occur in chronological time and physical space. Whereas maps have been the traditional manner by which geographers represent things in physical space, a new definition of mapping suggests that it allows for more than just placing things on maps; more basically, it allows for understanding the spatial nature of things (Edson, 2001). Spatial analysis is the term used to describe a set of tools and methods for examining the patterns of human activity or physical processes as well as movements across the Earth's surface (Hodgson and Cutter, 2001:50). In addition to statistical analysis and mathematical modeling, mapping (cartography) and GIS are the tools most commonly used in spatial analysis. Their use is equally relevant to pre-, trans-, and post-disaster investigations; they promote the development of standardized protocols on hazard vulnerability and disaster impacts; and they yield data that can be stored, merged, and disseminated electronically.

GIS is a rich set of tools that can be used for collecting, analyzing, storing, and displaying geographic data. All data must be georeferenced; that is, they must possess some locational attribute such as a coordinate (longitude/latitude) point, a polygon (such as a census tract), or a line (such as a road). Diverse data can then be combined by overlays to see the relationship between the two layers or the many layers that are included in the GIS. The simplest version is the construction of a data layer of housing properties overlain with a data layer depicting the 100-year flood zone to see which properties are inside or outside the zone for a given community. The Federal Emergency Management Agency's (FEMA's) HAZUS (NIBS-FEMA, 1999) is a GIS-based decision support tool that helps identify potential losses from a number of different scenarios.

GIS is widely used in some hazard-prone areas, among them the reverse 911 notification system (E-911) and wildfire hazards monitoring. Increasingly, GIS is becoming the preferred tool for vulnerability assessments and other hazards modeling applications. As noted earlier (Radke et al., 2000; Cutter, 2003a) there are a number of areas in which geographic vulnerability science can enhance the hazards and disasters research community. These include better temporal and spatial estimates of tourists, homeless people, and undocumented workers; better integration of physical processes and social data to predict hazard impacts; and interoperability where data in a variety of formats can be easily shared and exchanged by various systems in a highly decentralized and distributed system.

The advent of GIS has increased the ability of researchers to study the spatial nature of hazards and their relationship to human populations. Spatial data can be gathered through other means, however. Aerial photography has been used in post-disaster situations to visualize changes in topography and geography from pre- to post-disaster. An example is the pre- and post-impact comparisons of aerial photography to measure impacts such as those from Hurricane Andrew (Hodgson and Cutter, 2001; Ramsey et al., 2001) or the monitoring of heat from the debris pile at the World Trade Center using thermal sensors (Greene, 2002).

A newer trend is to utilize satellite-based technology to visualize changes associated with disasters impacts. Such technology as light detection and ranging (LIDAR) and RADAR can be utilized to visualize impacts and provide spatial data. Probably the most important attribute includes an "ability to quickly gain an overview understanding of the extent of damage" (EERI 2005:20) so that social scientists can identify geographic areas of interest or types of damages that probably resulted in significant social impacts that would be worthy of study. In recent articles (Adams et al., 2004a, 2004b, 2004c), several examples were provided, including the Bam, Iran, earthquake; Niigata, Japan, earthquake; Hurricane Charley; the catastrophic Indian Basin earthquake and tsunami, the World Trade Center, and the search for the *Columbia* Space Shuttle wreckage. Moreover, such remote sensing technologies are enhancing emergency preparedness activities and related loss estimation and decision support software tools.

Remote sensing technologies have been used in studies of many of the foreign earthquake events of the past five years. In order to be most effective, it has been noted that standardized damage scales need to be developed to ensure consistent interpretation of remotely sensed data and images. With standardized scales, these technologies can be used to identify areas of significant damage, collapsed structures, estimation of mortality (based on building damage), inundation zones, and areas of utility outage (Eguchi, 2005). Post-disaster investigations, especially those occurring during the reconnaissance stage, therefore benefit from state-of-the-art spatial technologies and methods. Real-time data from earthquakes, for example, when translated into ground-shaking maps can allow identification of the most likely area of serious impacts.

As illustrated in Box 7.4, GIS and remote sensing technologies and methods have powerful applications. However, with increased access to and usability of these tools, technologies, and methods, the risk exists of inaccuracies being promulgated. One significant constraint on the effective use of maps is the availability and quality of data being utilized. Two characteristics of data inputs are required in analyzing hazardous conditions and disaster events: a temporal dimension and a geographic or spatial dimension. The type of spatial data required depends on the chronological time phase of the application (e.g., an immediate post-impact response versus a longer-term reconstruction response).

In looking at the spatial nature of hazards, the scale, resolution, and extent of data are equally important. Map scale is the relationship between

---

**BOX 7.4**
**GIS and Remote Sensing Technologies**

Using an innovative approach with geographic information system (GIS) and remote sensing technology, the LandScan global population project has developed a population distribution model that produces the finest resolution population distribution data available for the entire world and the continental United States (Bhaduri et al., 2002). LandScan global at 1 km resolution represents an "ambient population" (average over 24 hours) and is 2,400 times more spatially refined than the previous standard. The LandScan population distribution model involves collection of the best available census counts (usually at subprovince level) for each country and four primary geospatial input data sets—namely, land cover, roads, slope, and nighttime lights—that are key indicators of population distribution. Relationships between any of these datasets and population distribution are not globally uniform. For each region, the population distribution model calculates a "likelihood" coefficient for each LandScan cell, and applies the coefficients to the census counts, which are employed as control totals for appropriate areas. Census tracts are divided into finer grid cells (1 km), and each cell is evaluated for the likelihood of being populated based on the four geospatial characteristics. The total population for that tract is then allocated to each cell weighted to the calculated likelihood (population coefficient) of being populated.

As an expansion of global LandScan, very high-resolution (90 m cell) population distribution data (LandScan USA) are being developed for the United States. LandScan USA includes nighttime (residential) as well as daytime population distributions. LandScan USA is more spatially refined than the resolution of block-level census data and includes demographic attributes (age, sex, race). Locating daytime populations uses a modeling approach that involves not only census data, but also other socioeconomic data including places of work, journey to work, and other mobility factors. Hourly population distribution at the 90 m cell have been developed for several major metropolitan areas The combination of both residential and daytime populations will provide significant enhancements to geospatial applications ranging from homeland security to socioenvironmental studies.

---

SOURCE: http://www.ornl.gov/sci/gist/; Bhaduri et al. (2002).

---

the length of a feature on the map and its length on Earth (Hodgson and Cutter, 2001). Many people mistake spatial scale for another dimension of data, spatial resolution. Spatial resolution is the observational or collection unit (e.g. county, census tract, individual). Finally, spatial extent is the area covered by the study (e.g., city, entire nation, world). The choice of spatial characteristics is driven by the research problem—for some, more detailed and fine-grained analyses (based on individual observations within one community) are more appropriate than larger, more generalized analyses such as those that concentrate on the hazard characteristics of counties (resolution) for the entire United States (spatial extent).

The chronological nature of spatial data involves two important concepts: frequency and lag time. Data often become "old" as the time increases between when they were first collected and ultimately used. Real time or near real time refers to data that has no discernible lag time, that is, receipt of the data is almost instantaneous to its collection. The use of sensors to provide data on traffic flows or Doppler radar that is used to identify tornado winds is a good example of real-time or near-real-time application. The frequency of data is another characteristic that has implications for social science research. For example, surveys about hazards and disaster experiences and expectations, as both relate to mitigation activities, heretofore have been done infrequently. While post-disaster field surveys are done in greater numbers, their frequency depends on the uncertainties of event frequencies. Post-hurricane evacuation behavior surveys are normally, but not always, conducted after major landfalls of hurricanes.

An example of a frequency concern is the decennial census. Population and housing data are essential for modeling populations and infrastructures at risk from hazards, yet these data are only collected every 10 years (frequency). At the same time, there is often lag time between when they were collected (e.g., 2000) and when they become available for use (e.g., 2002). Thus, data that represent the social or demographic situation in 2000 (the census year) may or may not be applicable to a community in 2005, especially in areas that have experienced rapid growth. Given this time lag, communities often resort to population projections in producing demographic profiles.

The temporal characteristics of data influence the types of research questions that can be addressed. A good example is data production with remote sensing technologies. Remotely sensed data are most often used for purposes of pre-event threat identification (e.g., identification of hurricanes in the mid-Atlantic) and post-event rescue and relief operations. While the collection of remote sensing data can be scheduled on demand, the lag time required for processing such data may negate their utility in immediate emergency response situations such as the attack on the World Trade Center (Thomas, 2001; Bruzewicz, 2003). Thus, both the frequency of data collection and the lag time between the collection of data and their availability influence what hazards and disasters researchers study and how the research questions are framed.

## Modeling and Simulation

Models are abstractions of reality, and modeling is the process of creating these abstractions. Because reality is nearly infinitely complex and all empirical data are processed with reference to that complexity, model building involves the simplification of reality as data are transformed into knowledge. The models created are essentially forms of codified knowledge and used to represent the "reality" of things not known from things that are known (Waisel et al., 1998). Modeling is the sine qua non of science. Virtually all scientific activities require modeling in some sense, and any scientific theory requires this kind of representational system (Neressian, 1992). The structure of a model can be symbolic (i.e., equations), analog (i.e., graphs to model physical networks), or iconic (physical representations such as scale models). Models are usually thought of as quantitative, and able to be represented mathematically. However, qualitative models are no less, and arguably more, common. For example, mental models play a very important role in our conceptualization of a situation (Crapo et al., 2000), and verbal and textual models are used in the process of communicating mental models.

Science can be seen as a model-building enterprise because it attempts to create abstractions of reality that help scientists understand how the world works. Technological advances in computing allow the development of complex computer-based models in a wide range of fields. These models can be used to describe and explain phenomena observed in physical systems from micro- to macrolevels, or to provide similar representations of real or hypothetical experiences of individuals and social systems. Models play an essential function in formalizing and integrating theoretical principles that pertain to whatever phenomena are being studied. For example, the computational models used for weather forecasting integrate scientific principles from a variety of natural science and engineering fields. In similar fashion, computational models used for social forecasting integrate theories from a variety of social science as well as interdisciplinary fields such as urban and regional planning, public policy and administration, and public health management.

Computational modeling provides an opportunity for social scientists conducting studies of hazards and disasters to integrate theories and empirical findings from the natural sciences, engineering, and social sciences into models that can be used for decision making. For example, one of the most widely used models in emergency management is that of loss estimation. Loss estimation modeling for disasters has grown in the last decade. Early loss estimation methods were grounded in deterministic models, based on scenarios. Scenario events were chosen and estimates of impacts were based on those events. During the 1970s, for example, NOAA scenarios (NOAA, 1972, 1973) estimated regional physical and social impacts for large earthquakes in the San Francisco and Los Angeles, California, areas and were intended to provide a rational foundation for planning earthquake disaster relief and recovery activities. By the 1990s, technological advances in personal computing technology, relational database management systems, and the above GIS and remote sensing systems had rendered the development of automated loss estimation tools feasible.

As noted above, HAZUS (NIBS-FEMA, 1999) was developed by FEMA and the National Institute of Building Sciences (NIBS). It is a standardized, nationally applicable earthquake loss estimation methodology, implemented through PC-based GIS software. HAZUS methodology estimates damage expressed in terms of the probability of a building being in any of four damage states: slight, moderate, extensive, or complete. A range of damage factors (repair cost divided by replacement cost) is associated with each damage state. While the front-end of the loss estimation methodology is clearly driven by the earth sciences and engineering, the outputs of the model are much more social science driven. The outputs of interest to urban and regional planners and emergency management professionals are not ground motions, but rather the impacts of ground motion at community, regional, and societal levels. Researchers from the Pacific Earthquake Engineering Center have developed a performance-based earthquake engineering model that describes these outputs as the "decision variables" and often refers to them as "death, dollars, and downtime."

Other far less used computational models have the potential for significant use in social science hazards and disaster research. For example, what has come to be known as agent-based modeling is a set of computational methods that allows analysts to engage in thought experiments about real or hypothetical worlds populated by "agents" (i.e., individuals, groups, organizations, communities, societies) who interact with each other to create structural forms that range from relatively simple to enormously complex (Cederman, 2005). Such modeling, which has grown out of work on distributed artificial intelligence, can be used to simulate mental processes and behaviors in exploring how structural forms operate under various conditions (Cohen, 1986; Bond and Gasser, 1988; Gasser and Huhns, 1989). A major strength of agent-based modeling is its focus on decision making as search behavior. Model applications have been used to address issues of communication, coordination, planning, or problem solving, often with the intent of using models as the "brains" of real or artificial agents in interactions with each other. These models can facilitate descriptions and explanations of many social phenomena and test the adequacy and efficiency of various definitions or representation schemes (Carley and Wallace, 2001). The earlier example (Box 7.1) of planned and improvised post-disaster responses illustrates the kind of research topic in hazards and disaster research that can be advanced through use of agent-based modeling techniques.

In that example, conventional and improvised roles are nested within different types of organizations and social networks, which connect roles and organizations. The networks themselves represent more inclusive structural (i.e., relational) aspects of agent-based modeling and inform knowledge of when, where, how, and why role behaviors and organizational adaptations occur following a disaster (Mendonca and Wallace, 2004). It is important in this regard to develop representations of both network adaptation and how "agent" knowledge, behaviors, and actions affect and are affected by their respective position within the network. Network models have been used successfully to examine issues such as power and performance, information diffusion, innovation, and turnover. The adequacy of these models is determined using nonparametric statistical techniques (Carley and Wallace, 2001).

From the perspective of a researcher concerned with social phenomena in disaster contexts, two issues stand out (Carley and Wallace, 2001). First, how scalable are agent-based models and representation schemes? That is, can the results from analyses of social networks from two to a relatively small number of members (agents) be generalized to larger more complex response systems that are so characteristic of events having high magnitude and scope of impact? Second, are cognitively simple characterizations of individuals as "agents" adequate or valid representations of agents when the actions of groups, organizations, communities and societies are at issue? Answers to these questions are not possible at this point in knowledge development. However, agent-based modeling techniques are developing rapidly (Gilbert and Abbot, 2005), their development is unambiguously interdisciplinary (Cederman, 2005), and their twin focus on human decision making and structural adaptation (Eguiluz et al., 2005) is a core feature of what has been termed the hazards and disaster management system. Decision support tools are needed in this system, and agent-based modeling techniques can facilitate their development and dissemination (Mendonca and Wallace, 2004).

Perhaps the most familiar computational modeling tool to social scientists is simulation. Simulation models often represent an organization or various processes as a set of nonlinear equations and/or a set of interacting agents. In these models, the focus is on theorizing about a particular aspect of social action and structure. Accordingly, reality is often simplified by showing only the entities and relations essential to the theory that underlies them. Models embody theory about how an individual, household, small group to larger organization, community, or society will act. With a model structure in place, a series of simulations or virtual experiments can be run to test the effect of a change in a particular process, action, policy, or whatever. In so doing, models are used to illustrate a theory's story about how some agent will act under specified conditions. Cumulative theory building evolves as multiple researchers assess, augment, reconstruct, and add variations to existing models (Carley and Wallace, 2001).

The dominant use of computing in the natural sciences, social sciences, and engineering continues to involve statistical models of existing data.

These statistical models range from relatively simple to highly complex configurations of variables, but over time the increasing capacity of computers to process enormous volumes of data has allowed the development of the kinds of computational techniques discussed above. Computational models are more powerful to the extent that simulated data are informed by real data. The ready accessibility of those doing computational modeling to empirical data previously collected on the same topics, and hopefully archived for secondary data analysis, is therefore essential.

## Laboratory and Field Gaming Experiments

It is certainly possible for any research program or center to include both field studies and laboratory simulations of responses to hazards and disasters. When the Disaster Research Center (DRC) was established during the mid-1960s, for example, its research program included both field studies and laboratory gaming experiments. The field studies have continued for decades. However, after a very creative early application (see Drabek and Haas, 1969), the simulation work was largely suspended because of its related cost and complexity. No formal program in social science hazards and disaster research involving laboratory or field gaming experiments has been sustained since the early 1970s. Certainly, emergency management professionals engage routinely in realistic simulations, either at their own local or regional emergency operations centers or perhaps at FEMA's Emergency Management Institute in Emittsburg, Maryland. However, these simulations are designed as training exercises not as research opportunities for assessment of their effectiveness or realistic foundation in disaster field studies. For the purposes of this chapter, the early combination at the DRC of field studies, data archiving, and simulations continues to serve as a template for future hazards and disaster research.

The use of experimentation has been both touted and criticized by researchers in the social sciences (Drabek and Haas, 1969; Hammond, 2000). Of particular concern is the need to ensure proper scientific conduct of experiments. Increasing realism in experimental situations leads potentially to problems of generalizability. However, the generalizability of "realistic" laboratory or field experiments may be compromised if participants are not experienced in the domain—the result being that the hypotheses postulated may not correspond to the phenomena actually encountered in a real decision environment. Moreover, the events or activities that are controlled in experiments may not be controllable in a real world. Gaming simulations are quasi-experimental designs that can provide both statistical power and the ability to generalize results to a variety of crisis situations.

The advent of computational modeling, as discussed above, has provided another application for gaming simulations (i.e., the testing and validation of computational models of social phenomena). In these simulations, "agents" in the computational model "play" the same roles as human participants. The actions taken by both human and artificial agents can then be compared. Also, agent-based models, when informed by empirical data, can be used to create a realistic setting for the gaming simulation and, in effect, "play against" the participants, again providing opportunities to investigate cognitive and behavioral phenomena of individuals in social entities of various types.

Both research and practical experience have shown that written plans and procedures serve the valuable purposes of training and familiarization with the role of incumbents such as public officials in crisis-relevant organizations (Salas and Cannon Bowers, 2000). These plans and procedures serve as a normative model for education and training activities. Gaming simulations can provide a means for evaluating the plans and procedures in laboratory settings or in the field (e.g., emergency operations centers). An additional and equally important benefit of these simulations is that they can provide a field laboratory or field venue for experimentation on multiple types of circumstances. Thus, experiments on responses to terrorist events can readily be compared with those related to natural and technological disasters.

A variety of data can be collected prior to a gaming simulation, subject only to the patience of the participants. Biographical data are certainly available—and they may be needed for designing the experiment (Grabowski and Wallace, 1993). For example, data could be collected on cognitive style prior to the exercise and the results used to design the experiment. However, it is important not to deluge participants with an extensive battery of questionnaires because they may create apprehension, alter behavior, or magnify the lack of realism of the simulation. Unobtrusive measures for data collection can also be devised in laboratory or field experiments to record the activities engaged in by the participants. All communications can be recorded and a digital record kept of phone messages, including recording sender, receiver, length of message, and content. These data can be collected for each sample run and categorized in a variety of ways. It must be recognized that participants may communicate with outsiders or with insiders who are not part of the experiment but are with the training group.

Unobtrusive measurements can be built into the exercise, such as recording time and measuring the difference between the time that an event was initiated and the appropriate responses were made. To measure the degree of correctness, every initiated event can have a set of appropriate decisions. In addition to maintaining a record of the activities of participants in the game, many times simulations lend themselves to observation. Participants in the exercise can be observed in a very structured manner with pre-designed instruments to be completed by trained observers. Video-taping can be also used, but usually needs to be electronically transcribed for analysis—resulting in a great deal of qualitative data that require extensive effort to analyze. Various techniques, such as protocol analysis, have been found useful for research purposes, but the benefits of their use must outweigh the costs because they are so time consuming. Behavioral coding of group interactions can be done both in real-time or from the videotapes. Here training of the coders is crucial (Fleiss, 1971).

After a gaming experiment is run, participants can complete self-reporting questionnaires. These can be done as part of, or immediately after the activity, and digitally recorded (Litynski et al., 1997). Participants can also be asked to describe and rate each other's behavior on a variety of dimensions, and to record their interactions with each other during the course of the exercise. Both preceding and following the exercise, interviews can be conducted with each of the participants.

The foregoing activities will create a wealth of data. Analysis of the data generated by experimentation using gaming experiments can usually be assessed by standard statistical techniques (Cohen, 1977). The degree of realism of a game is extremely important, not only from the point of view of evaluating the decision aid per se, but in maintaining the interest in and enhancing the educational benefits of the simulation. Such validity can be easily ascertained by having experienced field researchers and emergency management professionals walk through the simulation prior to the actual exercises.

Perhaps the most complex issue with gaming experiments is as follows: Do participants treat the simulation as realistic? This was certainly the case in the seminal work by Drabek and Haas (1969). Box 7.5 provides a case where the realism of the gaming simulation could be compared to an actual event that followed shortly after a gaming experiment was run. In this case, it was found that there was some in-game playing because the recovery activity in the simulation did taper off in comparison with the actual event; in fact it ended dramatically at 4:00 p.m. (Belardo et al., 1983). This suspension was obviously not the case with the actual event. However, gaming simulations can be designed in the laboratory or field as learning experiences, and the participants usually understand that training is very important as a precursor to the need to prepare for dealing with incidents with the potential to escalate to a disaster.

In conclusion, gaming simulations with hazards and disaster management professionals as participants have an important role in social science research on disasters. The core idea here is to build gaming simulations with an eye toward realism. Such realism can be captured through standardized data from previous field studies that are maintained in effectively managed data archives, accessible to multiple researchers, and used to every

---

**BOX 7.5**
**Realism in Gaming Simulation**

A serendipitous evaluation of a gaming simulation yielded the observation that realism in the crisis environment was replicated in the simulation environment in terms of both organizational- and individual-level responses. The evaluation entailed data collection during a training exercise held by the U.S. Nuclear Regulatory Commission and the Federal Emergency Management Agency (FEMA) at the Robert A.F. Genet Nuclear Facility in New York. Four days after the simulation an actual incident occurred that involved the activation of all emergency response activities throughout the State of New York. This provided an opportunity to evaluate the benefit of simulations. The realism of the crisis environment was well replicated, both organizationally and its impact on individuals. Stress levels were found to be similar between the simulation and the actual event. Communications were similar during the beginning of the crisis, but there were some differences during the latter stages of the exercise, particularly with respect to decisions concerning recovery operations. This may have been due to participants in the gaming simulation being aware of the need to end the exercise before the end of the working day.

SOURCE: Belardo et al. (1983).

---

extent possible in the development of computational models such as those summarized above. The hazards and disaster research community has developed knowledge to the point at which it is feasible to integrate these core informatics activities.

## RECOMMENDATIONS

The research findings and recommendations from Chapters 3 to 6 of this report summarize what has been done in the past under NEHRP support and what the committee feels should be done in the future. The discussions of technologies, methods, and informatics issues in this chapter relate the substance of past and future hazards and disaster research to its actual implementation. Thus, regardless of the topics discussed in previous chapters, social science studies in the next several decades must be responsive to the changing environment of hazards and disaster research. By whatever available technological and methodological means available, they must capture data that are more highly structured and standardized across natural, technological, and willful hazards and disasters. They must analyze, store, and manage data with dissemination and formal rules of data sharing in mind.

Recommendation 7.1: *The National Science Foundation and Department of Homeland Security should jointly support the establishment of a nongovernmental Panel on Hazards and Disaster Informatics. The panel should be interdisciplinary and include social scientists and engineers from hazards and disaster research as well as experts on informatics issues from cognitive science, computational science, and applied science. The panel's mission should be (1) to assess issues of data standardization, data management and archiving, and data sharing as they relate to natural, technological, and willful hazards and disasters, and (2) to develop a formal plan for resolving these issues to every extent possible within the next decade.*

As summarized in this chapter, there are continuing issues in the following areas: (1) standardizing data on hazardous conditions, disaster losses, and pre-, trans-, and post-impact responses at multiple levels of analysis; (2) improving metrics in all of these same research areas; (3) developing formal data standards for storing, aggregating, disaggregating, and distributing data sets among researchers; and (4) using computing and communications technologies to enhance quantitative and qualitative data collection and data management. Addressing these issues systematically can, and the committee believes should, lead ultimately to the establishment of both centralized (virtual) and distributed data repositories on hazards and disasters.

The range and depth of research inquiries and approaches in hazards and disaster research will perforce result in major increases of data. Thus, the status quo ante of continuing inattention to data management issues is no longer acceptable. Resolving what the committee has termed globally the "hazards and disasters informatics problem" will require careful consideration and planning. This research community is not in a position to simply adopt informatics solutions from other fields of inquiry because such solutions are only now in the process of being developed. Like other research domains, hazards and disaster research has its own unique theories, models, and findings. Yet informatics issues and their resolution are not field specific; they are generic to basic and applied science. The committee believes that the first step in becoming a more active participant in the "science of informatics" is to create the interdisciplinary panel of experts specified in Recommendation 7.1.

The research domain of this community includes natural, technological, and willful hazards and disasters. Thus, the committee believes that it is quite appropriate for the National Science Foundation and the Department of Homeland Security to provide joint support for the work of the recommended interdisciplinary panel. The conceptual framework developed in Chapter 1 (see Figure 1.2 and its related discussion)—placed within the changing societal context described in Chapter 2, the research findings and recommendations summarized in Chapters 3 to 6, and discussions of research methods, techniques, and informatics issues in this chapter—provides the foundation for the recommended panel. The work of this panel should commence as soon as possible.

> **Recommendation 7.2:** *The National Science Foundation and Department of Homeland Security should fund a collaborative Center for Modeling, Simulation, and Visualization of Hazards and Disasters. The recommended center would be the locus of advanced computing and communications technologies that are used to support a distributed set of research methods and facilities. The center's capabilities would be accessible on a shared-use basis.*

There is an immediate need in social science hazards and disaster research to expand the use of state of the art modeling and simulation techniques for studies of willful as well as natural and technological hazards and disasters. The joint support of the National Science Foundation and Department of Homeland Security is therefore encouraged for purposes of implementing Recommendation 7.2. Three areas of research would be supported by the center, each of which would be developed and maintained at distributed research sites:

- *Modeling and simulation*: The center would act as both a repository for models constructed by social science researchers at distributed sites, and would work to ensure collaboration, (including experimentation using the Internet), maintenance, and refinement of models. Compatibility, which permits "docking" of computational models, would be a major responsibility of researchers and support staff of the Center.

- *Visualization*: Social science researchers are making ever-increasing use of digitized spatial and graphical information, such as global positioning system (GPS)-GIS displays. In addition, human-computer interface technologies are being investigated for possible use as decision tools for hazards management and emergency response. Research on the cognitive processes underlying visualization under conditions of stress and information overload typical of emergency response situations is just one potential topic for this visualization component of the recommended center.

- *Gaming experimentation*: The recommended center would have its own and distributed laboratory settings with data collection technologies for research on individual, small group/team, and "organizational" decision making using exercises, "games," and other interactive experimental media. Researchers could gather data and control treatment from distributed locations networked to the center.

As documented in this chapter, computational modeling, visualization, and gaming experiments are important tools for building on and applying knowledge gained from field studies. Heretofore the use of these technical tools has not been integrated, thus reducing their potential value. Such integrated use is best accomplished within a center established for that purpose. As noted above for example, the core idea of gaming and simulation is to build them with an eye toward realism. Such realism is enhanced through standard data production from previous field studies. As the resulting data from field studies become more effectively maintained in distributed data archives, they can be used systematically by the proposed center in the development of computational models and simulations, and the design of gaming experiments specifically for hazards and disaster management professionals. The hazards and disaster research community has developed to the point at which the sustained integration of field research, modeling, and experimentation can be accomplished.

**Recommendation 7.3:** *The hazards and disaster research community should educate university Institutional Review Boards (IRBs) about the unique benefits of, in particular, post-disaster investigations and the unique constraints under which this research community performs research on human subjects.*

The committee has noted above the difficulties involved in harmonizing the actual practice of research with the demands placed on researchers during field studies by the fluid situations that inevitably follow disasters. In particular, the fine points of consent forms, detailed interview protocols, and other research infrastructure are often unachievable in the hours to weeks after a disaster. Furthermore, such requirements may violate cultural norms in the places studied. At the same time, IRB members may have real but sometimes misplaced concerns about the risks of psychological harm that they believe attach to research on hazards and disasters.

To the extent that they are not, hazards and disaster researchers must become familiar with federal (in particular, 45 CFR 46.101 et seq.) and local university regulations regarding human subjects research so that they can be knowledgeable resources for their respective IRBs and effective advocates for appropriate deviations from "standard" practices, while maintaining the personal privacy and dignity of research subjects. Members of the research community should seek to become members of human subjects review panels on IRBs or should assist in other policy-making roles.

# 8

# Knowledge Dissemination
# and Application

A s noted in previous chapters, much has been learned about the core
topics of hazards and disaster research. This accumulated body of
knowledge can serve as a foundation for science-based decision
making by individuals and households, policy makers in the legislative and
administrative branches of government, emergency managers at the community and state levels, and various stakeholders in the private sector.
Before such knowledge can be applied by potential users, however, they
must know of its existence and relevance for meeting the challenges they
face in coping with low-probability, high-consequence risks posed by natural, technological, or willful hazards. Knowledge utilization is also furthered when the information is demonstrably relevant to stakeholders, when
it is disseminated effectively, and when stakeholders are motivated to use it.
The absence of any of these conditions can contribute to the underutilization
of knowledge, the so-called implementation gap. More systematic research
is needed on the dissemination and application of hazards and disaster
information generated by the social sciences and other disciplines. This
research will provide a clearer understanding of what can be done to further
the implementation process, thereby advancing sound mitigation, preparedness, response, and recovery practices.

In response to the committee's statement of task, this chapter focuses
on the challenges to increasing the application of social science research
results on hazards and disasters. One of the most important challenges is
the lack of systematic and recent research on this topic, resulting in an undue
reliance on anecdotally derived insights. The chapter briefly discusses some

of the research on the knowledge utilization process in this field and the relevant literature on dissemination and utilization that has been produced by social scientists outside the field. This is followed by a discussion of several examples of knowledge diffusion and utilization efforts in hazards and disaster research that are at least anecdotally known to have experienced some degree of success. For analytical purposes, a simple matrix is used to categorize these efforts according to principles derived from the extant research utilization literature. The chapter concludes with a discussion on research needed to enhance future utilization in the hazards and disaster field.

As previously noted, social science hazards and disaster research emerged with a problem-focused orientation, which continues to this day even while researchers also give considerable attention to basic research and theoretical issues. Thus, much of the research described in Chapters 3 and 4, as well as elsewhere in this report, has been undertaken to advance social science theory and to further the reduction of disaster losses and social disruption, enhance emergency response, and speed disaster recovery. More specifically, much of the research conducted on hazards and disasters is geared toward providing a more informed basis for actions by policy makers and practitioners. Thus this body of work has implications for various types of applications, including disaster education and training, hazards reduction legislation and regulations, and emergency and recovery preparedness practices. Nevertheless, it is unclear to what extent stakeholders know about and use social science knowledge relevant to such applications and, when such knowledge is applied, what difference this actually makes. Research is therefore needed.

## SOCIAL SCIENCE RESEARCH ON THE UTILIZATION OF HAZARDS AND DISASTER INFORMATION

Very little research has been conducted on the utilization of social science knowledge of hazards and disasters. For example, prior research has not systematically addressed variations in utilization by different user communities. Most prior research, largely carried out in the 1980s, was qualitative in nature, and typically employed a case-study approach. Some of this work may not be as relevant today as it once was, especially given some of the societal changes discussed in Chapter 2. Anecdotal evidence about the way findings have been utilized by the practitioner community is fairly commonplace. Examples point to researchers who work with federal agencies to ensure that the results of their studies are incorporated into policies, planning guides, and training activities. Also noteworthy are examples of researchers who work with state and local governments to help translate research into practice. These and other examples of the promotion of knowledge application are discussed at length later in this chapter.

The most extensive study of the utilization of research on natural hazards and disasters was conducted by Robert Yin and his colleagues in the 1980s (Yin and Moore, 1985; Yin and Andranovitch, 1987). They analyzed the utilization of research in a variety of disciplines, including the social sciences and engineering. One study (Yin and Andranovitch, 1987) focused on the role of nine professional associations, including the American Planning Association (APA), the Association of American Geographers (AAG), and the American Society of Civil Engineers (ASCE), in stimulating the utilization of 14 innovations related to hazards such as earthquakes, landslides, and radon. A key finding from the research was that professional associations play the role of synthesizers of information from various sources. According to the researchers, these sources are not just limited to research projects, but also include insights derived from experience that represent "craft-based" knowledge. Part of the synthesizing role of professional associations involves the development of consensus among peers about how to tackle particular problems, and this consensus may result from insights derived from a combination of both research and experience.

Another research utilization investigation carried out by Yin and his colleagues involved case studies of nine applied projects in the hazards and disaster field dealing with earth science, engineering, and social science topics (Yin and Moore, 1985). One of the social science case studies concerned a project conducted by the National Academy of Sciences during 1974 and 1975 on the potential social, economic, political, behavioral, and legal consequences of earthquake prediction. The case study concluded that the project (1) influenced federal policy and federal agency research agendas, (2) helped shape federal legislation—the Earthquake Hazards Reduction Act which created NEHRP—and the implementation plan for that legislation, and (3) fostered a concern for the social and economic aspects of earthquake prediction within NEHRP.

The importance of social interaction between researchers and potential users came through strongly in these case studies in explaining the extent of research utilization. This interaction was important regardless of whether the project dealt with engineering, physical science, or social science issues (Yin and Moore, 1985:vi).

> The interactions led to a continued exchange of ideas, creating what might be called a "marketplace of ideas," in which investigators learn more about users' conditions, and users learn more about the ongoing array of research. In some cases, the exchange of ideas was facilitated by the activities sponsored by professional associations. In other cases, the exchange was the result of an active and communicative principal investigator. Overall, communications started earlier than and continued far beyond the ending of a specific project. Furthermore, the project design and conduct could be influenced by information from users, making the research more relevant to users' needs.

The above observations are consistent with comments made by practitioners at the committee's two workshops and with findings from studies outside the hazards and disaster field.

In addition to the case studies by Yin and his colleagues, several other social science research projects have examined research utilization. A study by Lambright (1984) considered the policy role played by the Southern California Earthquake Preparedness Project (SCEPP), a regional organization that emerged with government support to play a leadership role in earthquake preparedness in California. Lambright analyzed SCEPP's origins and development, and drew conclusions about the program's success in stimulating preparedness measures, including those that were science based, for a predicted or unpredicted earthquake. He concluded that SCEPP's mission was an extension of state and federal policy and that the organization, which no longer exists, had been successful in furthering research utilization. The reason for that success was straightforward. SCEPP was provided with necessary resources, had allies that championed its cause, and met with little external resistance.

Another example of studies carried out to better understand research utilization was part of the Second Assessment of Research on Natural Hazards. This involved a survey of 50 researchers and 28 practitioners (Mileti, 1999b; Fothergill, 2000). Researchers were asked how they disseminated their own work, the effectiveness of the different dissemination mechanisms they employed, and if their research was used. Practitioners were asked if they used research findings and, if so, how this came about. Findings from the survey indicated that:

- Local governments most frequently receive information from information dissemination organizations, through personal relationships, and at conferences and meetings. None reported using mainstream or specialty academic journals. Local practitioners noted that findings simply do not get disseminated to them and that they do not know where to go to obtain information. They believe that the federal-to-community dissemination process is flawed and that current federal dissemination practices favor large communities.
- State and federal practitioners reported using e-mail and Internet sources to obtain information. They also favored meetings and conferences as the most effective way of acquiring new research information.
- Practitioners and researchers described each other as having distinct cultures that preclude effective communication because of language barriers.
- Institutional barriers prevent the dissemination of knowledge from researchers to practitioners because academia does not reward

research faculty for such efforts. In fact, such service can be an impediment to obtaining academic tenure.

- There is a lack of formal means for bringing researchers and practitioners together. Not many practitioners have the opportunity and resources to attend conferences, even those who have research dissemination as a goal. There is a need to have better formal networks among the two groups that can act as translators of knowledge.
- There is a lack of meaningful interaction between researchers and practitioners to define research agendas and to interact during the research design and implementation phases. The concepts of community-based action research, or participatory action research (Huizer, 1997), and researcher-practitioner coalitions (Buika and Comfort, 2004) are two methods that potentially can make interactions more meaningful.

The general point is worth repeating: Little systematic research on information dissemination and implementation has been conducted in the hazards and disaster field. But the examples cited above indicate clearly that research utilization does take place under certain conditions, particularly when researchers and potential users interact in meaningful ways. The research also shows that a proactive response is needed by both researchers and potential users to further science-based decision making. Fortunately, additional insights can be acquired from research conducted outside the hazards and disaster field.

## GENERAL INSIGHTS ON KNOWLEDGE DISSEMINATION AND APPLICATION

Considerable research has been conducted in the social and management sciences generally on what has been variously described as research dissemination, knowledge utilization, research utilization, knowledge transfer, adoption of innovation, and technology transfer. A variety of such phrases have been used to characterize similar processes and the literature is filled with differing definitions and uses of them. At times the above phrases are carefully defined to characterize a narrow process, and at other times they are used interchangeably. The literature on dissemination and knowledge utilization spans a number of disciplines, including the fields of rehabilitation, education, sociology, psychology, and marketing. The committee concludes that the substance of this research is directly applicable to the transfer of social science knowledge on hazards and disasters. Some researchers have distinguished between a "push" process wherein providers of knowledge actively seek utilization versus a "pull" process wherein users actively seek knowledge from the research community. Other dissemination research has identified four functional types of dissemination:

1. *Spread,* which is defined as "the one-way diffusion or distribution of information;"
2. *Choice,* a process that "actively helps users seek and acquire alternative sources of information and learn about their options;"
3. *Exchange,* which "involves interactions between people and the multidirectional flow of information;" and
4. *Implementation,* which "includes technical assistance, training, or interpersonal activities designed to increase the use of knowledge or R&D or to change attitudes or behavior of organizations or individuals" (Southwest Educational Development Laboratory, 1996:5).

Much of the academic literature on knowledge dissemination concludes that lack of utilization results from fundamental differences in the world views of researchers and practitioners. Beyer and Trice (1982) concluded that "the most persistent observation . . . is that researchers and users belong to separate communities with very different values and ideologies and that these differences impede utilization." Similarly, Shrivastava and Mitroff (1984) suggested that academics and practitioners have fundamentally different frames of reference with respect to such things as the types of information believed to constitute valid bases for action, the ways in which information is ordered and arranged to make sense, the past experiences used to evaluate the validity of knowledge claims, and the metaphors used to symbolically construct the world in meaningful ways.

Practitioners attending the committee's workshops expressed similar perspectives on the dissemination problem, attributing the lack of use of research knowledge to factors such as the following:

- Information is not easy to digest and understand.
- Information is not relevant, or it takes too much time to sort the relevant from the irrelevant.
- The knowledge is targeted to the wrong end-user or consumer.
- Information is not concise, bulleted, and to the point.

In the hazards and disaster field, another major issue involves the saliency of emergency preparedness and disaster response to state and local political officials. Some years ago a national study found that disaster management is very low on the agenda of city officials (Rossi et al., 1982). Although natural, technological, and willful hazards are more prominent in the current political climate, the topic must still compete for attention in the face of a host of everyday concerns and scare resources.

While no all-encompassing theory or explanation of knowledge utilization has been described and tested, the broader literature includes many insights that can help strengthen the dissemination and application of hazards and disaster findings. Within the varied perspectives about knowledge dissemination and utilization, some combinations of the following four elements are considered in the literature:

- the dissemination *source*—that is, the agency, organization, or individual responsible for creating the new knowledge or product, and/or for conducting dissemination activities;
- the *content* or message that is disseminated—that is, the new knowledge or product itself, as well as any supporting information or materials;
- the dissemination *medium*—that is, the ways in which the knowledge or product is described, "packaged," and transmitted; and
- the *user,* or intended user, of the information or product to be disseminated (Southwest Educational Development Laboratory, 1996:12).

On a more practical note, the broader research literature enables some generalizations about the circumstances underlying successful knowledge utilization. Six general principles or strategies have emerged from prior research that accounts for successful knowledge utilization (Backer et al., 1995).

1. *Interpersonal contact.* For knowledge to be used in new settings there has to be direct, personal contact between those who will be using the knowledge and its developers or others with relevant scientific information. This principle was strongly confirmed by practitioners who participated in the committee's workshops. They reported frequently turning to local colleges and universities for technical support or to trusted consultants.
2. *Planning and conceptual foresight.* A well-developed strategic plan for how knowledge will be adopted in a new setting—including attention to implementation problems and how they will be addressed—is essential to meeting the challenges of adoption and sustained change. This approach has been institutionalized at the National Science Foundation (NSF) where proposals on disaster research and other topics must address how the work will provide societal benefits to potential user communities.
3. *Outside consultation on the change process.* Consultation can provide conceptual and practical assistance in designing the adoption or change effort efficiently and can offer useful objectivity about

the likelihood of success, costs, possible side effects, and so forth. The city and county participants in the committee's workshops reinforced this notion of outside consultant involvement to affect change; they commonly used expert consultants to bring the knowledge generated by researchers to bear on their issues.

4. *User-oriented transformation of information.* What is known about scientific information needs to be translated into language that potential users can readily understand, abbreviated so that attention spans are not exceeded, and made to concentrate on the key issues: Does it work? How can it be replicated? A key conclusion that emerged from the committee's workshops was the need to translate academic findings into understandable language and into products with practical application.

5. *Individual and organizational championship.* Chances for successful adoption of knowledge are much greater if influential potential adopters (opinion leaders) and organizational or community leaders express enthusiasm for its adoption. Again, participants at the committee's workshops affirmed the need for political support to implement new programs.

6. *Potential user involvement.* Stakeholders who will have to live with the results of the adoption process need to be involved in planning for adoption, both to obtain suggestions for how to undertake the adoption effectively and to facilitate ownership of the new program or activity, thus decreasing resistance to change.

## VIGNETTES FROM THE KNOWLEDGE DELIVERY SYSTEM

While much remains to be learned about research utilization in the social science hazards and disaster field, efforts to stimulate utilization have been carried out for many years by a variety of entities—especially in academia, government, and the nonprofit sector. The knowledge that they disseminate to spur science-based decision making and implementation cuts across all of the core topics of hazards and disaster research depicted in Figure 1.1 and Figure 1.2. A subset of such efforts, both past and present, is described in this section. These activities or programs were selected by the committee mainly because they involve rather significant attempts to further the dissemination and application of knowledge developed by social scientists. Sometimes this is the principal type of knowledge disseminated by an entity, while in other instances, knowledge from other relevant disciplines is also promoted to further disaster reduction.

Information on these activities came from various sources. The committee had first-hand knowledge of many of them. This experience was supplemented by information provided directly by some of the entities and by their public documents, including those on Web sites. Without systematic assessment data on the efforts and initiatives discussed here, it is not possible to be very precise about how successful many of them have been, which again reflects the great need for more research on knowledge utilization. In trying to understand how successful many of these programs and activities have been, the committee has had to rely principally on anecdotal or experiential insights and reputation, rather than on research-based evidence. In the future, it would be worthwhile not only to know the relative degree of success that such programs and activities have achieved over time, but also to determine such things as what core hazards and disaster research topics are the most challenging in terms of meeting research utilization goals, and the impact of new technologies such as geographic information systems (GIS) and the Internet on effectiveness. It would also be useful to make comparisons across hazard types to determine the degree to which research dissemination and application efforts need to be tailored to particular natural, technological, and willful threats.

The matrix in Table 8.1 is used to organize the committee's discussion of a set of 18 efforts and programs chosen because of their commitment to furthering the dissemination and application of social science knowledge on hazards and disasters. This list is not intended to be exhaustive; rather, it is a capsule of the larger knowledge delivery system in the hazards and disaster field. However, enough of these types of activities and programs have been selected to demonstrate the variability in the strategies employed to further research application.

The matrix combines the four approaches to knowledge dissemination and the six factors for success previously identified in the review of the broader research literature on knowledge utilization. Each of the 18 activities is placed in the cell that best characterizes it. Some of the cells remain blank, but certainly, examples of activities might exist for every cell. And some of the 18 selected activities included here can apply to more than one cell. In such cases, placement was decided on the basis of the major characteristics of the programs or activities. For example, while Thomas Drabek's efforts are placed in only one cell, they actually spill over into several cells in the matrix.

This discussion is intended to illustrate the range of activities and programs that comprise the hazards and disasters research utilization infrastructure and to demonstrate the principles of information dissemination and application derived from the broader research literature. It also sets the stage for the discussion on needed research at the end of this chapter.

**TABLE 8.1** Examples of Knowledge Diffusion Efforts

| Characteristics for Success | Approaches to Knowledge Dissemination | | | |
| --- | --- | --- | --- | --- |
| | Spread | Choice | Exchange | Implementation |
| Interpersonal contact | Drabek's dissemination efforts | State Attorney General's Office | Boulder, CO Floodplain Management | |
| Planning and conceptual foresight | Red Cross information brochures | Texas A&M Hurricane Planning | FEMA's Higher Education Program | NEHRP |
| Outside consultation on the change process | FEMA's planning and mitigation guides | University of South Carolina Hazards Research Lab | | Indian Point Expert Task Force |
| User-oriented transformation of information | University of Colorado Natural Hazards Center | | Latin America Vulnerability Project | NWS's Warning Programs |
| Individual and organizational championship | | | Tulsa, Oklahoma Floodplain Management | |
| Potential user involvement | FEMA's EMI | FEMA's CSEPP Training | Association of Bay Area Governments | Association of State Floodplain Managers |

NOTE: CSEPP = Chemical Stockpile Emergency Preparedness Program; EMI = Emergency Management Institute; FEMA = Federal Emergency Management Agency; NWS = National Weather Service.

## INTERPERSONAL CONTACT

### Spread: Thomas Drabek's Dissemination Efforts

Thomas E. Drabek (John Evans Professor Emeritus, University of Denver) has conducted disaster research studies during the past four decades. Committed to the premise that research findings and conclusions should do more than gather dust in academic libraries, Drabek has employed a variety of dissemination strategies that have brought his work to thousands of emergency management professionals. Many of his previous projects were guided by active advisory committees who performed six key functions: (1) facilitation of field work, (2) assistance with field site selection, (3) review of data collection instruments, (4) review of working papers, (5) review of drafts of project books, and (6) assistance with dissemination of project results by arranging for conference and workshop presentations, newsletter and journal publication suggestions, and informal contacts with key government agency officials. The expertise of committee members reflected both academic and practitioner experiences, and all members had high name recognition within their respective reference groups. Often, they began to use and distribute preliminary findings before project completion.

Like the relationships with his advisory committee members, Professor Drabek has developed bonds of trust with numerous practitioners. Reflecting his stated respect for those who do "the real work of emergency management," he has maintained membership in the core organizations of both his discipline and the emergency management profession. He has spanned successfully boundaries that few others were willing to traverse. Upon the completion of each previous research project, he wrote a summary book that was distributed by the Natural Hazards Research and Applications Information Center (NHRAIC). With the assistance of the center's staff, he was able to state his findings and conclusions in a crisp and clear style that communicated well to both practitioners and academics. All of his books were distributed with purchase prices that reflected only NHRAIC production and printing costs. This facilitated their widespread circulation and frequent use in educational workshops.

While Drabek frequently presented his research conclusions at both professional sociological and social science association meetings, he also made presentations at national, regional and state emergency management conferences. Additionally, he accepted lecture invitations extended by emergency managers in Italy, Thailand, Switzerland, Australia, New Zealand, Canada, and Mexico. Many of these reflected his comprehensive inventory of the sociological literature, i.e., *Human System Responses to Disaster* (Drabek, 1986). While the specific content of his presentations varied so as to reflect his work at particular points in time, Drabek consistently carried a singular message: Emergency management can be practiced best if it reflects actions rooted in scientific knowledge rather than myth. This theme was brought into hundreds of classrooms where students have reviewed the conclusions summarized in *Emergency Management: Principles and Practice for Local Government* (1991). Drabek coedited this volume with Hoetmer, which was published by the International City Management Association in its distinguished Green Book series.

During the 1990s, Drabek conducted three major studies that documented a catastrophic vulnerability in the tourism industry. His projects underscored the wide gaps between the expectations of business managers and their customers regarding disaster preparedness, behavioral responses, and approaches to mitigation. Employees caught between the directives of their bosses and fears and desires of family members during numerous large-scale evacuations revealed portraits that required action. Federal Emergency Management Agency (FEMA) staff agreed and asked Drabek to tackle the problem. He recruited Chuck Gee, a former classmate from the University of Denver who had long held the post of dean at the School of Travel Industry Management (STIM), University of Hawaii at Manoa. Together, with the assistance of Ruth Drabek and two STIM staff, George Ikeda and Russell Uyeno, they prepared a guide for university faculty in departments of tourism, hospitality, and travel management. This resource, like the other FEMA-sponsored Instructor Guides, was made available free of charge through the Internet. It facilitated the rapid dissemination of research findings. As in the past, Drabek also published study results in academic journals, practitioner publications, and other outlets that are received routinely by both emergency managers and tourism executives and faculty.

Most recently Drabek has produced a revised edition of the *Social Dimensions of Disaster* (Drabek, 2004). This volume summarizes key findings and conclusions from sociological studies completed during the past 10 years. Like his many other efforts, this has brought the work of disaster sociologists to large audiences who might otherwise have never learned of them.

## Choice: Warning Research Utilization by the New York State Attorney General

Interactions between researchers and policy makers can facilitate disaster management. While such interactions are common, they are rarely documented. Social science researchers at Oak Ridge National Laboratory conducted studies of emergency warnings for hazardous materials incidents (Rogers and Sorensen, 1988). This work simulated the speed at which warnings could be disseminated to a population at risk. Researchers worked with the state attorney general's office in connection with legislation on chemical hazard mitigation requirements for fixed facilities. Contact between the two groups was made at an Environmental Protection Agency (EPA) Hazardous Material Spills Conference. The research presented at that conference showed that it would be extremely difficult to warn residents on the borders of facilities storing hazardous chemicals in a timely manner after an accidental release (Sorensen et al., 1988). The state attorney general's office used the research to justify the need for legislation that required facilities to systematically identify potential accidents and develop the means to prevent the accidents or mitigate their effects (Skinner et al., 1991).

### Exchange: Floodplain Management in Boulder

Some communities can benefit from access to national experts who reside locally. The City of Boulder, Colorado has a major potential for serious flash flooding. It is also the location of the Natural Hazards Research and Applications Information Center, founded by Gilbert White, one of the nation's leading floodplain experts in the post-World War II era. Boulder has a flood problem similar to the Big Thompson Canyon below Estes Park, Colorado. White and one of his former students, Eve Gruntfest, have worked over the years with city officials to develop a comprehensive floodplain management plan for Boulder Creek and its tributaries. Many hours of professional community service were provided in meetings with various city and county officials. As part of the effort, a comprehensive survey was conducted for two populations living in the Boulder Creek 100-year floodplain. Population A included year-round, non-student residents, and population B included residents of the University of Colorado Student Family Housing. Residents were surveyed about their knowledge of the 100-year floodplain, flood risk awareness, preferred warning methods, perceived response, impacts of false alarms, and flood and weather information (Gruntfest et al., 2002). This survey provided an important database for city officials.

## PLANNING AND CONCEPTUAL FORESIGHT

### Spread: Red Cross Disaster Education

The Red Cross has been a major user of social science disaster research. The agency has reflected research in a series of public information brochures on disaster preparedness that are made available to the public through the Web and local Red Cross chapters. Brochures are available in English and 14 other languages. These brochures make recommendations on such topics as how to develop a family disaster plan, which is based on research findings from studies of evacuation behavior during disasters conducted by Perry and colleagues (1980). This application is significant because families with written emergency plans are more likely to engage in protective behaviors when confronted by a disaster. Other brochures developed on the basis of social science research include such topics as planning for special population groups such as the disabled, elderly, or children, and assembling an emergency supply kit.

## Choice: Hurricane Planning and Research in Texas

Several states actively engage social scientists while developing emergency planning strategies. The State of Texas has been developing hurricane evacuation plans based on social science research carried out by the Hazard Reduction and Recovery Center (HRRC), which was established at Texas A&M University in 1988. HRRC researchers focus on hazards analysis, emergency preparedness and response, disaster recovery, and hazards mitigation. Researchers study the full range of natural and technological hazards and disasters, including hurricanes, floods, earthquakes, and chemical plant and transportation accidents. Two core missions of the center relate directly to knowledge transfer:

1. To disseminate findings to the research community and to practitioners so they can use this knowledge to mitigate, prepare for, respond to, and recover from disasters.
2. To provide assistance and consultation to those state, national, and international agencies charged with responsibility for hazard analysis, emergency preparedness and response, disaster recovery, and hazard mitigation.

HRRC's hurricane-related research has focused, first, on behavioral responses to hurricanes that have impacted Texas and, second, on coastal residents' perceptions of disaster risk. One goal of this research is to examine the correlation between what people think they would do during a hurricane evacuation and what they actually do. This research has been used in predicting evacuation times and developing evacuation plans for coastal counties (see http://hrrc.tamu.edu/).

## Exchange: FEMA's Higher Education Program

FEMA has developed long-term programs to help professionalize emergency management. One of the goals of its Higher Education Program is to encourage and support the dissemination of information on hazards, disasters, and emergency management in colleges and universities across the United States. This goal is based on the anticipation that in the future more and more emergency managers in government as well as in business and industry will need to come to the job with a college degree in emergency management. Through the Higher Education Program, FEMA works closely with the research community to develop standardized curricula on hazards and disasters. At least four of the courses developed by the higher education program are social science related:

1. Social Dimensions of Disaster
2. Sociology of Disaster
3. Social Vulnerability Approach to Disasters
4. Public Administration and Emergency Management

Three of these courses were developed by Thomas Drabek, who as noted above, has been actively engaged with professionals and practitioners in the emergency management field, including those at the local, state, and national levels. Much of Drabek's research, including his work on emergency planning related to tourism, has been NEHRP supported. Drabek's strategic information sharing activities have been supported by both NSF, such as his project advisory committees, and FEMA, including his work involving the development of courses for its Higher Education Program. In developing course material for this program, Drabek has relied not only upon his own disaster research results but also on those produced by a host of other NEHRP-funded researchers in the social sciences. This encyclopedic combining of research and training activities by social scientists has resulted in a more informed emergency management community, especially in the case of those emergency managers who graduate from university and college programs that offer courses based on the material prepared in collaboration with FEMA's Higher Education Program.

An activity called the Practitioner's Corner was launched as part of the Higher Education Program to create another way for emergency management practitioners to communicate their thoughts and ideas concerning college-level hazard, disaster, and emergency management courses and programs to the educators responsible for them (see http://training.fema.gov/EMIWeb/edu/practitioner.asp). Volunteers are solicited for papers on such subjects as competencies, knowledge, skills, and abilities that emergency management educators should develop or bring out in their students and philosophical perspectives on the different ways to look at or approach the emergency management position, for example,

- the most appropriate organizational placement of emergency management responsibilities at the local government level;
- lessons learned in disasters;
- lessons learned in bureaucratic politics;
- success stories, obstacles overcome, and challenges met; and
- emergency management public policy issues.

### Implementation: National Earthquake Hazards Reduction Program

NEHRP pursues the objective of transferring knowledge, including that derived through the social sciences, on a sustained basis to reduce risks to life and property from earthquakes. The four agencies in the program—National Institute of Standards and Technology (NIST), which is currently the lead agency, FEMA, the U.S. Geological Suryve (USGS), and NSF—are expected to work collaboratively with each other as well as with other stakeholders to achieve this objective. Underpinning earthquake risk-reduction efforts through NEHRP is the provision of technical assistance and research that develops new knowledge about (1) earthquake hazards; (2) the response of the natural, built, and social environments to those hazards; and (3) techniques to mitigate the hazards. A major challenge facing NEHRP is furthering the use by local, state, and private stakeholders of the science-based knowledge it generates. As discussed in other chapters of this report, NEHRP has fostered social science research on disasters and has championed the application of knowledge generated by this research.

## OUTSIDE CONSULTATION ON THE CHANGE PROCESS

### Spread: FEMA Planning Guides

Federal agencies have incorporated social science knowledge, albeit not always systematically, into guidance documents for local emergency management agencies and, in doing so, have engaged social scientists to help prepare these guides. FEMA produces some planning guides that are knowledge based and rooted in social science research, such as its planning guidance for the Chemical Stockpile Emergency Preparedness Program (CSEPP). For example, the guide's recommendation for planning community shelter capacities were based on social science research on shelter use in emergencies (Mileti et al., 1992). FEMA's mitigation planning guide series also contains good examples of research-based guidance.

### Choice: Hazards Research Lab

The Hazards Research Lab (HRL) at the University of South Carolina (USC) was established in 1995. The HRL specializes in the application of geographic information science to environmental hazards analysis and management. In addition to its basic research and training mission, the HRL facilitates federal, state, and local efforts to improve emergency preparedness, planning, and response through its outreach activities. For example, the HRL maintains the most comprehensive database in the nation on hazard events and losses in the United States (http://sheldus.org). In partnership with the South Carolina Emergency Management Division, the HRL provided the methodology and baseline information for conducting hazard vulnerability assessments under the Disaster Mitigation Act (DMA

2000) through its South Carolina Hazards Mapping Interface (a Web-based interactive product). The HRL also conducts post-disaster studies (e.g., on Hurricane Floyd and the Graniteville, South Carolina, train derailment and chlorine release) and these findings are given to the state and local emergency responders to help improve disaster preparedness. The partnership between the academic and practitioner communities is realized in all the activities of the HRL. As noted by John Knight, director of risk assessment for the South Carolina Emergency Management Division (University of South Carolina Research, 2005):

> It's been helpful to have a reliable source like USC for so much of the information we use," Knight said. "The natural hazards mapping and analysis has been a very useful tool for us at the state level, and we continue to work very closely with Susan Cutter (HRL director) and her colleagues.

### Implementation: Indian Point Expert Task Force

Social scientists have worked with the private sector to improve their implementation of regulatory requirements. In the fall of 2002, the governor of New York hired the consulting firm James Lee Witt Associates to review the status of emergency planning at the Indian Point Nuclear Power Plant, located in Westchester County. The draft Witt report was very critical of the status of planning at the plant and surrounding communities. As a result, Entergy, the company that operates both of the reactors at the site, formed an advisory group called the Indian Point Expert Task Force to advise on which of the Witt report's criticisms were valid and what to do to improve emergency systems and plans. The nine-member team consisted of nuclear engineers, health specialists, planners, and social scientists. The goal of Entergy was to develop the best emergency plans and response in the nuclear industry. The task force reviewed the Witt report and dismissed some of its findings as scientifically invalid, endorsing others. The task force also had access to plant and community personnel, and it observed and evaluated exercises and drills. One of the drills evaluated was the functioning of the joint news center located at Westchester County airport. The social science evaluators noted that the effectiveness of the operation was hampered by the physical layout of the building, which interfered with social interactions, and also by the reliance on out-of-date communications protocols and equipment. Based on the unanimous recommendations of task force observers, Entergy committed to opening a new joint news facility that would be co-located with the county 911 center and to develop a new concept of operations that would improve interactions and communications. Other functional areas that have been influenced by social scientists at Indian Point include revisions of strategies for providing public information, implementing emergency communications, and issuing warning messages.

## USER-ORIENTED TRANSFORMATION OF INFORMATION

### Spread: Natural Hazards Center

The Natural Hazards Research and Applications Information Center (NHRAIC) at the University of Colorado at Boulder was founded in 1976 by Gilbert White and J. Eugene Haas as a direct outgrowth of the First Assessment of Research on Natural Hazards. NHRAIC is funded by grants from NSF and annual contributions from other agencies, including FEMA in the Department of Homeland Security (DHS), USGS, the National Oceanic and Atmospheric Administration (NOAA), the National Aeronautics and Space Administration (NASA), U.S. Army Corps of Engineers (USACE), U.S. Forest Service, National Weather Service, Department of Transportation, Environmental Protection Agency, and the Centers for Disease Control and Prevention. NHRAIC is affiliated with the University of Colorado's Institute of Behavioral Science and the Institute's Environment and Behavior Program. NHRAIC's mission is to disseminate information on the societal dimensions of hazards and disasters and to foster linkages between the research community and public and private sector users of research. The center also sponsors quick-response research following disasters, conducts research projects, and engages in activities aimed at enhancing the hazards research workforce.

As part of its information dissemination program, NHRAIC maintains a library totaling approximately 28,000 items, which is available to students, visiting scholars, and practitioners. *Hazlit*, the library database, can be searched on the Web, and the library staff is also available to conduct customized searches. NHRAIC's newsletter, the *Natural Hazards Observer*, is published six times a year and distributed to approximately 15,000 readers in the United States and abroad. Typically 28 pages in length, the *Observer* features invited comments from disaster experts and practitioners, as well as timely information on meetings, conferences, web resources, pending legislation, research and government reports, and grant awards. A shorter publication, the *Natural Hazards Informer*, contains research-based guidance geared specifically to practitioners. NHRAIC maintains two listservs: *Disaster Research*, which provides a forum for research-related queries and discussions, and *Disaster Grads,* which is tailored to the needs of graduate students and young professionals.

NHRAIC also organizes and conducts an invitational workshop on hazards, disasters, and, more recently, homeland security, which has been held annually in July since 1976. The goal of the workshop is to bring together researchers, public and private sector practitioners, agency officials, and students for discussions of research, educational, and policy issues. The workshop program is designed to be less formal than a professional conference and is intentionally organized to span research-practitioner boundaries and to encourage networking and information sharing. Over the years, workshop attendance has grown to well over 300 participants.

NHRAIC approves small grant proposals for quick-response research on an annual basis. This pre-approval process enables researchers to go into the field rapidly if a disaster event occurs that falls within the parameters of their proposals. Grantees are required to prepare reports based on their quick-response studies, which are then disseminated by the center. NHRAIC also publishes monographs and special reports based on social science research on hazards and disasters, including many NEHRP-sponsored studies. Its most recent special report *Beyond September 11* (NHRAIC, 2003) consists of a compilation of quick-response studies that were carried out following the terrorist attacks of September 11, 2001. Faculty affiliated with the center also carry out their own research projects, funded by agencies such as NSF and FEMA, which provide training and educational opportunities for graduate students and postdoctoral scholars.

The NHRAIC Website, http://www.colorado.edu/hazards, is among the most visited sites in the hazards and disasters field. Users of the site can search the *Hazlit* database, find links to other information sources, and gain access to online versions of the *Observer*, quick-response reports, and other center publications, as well as programs and session summaries from past workshops, a directory of academic centers and government programs focusing on hazards and disasters, and other relevant information. Popkins and Rubin (2000) assessed user views of the Natural Hazards Center and concluded that it has been a vital information resource to both academic researchers and practitioners in the emergency management field, making information easily accessible to them.

### Exchange: Latin America Vulnerability Project

Some social science researchers have actively pursued participatory action research to help reduce disaster vulnerability. One such project produced *Working with Women at Risk—Practical Guidelines for Assessing Local Disaster Risk* (Enarson et al., 2003). This project is an example of social science researchers working together internationally to develop new ways of studying community vulnerability and improving local capabilities for response to hazards and disasters in rural areas. The approach builds on local women's knowledge and understanding of risk and vulnerability developed from their social roles, economic activities, and family and community networks. The research led to a step-by step guide developed by the researchers based on an integrated gender-based model. The guide suggests methods to (1) identify women's groups who might take on a vulnerability project, (2) train the women to be community researchers, (3) develop strategies for collecting information about hazards, and (4) utilize resulting knowledge to reduce risk through sharing of the work with community members, officials, and the media. The guide takes into consideration both the social and the behavioral context of the lives of those at risk. It was developed over a two-year period during which the research team worked with villages in rural El Salvador, St. Lucia, Dominica, and the Dominican Republic.

### Implementation: National Weather Service (NWS) Warning Programs

As other examples have shown, federal agencies sometimes actively engage social science researchers when developing disaster reduction programs. An early example of the direct utilization of social science input by the National Weather Service (NWS) involved its Southern Region Headquarters initiating an emphasis on calls to action (CTAs) as part of its warning process in the early 1970s (Troutman et al., 2001). Benjamin McLuckie of the Disaster Research Center was asked to study how to improve the effectiveness of written warnings. McLuckie (1974) developed a workbook and self-study course titled "Warning—A Call to Action," which became an important tool for forecasters to improve the effectiveness of their warnings. The goal of this effort was to convey specific information as concisely as possible. Individual weather offices were therefore encouraged to develop a set of CTA statements that were specific to their local regimes.

The NWS has continued to utilize social science research on warnings in designing and implementing warning systems. In the early 1990s the NWS adopted a systems approach to issuing warnings based on the work of social scientists which involved addressing four aspects of the problem: (1) detection and forecasting; (2) developing the warning message; (3) disseminating the warning; (4) and getting people to respond. This approach is disseminated to communities through the NWS Storm Ready Program. To become "storm ready," a community or county must

- establish a 24-hour warning point and emergency operations center,
- have more than one way to receive severe weather warnings and forecasts and to alert the public,
- create a system that monitors weather conditions locally,
- promote the importance of public readiness through community seminars, and
- develop a formal hazardous weather plan, which includes training severe-weather spotters and holding emergency exercises.

A frequently cited example of the success of this program comes from the experience of Van Wert County, Ohio, which experienced an outbreak of tornadoes in 2002. As part of the program, the county placed a series of warning alert systems in public locations, including retail stores and movie theaters. During the outbreak, the Van Wert County emergency operations center received a NWS Tornado Warning via a NOAA Weather Radio receiver. The Van Wert County emergency manager immediately activated the City of Van Wert's siren warning system and broadcasted the NWS tornado warning and action statement live. Quick action by the manager of Van Wert Cinemas and his staff got more than 50 adults and children out of theaters in the multiplex and into safer conditions in a hallway and restrooms. Minutes later, a tornado tore off the building's roof and tossed cars into the screen and front seats where minutes earlier children and their parents had been watching a popular holiday movie.

## INDIVIDUAL AND ORGANIZATIONAL CHAMPIONSHIP

### Exchange: Tulsa, Oklahoma, Floodplain Management

Persistent efforts by individuals can result in the adoption of programs that involve the application of hazards and disaster research knowledge. Through the leadership of resident and mitigation champion Ann Patton, Tulsa, Oklahoma, developed one of the premier floodplain management programs in the country. Tulsa's frequent flooding led to recurring losses and hardship. The ultimate response of Tulsa's local authorities was to create a flood mitigation regulatory climate that encouraged private participation. Ann Patton worked to convince people to understand that "everyone contributes to flooding in Tulsa, so everyone should pay something to prevent it." To do so, she brought in two eminent social scientists to convince the city's leadership to use nonstructural measures to reduce flood losses (Meo et al., 2004). The two were planner Ian McHarg and floodplain expert Gilbert White. Based on their encouragement and leadership, Tulsa developed a floodplain management strategy that would win recognition as one of the leading hazard reduction efforts in the country. One of the innovative features of the program is that Tulsa charged a $4 per month drainage fee, collected with water bills, to support land management and maintenance of the stormwater drainage system for land acquired by the city. Once the city owns the land, it is used for a wide range of flood-compatible activities and thus taken off the market for potential development.

Public participation was a major component in Tulsa's planning efforts. Citizen advocates, including Ann Patton, played a critical role in pressing for tough flood mitigation actions. For example, flood channel and river bank cleanups became commonplace. Properties highly vulnerable to flooding were bought out or donated in an effort to eliminate structures in flood hazard zones. Experts in policy, planning, and mitigation stepped forward and volunteered in the flood mitigation effort. Open space adjacent to rivers and streams was preserved for public parks, recreation sites, and gardens. Detention basins, which are now local amenities and instruments of flood management, were built and old ones cleaned up. Frequent flooding of the Arkansas River prompted Tulsa officials to develop a system of river parks to minimize the effects of recurring floods. The area now boasts 50 miles of scenic trails along the banks of the Arkansas River. Today floods pose fewer dangers to citizens of the City of Tulsa because of these mitigation activities (NRC, 2004a).

## Spread: FEMA's Emergency Management Institute

Through its courses and programs, FEMA's Emergency Management Institute (EMI) serves as the national focal point for the development and delivery of emergency management training to enhance the capabilities of federal, state, local, and tribal government officials, volunteer organizations, and private sector organizations. EMI's curricula are structured to meet the needs of this diverse audience with an emphasis on how the various stakeholders can work together to save lives and protect property. Instruction focuses on four phases of emergency management: mitigation, preparedness, response, and recovery. EMI develops courses and administers resident and nonresident training programs on coping with natural hazards, technological hazards, and terrorist threats. EMI has regularly engaged social science hazards and disaster researchers to help develop curricula and serve on its advisory board. The library at EMI, with more than 100,000 publications, represents one of the major repositories of disaster research documents. The Disaster Research Center at the University of Delaware and the Natural Hazards Center at the University of Colorado are the other two major repositories of social science hazards and disaster research publications in the country. Approximately 5,500 participants attend resident EMI courses each year, while 100,000 individuals participate in nonresident programs sponsored by EMI and conducted by state emergency management agencies under cooperative agreements with FEMA. Another 150,000 individuals participate in EMI-supported exercises, and approximately 1,000 individuals participate in the Chemical Stockpile Emergency Preparedness Program (CSEPP). Additionally, hundreds of thousands of individuals use EMI distance learning programs, such as the Independent Study Program.

## USER INVOLVEMENT

### Choice: FEMA's CSEPP Training

FEMA's CSEEP develops training for planners and first responders at sites that store dangerous chemical weapons. When a state in the program identifies a training need, FEMA assembles a training development team consisting of a user from local government, a state emergency management representative, a FEMA representative, subject matter experts, a social science disaster research expert, and a production expert. This team works together over the course of training development. The role of the social scientist is to ensure that the training material applies relevant social science knowledge on hazards and disasters and that this is done accurately. The final product is expected to meet user's needs. CSEPP training products, because of the social science involvement, reflect current knowledge derived from social science research. For example, much of the information presented in a CSEPP training course on "Public Information and Education" was develop from social science research on disaster education and warning. Likewise, a training course on "Command and Control" reflects the extensive research on organizational behavior in disasters carried out by the Disaster Research Center.

### Exchange: Association of Bay Area Governments

The Association of Bay Area Governments (ABAG) serves the nine counties in the San Francisco Bay area through a variety of preparedness and mitigation projects focusing on earthquakes and a host of other environmental risks. Its Earthquake Program is a major activity at ABAG. For many years the program has been managed by planner Jeanne Perkins. With funding from and collaboration with such NEHRP agencies as NSF, USGS, and FEMA, Perkins has over the years carried out many research and related projects on such subjects as the legal aspects of earthquake management and housing vulnerability, in the process building a science-based approach for helping stakeholders reduce disaster risks in the region. An aggressive strategy has been used to communicate the results of these efforts to potential local users in the public and private sectors based on an understanding of what motivates them to take the needed action. The strategy has included the dissemination of important hazard-related information to potential users through such means as workshops and conferences, the organization's Web site, and various publications.

### Implementation: Association of State Floodplain Managers

The Association of State Floodplain Managers (ASFPM), a member organization representing flood hazards specialists in government, academe, and the private sector, is heavily engaged in the transfer of knowledge to potential users and has the reputation of doing it successfully. The floodplain management and policy issues that come under its purview include flood mitigation, preparedness, warning, and recovery, with particular attention given to the National Flood Insurance Program administered by FEMA (www.floods.org). One of the organization's most significant contributions has been the development of the No Adverse Impact campaign, designed to promote an approach to community development in which the actions of one property owner or of the community do not adversely affect the flood risk of another property owner or community. Instead of structural approaches to floodplain development, the No Adverse Impact campaign attempts to promote what ASFPM considers to be more sustainable strategies involving such nonstructural measures as land-use planning long advocated by such social scientists as Gilbert White (Larson and Plascencia, 2001).

## NONADOPTION OF SOCIAL SCIENCE KNOWLEDGE

The activities and programs discussed in the previous section exemplify a commitment to using social science knowledge to improve decisions and actions related to disaster management. However, as discussed earlier, many barriers must be overcome before relevant knowledge from the social sciences as well as other disciplines becomes an important factor in what individuals and organizations do about the risks they face or are responsible for managing in cooperation with others, including the public. It is not surprising then when some responsible organizations are late or nonadopters of extant social science knowledge relevant to disaster management. Two such examples are briefly discussed below as a contrast to the more successful examples noted earlier. It is unclear why progress has not been made in these two cases, which again point to the need for more studies of research utilization. If the principles of research utilization are to be fully elaborated, research on negative cases is just as important as research on the more successful ones.

### Homeland Security Threat Advisory System

The Homeland Security Advisory System (HSAS) was designed to provide a comprehensive means to disseminate information regarding the risk of terrorist acts to federal, state, and local authorities and to the American people. This system provides warnings in the form of a set of graduated "threat conditions" that increase as the risk of the threat advances. At each threat condition, the intention is for federal departments and agencies to implement a set of protective measures to reduce the nation's vulnerability during the heightened alert. Although the HSAS is binding on the executive branch, it is voluntary to other levels of government and the private sector. There are five threat conditions, each identified by a description and corresponding color (see Figure 8.1).

The higher the threat condition, the greater is the assumed risk of a terrorist attack. Risk includes both the probability of an attack occurring and its potential gravity. Threat conditions are assigned by the attorney general in consultation with the Secretary of Homeland Security. Threat conditions may be assigned for the entire nation, or they may be set for a particular geographic area or industrial sector. Assigned threat conditions are reviewed at regular intervals to determine whether adjustments are warranted.

The usefulness of the scale has been criticized by many media personalities and journalists. Even government investigators have indirectly criticized the design and implementation of the scale. The Government Accountability Office (GAO) (2004) notes that the development of the scale largely fails to reflect the expertise derived from risk communications and disaster warning research. While most disaster researchers would agree that the scale is not a warning system, much of what has been learned by disaster researchers on effective risk communication practices is largely ignored in the development of the system (NRC, 2002a). As Aquirre (2004:13) has observed:

> Summarizing some of the most important problems with HSAS, the hazards it addresses are unspecific as to their origin, the nature of the threats, their time and place configurations, and what to do about them; the likely victims are unknown; the local government and emergency management response networks as well as the local and state political systems do not participate in preparing and mitigating their effects, although they are liable for the costs of reacting to the warnings; and it lacks an accurate understanding of the social psychology of people's response to warnings, assuming an undifferentiated public that automatically behaves as it is told by the authorities. Moreover, it confuses warnings with mitigation and public relations and is too closely linked to partisan political processes.

### National Incident Management System

In Homeland Security Presidential Directive-5 (HSPD-5), the president called on the Secretary of Homeland Security to develop a single incident management system to provide a consistent nationwide approach for fed-

FIGURE 8.1 Homeland Security Advisory System (available at http://www.dhs.gov/dhspublic/display?theme=29).

eral, state, tribal, and local governments to work together to prepare for, prevent, respond to, and recover from domestic incidents, regardless of cause, size, or complexity. This resulted in the establishment of the National Incident Management System (NIMS). The basis for NIMS centers around the Incident Command System (ICS). One of the first steps in becoming compliant with NIMS requires states and local governments to institutionalize the use of ICS (as taught by DHS) across the entire response system. This means that ICS training must be consistent with the concepts, principles and characteristics of the ICS training offered by various DHS training entities.

The concept of ICS was developed more than 30 years ago in the aftermath of a devastating wildfire in California. During 13 days in 1970, 16 lives were lost, 700 structures were destroyed and more than a half million acres burned. The overall cost and losses associated with these fires totaled $18 million per day. Although all of the responding agencies cooperated to the best of their ability, numerous problems with communication and coordination hampered their effectiveness. Consequently, Congress mandated the creation of a system by the U.S. Forest Service that would "make a quantum jump in the capabilities of Southern California wild land fire protection agencies to effectively coordinate interagency action and to allocate suppression resources in dynamic, multiple-fire situations" (NIMS, 2004). The California Department of Forestry and Fire Protection; the Governor's Office of Emergency Services; the Los Angeles, Ventura and Santa Barbara County fire departments; and the Los Angeles City Fire department joined with the U.S. Forest Service to develop the system. This system became known as FIRESCOPE (Firefighting Resources of California Organized for Potential Emergencies).

Organizations are crucial in planning for, managing, responding to, and recovering from emergencies. Disaster research has given major attention to the behavior of organizations. Researchers have identified a number of key factors in promoting organizational effectiveness during the immediate emergency period of disasters. These include a flexible structure suited to activity coordination, good interpersonal relationships, frequent and open communication, adaptability, and shared responsibilities. One general finding is that civilian organizations in the United States do not function well under military models of command and control. However, much of this research dates back to the 1950s-1970s era and needs to be updated because of the changes that have taken place in emergency management as well as the broader range of threats that must now be confronted. Currently, for example, little research exists on the effectiveness of emergency operations centers (EOCs), let alone ICS operations.

Disaster researchers have long argued for the utilization of a different management model that sees the emergency manager more as a "broker" or "emergency resource manager" than as an "incident commander." As opposed to the hierarchical, top-down, command-and-control management model inherent in ICS, researchers suggest that effective emergency management is decentralized, organizationally flexible, adaptable, and resilient. Future research in this area should address the strengths, weaknesses, and overall effectiveness of command-and-control versus resource management models.

## DISASTER RESEARCH AND APPLICATION
## AND HURRICANE KATRINA

In 2004, the *Natural Hazard Observer,* published by the Natural Hazard Center at the University of Colorado, featured a series of articles that examined Disasters Waiting to Happen with the intent of generating a discussion about creative approaches to mitigation. In one of the articles, sociologist Shirley Laska (2004) discussed the impact of a major hurricane striking New Orleans. The article suggested that:

- flooding of New Orleans to depths as great as 20 feet would occur;
- that up to 80 percent of the structures in flooded areas would be severely damaged;
- there is a need to develop a plan to evacuate the estimated 120,000 residents without the means of evacuation;
- early evacuation of the city is needed due to limited routes of egress;
- all modes of transportation should be utilized in an evacuation;
- there is a lack of mass care centers to house those not evacuating;
- problems will occur with search and rescue due to hazardous conditions;
- it would take an estimated 10 days to complete search and rescue;
- there will be a need to house hundreds of thousands of displaced citizens unable to return to their residences; and
- that survivors would have to endure conditions never before experienced in a disaster in the United States.

These predictions became reality when flooding caused by Hurricane Katrina devastated New Orleans. Researchers have understood the consequences of a major hurricane hitting New Orleans, not just in a broad sense, but in a fairly detailed understanding of planning and response needs. So why was this knowledge ignored at all levels of government? While many have speculated on the answer to this question, it will take careful investigation to determine the root causes of why the country was not prepared for this event.

## RECOMMENDATIONS

Previous chapters of this report have documented the contributions that social scientists have made to understanding hazards and disasters of various types. Such knowledge is important not only because it sheds light on human behavior in very challenging and dynamic situations, but also because it provides a foundation for science-based decision making by at-risk populations seeking to manage willful, natural, and technological hazards and disasters. The utilization of social science information on hazards and disasters, when combined with relevant knowledge derived through the efforts of researchers in other disciplines, has the potential to significantly reduce the societal impacts of disasters. As noted in this chapter, however, there are many barriers that must be overcome before potential users will adopt information produced by social scientists and other researchers.

Evaluations of disaster related policies and programs are rare. The few case studies that were conducted many years ago may have limited application today; thus, much work needs to be done. More studies are needed on the utilization of research results in the hazards and disaster field, an area of investigation that has suffered from major neglect in recent years. Such studies could provide a basis for overcoming barriers to more effective dissemination and application of extant knowledge. While much anecdotal information, including that on the 18 dissemination activities discussed above, conforms to general theories about successful knowledge dissemination and utilization, there is a clear need to proceed with studies that use rigorous research methods to determine where and how improvements should be made.

Recommendation 8.1: *Renewed attention should be given by the social science hazards and disaster research community to the need for formal evaluation research on knowledge utilization in the field. New research should be carried out using all of the relevant methodologies and technologies available to the social sciences today.*

As part of its NEHRP role, NSF supported early social science studies on research utilization in the hazards and disasters field. As noted, Yin and his colleagues (Yin and Moore, 1985; Yin and Andranovitch, 1987) carried out their important work on research utilization in the 1980s. However, much has changed since that time. The knowledge base generated by the social sciences on hazards and disasters has grown significantly, as discussed in previous chapters of this report. This increased output has important implications for what practitioners such as urban and regional planners and emergency managers can conceivably do to decrease society's vulnerability and to enhance its capacity to mitigate, prepare for, and respond to disasters. Furthermore, as illustrated in Table 8.1, there are now numerous brokers of social science hazards and disaster information, including individual researchers as well as outreach and dissemination programs in government, academia and civil society. Many of these brokers did not exist when Yin and other researchers (Yin and Moore, 1985; Yin and Andranovitch, 1987) conducted their studies. Their existence now offers a real opportunity to better understand the research utilization process. Given these and other important societal changes, this is an opportune time to revisit the issue of hazards and disaster research utilization after so many years of neglect. And unlike the earlier era of research utilization studies, the next round of studies should see social scientists taking advantage of the full arsenal of methodologies and tools now available to them (see Chapter 7). Future evaluation studies should augment the case study and qualitative approaches favored by Yin and his colleagues (Yin and Moore, 1985; Yin and Andranovitch, 1987) with those approaches that allow for quantitative and other kinds of analyses. Moreover, statistical and computational modeling of the research utilization process could lead to greater theoretical understanding and provide a firmer basis for improving future efforts.

Table 8.1 suggests a number of areas in which research on knowledge utilization might address important issues from a comparative standpoint. One topic would be to document variations in the accuracy of the information being provided by different types of information brokers. Accuracy is obviously important because the information disseminated should be based on valid social science input if users are to become positioned to make the most effective risk adjustments. Another comparison that Table 8.1 suggests involves determining if some approaches work better than others with different users, such as land-use planners, emergency managers, and public health officials. For example, are strategies such as participatory action research and FEMA's training program effective with different potential user groups? Finally, some of the 18 activities and programs shown in the matrix use new technologies such as GIS and the Internet as part of their strategies for disseminating information. These tools did not exist when earlier research utilization studies were conducted. A fruitful line of research would be to compare the use of such technologies among information providers and to measure their value in stimulating research utilization.

Finally, future research utilization studies should focus not only on the ways information is introduced to potential users, (i.e., process issues), but also on the actual results of such efforts. This requires a "soup-to-nuts" research strategy. For example, it is crucial to have an understanding of just how much practitioners such as urban planners, emergency managers, and public health officials know about social science knowledge on hazards and disasters, the source of their information, and whether or not they ever apply it when making decisions about risk reduction. Here is perhaps the most challenging part of the process: In those documented cases where stakeholders have actually applied such knowledge, research should be focused on determining the extent to which this knowledge has made a difference. More than anecdotal information is needed about this outcome if social scientists are to be in the best position to help practitioners.

Recommendation 8.2: *Building on earlier practice, social scientists should conduct research utilization studies involving knowledge on hazards and disasters produced by other research disciplines.*

In their 1980s studies, Yin and his colleagues (Yin and Moore, 1985; Yin and Andranovitch, 1987) examined the utilization not only of social science knowledge on hazards and disasters, but also knowledge generated by physical scientists and engineers. The committee feels that it is essential for social scientists to continue this practice. First, social scientists have the methodological tools to carry out such research, perhaps even more so now than a generation ago. Second, more can be understood about the challenges of social science research utilization when they can be compared with the challenges facing disciplines such as earthquake engineering and earth science. Third, this practice could create opportunities for social scientists to engage in fruitful multidisciplinary and interdisciplinary research, as discussed in Chapter 5 of this report.

Recommendation 8.3: *Cross-cultural research utilization studies should be pursued by social scientists. Such research could contribute to global understanding of knowledge dissemination and application.*

As discussed in Chapter 6, disasters impact developed and developing countries alike. Stakeholders in all nations exposed to natural, technological, and willful hazards must make decisions about how to manage them. As in the United States, some of the decisions made in other countries are science based, while others are not.

With so many nations having variations in exposure to disaster risks, cross-cultural research on knowledge utilization is a promising area of inquiry, one that social scientists in the United States should pursue aggressively. Such research would provide an opportunity to test cross-culturally the principles of research utilization discussed earlier, determining their degree of universality. Also, through comparative analyses, approaches to research utilization in one country might be identified as relevant for consideration in another. Finally, cross-cultural studies on research utilization, involving social science or other kinds of knowledge related to hazards and disasters, would provide an opportunity for the collaborative international research called for in Chapter 6.

# 9

# The Present and Future Hazards and Disaster Research Workforce

T his chapter provides an overview of the social science hazards and disaster research workforce in the United States and discusses how it should be shaped to meet future societal needs. The needs relate to the broad range of hazards and disasters facing the nation and world in the twenty-first century (see Chapters 1, 2, and 6). Responding to them will require proactive steps to expand the relatively small research workforce in this field. Simply stated, the size and composition of the hazards and disaster workforce will significantly determine the extent to which the social sciences, in general, can respond forcefully to twenty-first century demands for basic social science knowledge and its application. This chapter therefore concludes with several recommendations to achieve that objective.

As is in all areas of scientific research, the future of social science hazards and disaster research is highly dependent on its human resources. This research specialty offers many rewards, not the least of which is the opportunity to make significant contributions to advancing theory and knowledge in the social sciences (see Chapters 1-4, 6, and 7). Social scientists have opportunities for many rewarding disciplinary, multidisciplinary, and interdisciplinary research experiences (see Chapters 5). Researchers also receive significant satisfaction from knowing that findings from their work have implications for reducing vulnerability nationally and internationally to multiple types of hazards and disasters (see Chapters 8). In spite of these intrinsic rewards, moving beyond the historically small supply of talented social science investigators described below is a major challenge for the hazards and disaster research community.

This challenge results, in part, from reliance on traditional recruitment strategies and the relatively modest funding that has been available for research and education in the social sciences (in comparison with the natural sciences and engineering). Traditional recruitment strategies are not likely to yield the number of new researchers that will be needed in hazards and disaster research. Academically based researchers have been the mainstay of this research specialty within the social sciences for decades. For these professionals, issues of funding and publication in mainstream as well as specialty journals are crucial considerations for achieving tenure and promotion. Given the plethora of research specialties in all of the social sciences, the competition for space in mainstream journals is very intense, requiring major efforts to link respective specialty research interests and findings with mainstream theoretical developments and issues. While hazards and disaster researchers in the social sciences have had notable successes in this regard, the trade-offs of publishing in specialty versus mainstream journals are particularly pointed for junior scholars. Another related and major challenge is changing the composition of the hazards and disaster workforce, which quite frankly has never been very diverse.

New opportunities and challenges have emerged, however, that may facilitate expansion of the hazards and disaster research workforce within the social sciences. The tragedy of September 11, 2001, for example, which involved the attacks on the World Trade Center and the Pentagon, and the crash of the terrorist-held plane in Pennsylvania, has generated significant interest in hazards and disaster research related to terrorism. As a result, the Department of Homeland Security (DHS) is now funding major research and educational initiatives that, consistent with its congressional mandate (see HR 5005 as amended, November 25, 2002), have implications for both terrorism and other types of hazards and disasters. Specifically, DHS has established a fellowship and scholarship program to produce a new generation of researchers, including social scientists. It also has established a Centers of Excellence Program, one that includes the Center of Excellence for Behavioral and Social Research on Terrorism and Counter-Terrorism. An even more direct recruitment approach is the Enabling Project funded by the National Science Foundation (NSF), which offers an innovative and promising strategy for mentoring junior faculty in the social sciences interested in research on natural, technological, and human-induced hazards and disasters. Another positive trend is the establishment of new homeland security journals, some of which are online, that can provide additional specialty publication outlets for young as well as established hazards and disaster researchers. Finally, NSF continues to avow its commitment to increase the participation of underrepresented groups, a commitment that is now shared by DHS.

These and other developments to be discussed in this chapter offer possible means to maintain and hopefully expand the talented research workforce that will be needed to address the research gaps and opportunities identified in previous chapters of this report. Ideally such a workforce will be of adequate size, reflect the diversity of the nation, and include researchers who have both basic and applied research interests and are capable of carrying out disciplinary, multidisciplinary, and interdisciplinary research. The natural, technological, and willful disasters that confront humankind in the foreseeable future require such a research workforce within the social sciences.

## WORKFORCE STRUCTURE

Although scant data exist on the infrastructure of social science hazards and disaster research, it is safe to conclude that this community has remained relatively small throughout its history, with only a modest number of social scientists viewing hazards and disaster research as their principal focus (Quarantelli, 1987). A comparable research workforce that comes immediately to mind is the small, specialized field of volcanology in the earth sciences, thought to number approximately 200 (Applegate, 2004). As summarized in Chapters 3 through 7, despite its small size the social science hazards and disaster research community in the United States has been very productive over the years, contributing to a greater understanding of hazard vulnerability and helping to lay the groundwork for more effective mitigation, preparedness, and response and recovery efforts (Dynes and Drabek, 1994). And the field's size has certainly not hindered it from becoming the world leader in social science hazards and disaster research (Britton, 2004). This leadership claim can be made both in terms of knowledge production and the major role American scholars have played in the development of global institutions and collaborative networks, such as the International Sociological Association's (ISA) International Research Committee on Disasters (IRCD). Many of the key concepts and findings from hazards and disaster research have come from American investigators; they were the driving force behind the establishment of the IRCD in 1982, and they remain essential to its continuing success (Quarantelli, 1999).

An important feature of the hazards and disaster workforce is its fluidity. While funding for this fledgling research specialty in the immediate post-World War II period was motivated by very applied concerns (e.g., war-related preparedness and response, floodplain management), the intersection of basic and applied interests of academic researchers was inevitable (Quarantelli, 1994). Thus, over the years social scientists have come into the field pursuing mainstream theoretical interests, often have left the field to undertake other kinds of research efforts, and sometimes have returned because their interest in hazards and disasters has been rekindled. This means that the research workforce of the field has to be replenished continuously to sustain the knowledge production and world leadership that are needed. Current trends such as the increasing numbers and costs of peacetime disasters and the growing threat of terrorism will probably facilitate more sustained involvement in hazards and disaster research by senior people. For the workforce to be sustained at a desired level, however, specific strategies must be devised (1) to put the next generation of researchers in the pipeline and (2) to recruit new researchers from the existing pool of social scientists.

The committee does not have a precise accounting of the numbers of social scientists from respective disciplines currently engaged in hazards and disaster research. Neither government agencies nor professional associations systematically collect data on this research workforce, which as noted above resides primarily in academia. This imprecision also applies to students, both graduate and undergraduate, who might be included in the pipeline because they are working on hazards and disaster research projects or are being mentored by senior scholars. However, undoubtedly the social science hazards and disaster researcher community is relatively small, particularly when one considers the thousands of persons trained in the relevant social science disciplines. As a very conservative indication of the size of selective social science disciplines, for example, the committee obtained the following membership numbers from various staff members of professional associations for the year indicated: Association of American Geographers, 2003 (8,475); American Sociological Association, 2004 (13,246); American Economic Association, 2004 (approximately 18,000); and American Political Science Association, 2004 (13,597). If the size of the current hazards and disaster research workforce is approximately 200, this community comprises only a fraction of the total social science pool, a pattern that is similar to the subfield of volcanology relative to the larger discipline of earth science.

There are some clear differences among the social sciences regarding the number of researchers that each contributes to the hazards and disaster workforce. Such differences are at least partially attributable to historical circumstances. As noted in Chapters 1 and 2, sociology and geography played leading roles in the origins of hazards and disaster research during the early part of the last century and in its full emergence following World War II, giving them a head start on such disciplines as psychology, political science, economics, anthropology, and urban and regional planning (Quarantelli, 1987). It is important to note that most pioneer researchers had mainstream theoretical interests and published their findings in academic presses and mainstream journals, frankly because no specialty outlets existed. In any event, by the 1960s and 1970s small numbers of sociologists and geographers were clearly committed to this field, either as individual researchers or members of research centers. This disparity of respective involvement among social science disciplines continues to this day.

A simple typology is useful for characterizing the social science hazards and disaster research workforce. In broad terms, this workforce is comprised of three somewhat distinguishable types: (1) *core researchers*, (2) *periodic researchers*, and (3) *situational researchers*. Core researchers are the most committed to the field by virtue of their acknowledged interest in hazards and disaster research and the considerable amount of time they spend engaged in specific studies. Although core researchers may have multiple interests, they essentially see themselves as hazards or disaster researchers. This self-conception is reflected both in their research programs and their training of others, which typically involves sustained mentoring of junior scholars. Core researchers are more likely than not to have ties to larger networks of researchers and practitioners, as evidenced by collaborative work with other researchers, attendance at conferences and workshops with hazards and disaster themes, and interaction with disaster policy makers and managers. They are also more likely to be conversant with major paradigms in the field and have a thorough understanding of key application issues. Early pioneers of the field such as Harry Moore, Charles Fritz, Gilbert White, and Enrico Quarantelli are exemplars of this type.

Periodic researchers are scholars who do not see themselves as primarily hazards and disaster researchers, but focus on related topics from time to time throughout their professional careers. Their less frequent engagement in the field does not prevent them from making important contributions, but because of competing interests they are less likely to direct students toward careers in hazards and disaster research, attend specialty conferences and symposia, or interact on a regular basis with policy makers and practitioners. Within sociology, for example, scholars such as Allen Barton, Ralph Turner, Peter Rossi, Charles Perrow, and Kai Erikson have been periodic researchers who have made significant contributions to the field. Barton recently received the International Research Committee on Disasters (IRCD) 2002 E.L.Quarantelli Award for contributions to disaster theory, along with Russell Dynes who clearly became a core researcher early in his career. It is interesting to note also that at least five presidents of the American Sociological Association (Neil Smelser who is a member of the National Academy of Sciences, Ralph Turner, Peter Rossi, Kai Erikson, and James Short) have been involved in hazards and disaster research at some point during their careers. Similarly in geography, while such scholars as Gilbert White, Robert Kates, and Roger Kasperson have long been core hazards and disaster researchers, periodic researchers who have made significant contributions to the field include John Borchert, Walter Isard, M. Gordon Wolman, and Julian Wolpert (Mitchell, 1989). Like White, Kates, and Kasperson, the latter four geographers were elected to membership in the National Academy of Sciences.

Situational researchers are scholars who have not been involved previously in the field, but who become interested because of the opportunity to explore new and interesting phenomena. Eric Klinenberg, who studied the 1995 Chicago heat wave (Klinenberg, 2002) is a recent example. An earlier example is Ralph Ginsberg whose expertise in survey research and quantitative methods made him a valuable interdisciplinary team member in a study of market failure in earthquake and flood insurance (Kunreuther et al., 1978). Like core and periodic researchers, situational researchers often make significant contributions to the field. And like periodic researchers, situational researchers often are able to offer challenges to generally accepted theoretical formulations and frameworks in the field because they are not part of the core, thereby opening up new avenues of inquiry.

There is a healthy fluidity in the mix of situational, periodic, and core hazards and disaster researchers. A situational researcher may become so intrigued by hazards or disasters that he or she may develop a long-term interest in studying these phenomena. Indeed, all periodic and core researchers originally began their involvement in the field as situational researchers. That involvement then becomes intermittent for periodic researchers and sustained for core researchers. It is also the case that a core researcher may become a periodic researcher because of limited funding opportunities or changed career circumstances. History suggests that all three types are important for the advancement of this field. The committee concludes, however, that what is most needed for the future of hazards and disaster research *is a larger and more stable cadre of core researchers who are committed to disciplinary, multidisciplinary, and interdisciplinary research (as defined in Chapter 5); to training and mentoring students; and to furthering the sharing of data and the dissemination of findings (as discussed in Chapters 7 and 8).* Core researchers are also valuable because they provide needed intellectual and institutional leadership, serve as spokespersons for the field, project its identity as a community of scholars, and serve as links between succeeding generations of scholars.

## WORKFORCE PROFILE

Besides its relatively small size, it is difficult to be very precise about the demographic structure of hazards and disaster research due to the absence of good data. Mostly indirect measures will have to suffice in providing a general sense of this community's workforce profile. It is appropriate to begin by looking at sociology and geography because these disciplines historically have had the largest concentration of hazards and disaster researchers. Between the end of World War II and 1963, approximately 10 doctoral dissertations were published in sociology on hazards and disaster research topics. When the Disaster Research Center (DRC) was established at the Ohio State University in 1963 by three sociologists, only one of the cofounders (Enrico Quarantelli) had prior disaster research experience. The others (Russell Dynes and J. Eugene Haas) had related expertise and research experiences in studying small groups and larger organizations of various kinds. In terms of the above typology, Quarantelli was already a core disaster researcher at the time of the establishment of the DRC, while Dynes and Haas began as situational researchers. By the time the DRC moved from Ohio State to the University of Delaware in 1985, sociology doctoral candidates had completed 29 dissertations directly under DRC funding and three additional dissertations were completed using DRC data. Many of these Ohio State graduates are now part of the hazards and disaster research workforce, either as core or periodic researchers. Additionally during the Ohio State period, several other former DRC graduate students completed dissertations on nondisaster related topics, but then went on to very productive careers as core researchers in the field (Quarantelli, 2004).

Since its move to the University of Delaware 20 years ago, only six dissertations on hazards and disaster topics have been produced by students funded directly by the DRC. Another 16 graduate students along with 12 undergraduates have been funded under the National Earthquake Hazards Reduction Program (NEHRP); (through individual investigator awards and NSF's Research Experience for Undergraduates program) at William and Mary for disaster studies using DRC archival materials. Of the 16 graduate students, 12 completed M.A. theses, and 7 of these 12 ultimately were awarded Ph.D.s at other universities. Even including data from William and Mary, there is a clear shortfall when comparing the 1963–1985 and 1985–2004 periods at the DRC. This shortfall at what continues to be a major research center in the field could have important implications for hazards and disaster research in sociology when currently active core researchers from the pre-1985 graduating cohorts retire. This overall "graying" of the field in sociology has sparked some concern among core members. One of the responses to this concern has been the development of the NSF Enabling Projects (discussed below). Other institutions with sociologically oriented disaster research programs have also contributed to the disaster research workforce pool, though on a smaller scale than the DRC. For example, a small number of core or periodic disaster researchers, perhaps not much more than one or two dozen total, have emerged from programs at the University of Georgia, the University of Denver, the University of California at Los Angeles, Colorado State University, the University of Massachusetts, and the University of Colorado.

Turning to geography, Gilbert White built his highly distinguished hazards research career at the University of Chicago during the 1940s–1960s. White's numerous credits include paving the way for the emergence of floodplain management, putting geography in the forefront of hazards research, and mentoring the next generation of geography hazards researchers who today comprise a major part of the hazards and disaster research workforce core (Mitchell, 1989). One of the students White mentored was Robert Kates, who over the years contributed much to the field while at Clark University and helped make it a leading producer of geography hazards and disaster researchers. As in the case of sociology (and other contributing social science disciplines), the precise number of hazards and disaster researchers within geography is not known. In a 1989 article, Mitchell noted that between 1981 and 1986, 44 Ph.D. dissertations and 126 master's theses on hazards topics were completed in North American colleges and universities (Mitchell, 1989). In a more recent article, it was noted that 100 geography dissertation titles and abstracts dealt with hazards-related topics (Montz et al., 2003a). These data give at least some indication of the pool of academic geographers from which core, periodic, and situational types of researchers might have come.

In addition to the University of Chicago and Clark University, other key players are Rutgers University and the University of South Carolina, which now is the largest producer of Ph.D. hazards and disaster geographers in the country. In recent years, six Ph.D.s were completed at the Hazards Research Laboratory (HRL) at the University of South Carolina and five more are in the pipeline. Similar to sociology, overall the number of hazards and disaster geographers produced by the above and other institutions is small but nonetheless crucial to the development of the field. The contribution of geographers entering the hazards and disaster research arena after receiving graduate training is reflected in the fact that several have been elected to the National Academy of Sciences, as mentioned previously, and four have been elected president of the Association of American Geographers.

A major turning point in hazards and disaster research occurred when Gilbert White moved from the University of Chicago to the University of Colorado in 1968. At Colorado, White championed interdisciplinary research and established collaborative hazards research projects with colleagues from other disciplines. One of these projects, carried out jointly with J. Eugene Haas (who joined the University of Colorado faculty after leaving the Ohio State University) was the First Assessment of Research on

Natural Hazards. This assessment resulted in a landmark publication in hazards and disaster research (White and Haas, 1975). No less important, the University of Colorado served as the training ground for an interdisciplinary team of a dozen or more graduate students, including those from the social science disciplines of geography, sociology, economics, and psychology. Many of the First Assessment veterans became part of the core of the hazards and disaster research community, evidenced by the fact that several also took part in the more recently completed Second Assessment (Mileti, 1999b). Like sociology Ph.D.s from Ohio State and the University of Delaware, this group has also been in the field for nearly 30 years, and thus is part of the graying generation who must be replaced by junior colleagues.

Many researchers belong to professional organizations of specialists in the field. Membership in such organizations provides some additional clues about the size of the workforce in the social sciences. While multiple memberships can distort the picture, this distortion is perhaps balanced by the fact that some active researchers in the field choose not to join such organizations. Thus, there is some value in looking at organizational membership as a proxy or indirect indicator of workforce size. For example, Mitchell noted in 1989 that approximately 5 percent (about 120) of the 2,400 college faculty members who belong to the Association of American Geographers (AAG) identified hazards as one of their primary areas of specialization (Mitchell, 1989). More recently (June 2004), the Hazards Specialty Group within the AAG had 263 American members. Most of the Hazards Specialty Group members are geographers (Mitchell, 2004). Similarly, many sociologists with interests in hazards and disasters are members of IRCD. Some 46 of the IRCD's 250 members are American scholars (Phillips, 2004). The IRCD is an inclusive organization; thus, some of its American, as well as international, members are from social science disciplines other than sociology. To compare geography and sociology with anthropology, in 2005, seven members of the American Anthropological Association listed disaster research as an area of interest.

Examining social science membership in an association dominated by another discipline is also instructive for getting some sense of how social science disciplines other than sociology and geography fit into the workforce profile. The Earthquake Engineering Research Institute (EERI) is one such organization. EERI is the major earthquake engineering association based in the United States and membership is open to professionals from other disciplines. For the past few years, EERI's membership has remained around 2,400 (Tubbesing, 2004). In 2005, 38 members were listed as falling into four professional categories that are of interest: social science, public policy, urban planning, and public health (EERI, 2005). Of these 38 members, 23 are core hazards and disaster researchers from the United

States, including 6 who identify themselves as public policy experts (e.g., with political science and public administration backgrounds) and 8 who are self-identified as urban planners. Those listed as public policy specialists and urban planners arguably comprise a large portion of the researchers who make up the core of the workforce from those disciplines. Both the University of North Carolina at Chapel Hill and Cornell University have played major roles in training planners entering the hazards and disaster research workforce. No one from the disciplines of economics or psychology was listed as a member of EERI in 2005. The committee concludes that the number of economists and psychologists in the core hazards and disaster workforce is certainly no greater than those from the policy and planning disciplines.

Clearly, there is a need for more hazards and disaster researchers from disciplines such as economics and political science that have not had as much involvement in the field as geography and sociology. A workforce with greater disciplinary balance would further increase coverage of important and sometimes understudied topics and issues, such as the economics of disasters, intergovernmental dynamics during disasters, and cross-societal impacts.

In summary, and based on the above limited information, the core hazards and disaster research workforce is small and has its deepest roots in sociology and geography (Anderson and Mattingly, 1991). Estimating the number of periodic and situational researchers is inherently difficult because the involvement of these types is intermittent. Taking all of the above information into account, as noted earlier the committee concludes that the current supply of hazards and disaster researchers within the social sciences is comparable to the field of volcanology. As aforementioned, Applegate (2004) suggests that a first-order estimate of the number of volcanologists in the United States is about 200. His estimate is based on combining the 78 volcanologists employed by the U.S. Geological Survey (USGS) with 112 faculty members who identify themselves as volcanologists in the American Geological Institute's Directory of Geoscience Departments lists, and on assuming that some igneous petrologists and seismologists also study volcanoes. Hazards and disaster research in the social sciences is much more inclusive than volcanology; thus, its small pool of core, periodic, and situational researchers poses a major challenge for the future.

## Workforce Composition

Any workforce has compositional features such as age, gender, race, and ethnicity. Such features reflect societal forces and the related distribution of opportunities faced by particular individuals and groups. The committee's conclusion that this workforce has an aging core is particularly significant because of the small size of the community as a whole. Simply put, the core must be replenished as soon as possible. This fundamental workforce requirement creates an opportunity for junior prospects to play important roles in the future. Senior scholars can facilitate this succession, first, through their teaching and mentoring activities and, second, through actively promoting the importance of basic and applied research in their respective disciplines.

Like many scientific and professional fields, this research specialty was long dominated by men. The pattern began to change in the 1970s and 1980s, however, as more women entered the field after completing their graduate work. Today, women are increasingly represented in the core workforce, although again the committee does not have precise numbers to that effect. Women who have become a part of the core workforce have addressed both mainstream and specialty research issues. They have also encouraged and recruited other women to the field. In so doing, women have called attention to the importance of such previously neglected topics as gender and equity in relation to hazard vulnerability. In recent years, women have made significant gains in terms of their influence in the field, including assuming research center directorships and other key leadership posts. They also have moved closer to parity with male scholars in terms of research funding received from NSF (Anderson, 2000). The increased involvement of women social scientists in hazards and disaster research is reflected in other ways as well, including their active participation in multidisciplinary research settings such as the NSF-funded earthquake engineering research centers and in professional organizations. For example, a prominent female social scientist was a recent president of EERI and the organization's long-term executive director is a woman with social science training. Even with this progress, attention must remain on encouraging the full participation of women in hazards and disaster research.

While change in the status of women in the field has been a positive development, little has happened over the years to increase the involvement of racial and ethnic minorities. Although minorities such as blacks and Hispanics have doctoral degrees in larger numbers in the social sciences than in other research disciplines (National Science Board, 2004), they remain so underrepresented in hazards and disaster research as to be practically hidden from view. There are only a handful of African-American and Hispanic hazards and disaster researchers in the social sciences known to the committee. This group includes two senior Hispanic researchers at the Disaster Research Center, one of whom is the director. The underrepresentation of racial and ethnic minorities has existed for much, if not all, of the history of this field. Thus a continuing opportunity is being lost in a nation with an increasingly diverse population. As noted in Chapters 2 through 4, there is much irony in this underrepresentation because minorities are among the most at-risk population groups in the United States.

The benefits of an increasing number of women in hazards and disaster research apply equally to minorities. Thus, having additional minority researchers would add value on its own terms and also make it much easier to recruit and retain more minority students in the future. The mentoring of minority students by minority scholars is especially important. As in the case with women, a critical mass of minority researchers would likely compel studies of new research topics and issues such as some of those identified in Chapters 3, 4, and 6. Among the studies the committee has recommended are those that will help to reduce the vulnerability of minority and other population groups.

## WORK SETTINGS

As previously noted, the vast majority of social science hazards and disaster researchers work in academic settings, including both Ph.D.- and non-Ph.D.-granting institutions, where research and educational activities are combined. Far fewer researchers work in federal agencies and laboratories or in private sector organizations such as consulting firms. While some researchers work alone or with one or a few graduate students, the pattern for core researchers in this field is to work in teams. These teams can be project specific or created under the auspices of an established research center. Within the social sciences, the term "center" does not generally describe an entity with a large staff. To the contrary, these centers tend to be relatively small, often comprising only a few senior investigators and several graduate students. This characterization can be applied, for example, to the Disaster Research Center (University of Delaware) and the Hazards Research Laboratory (University of South Carolina). Most social science research centers, including the two just mentioned, are located in disciplinary departments even though their principal investigators, postdocs, and graduate students may from time to time team up with researchers in other disciplines. One center that is different in this regard is the Hazard Reduction and Recovery Center at Texas A&M University. This center, which is located in the College of Architecture, has been led by a social scientist since its inception. It carries out multidisciplinary and interdisciplinary research that combines social science with engineering, architecture, and planning disciplines.

Social scientists also work as team members in other types of centers, such as the University of California, Los Angeles' Center for Public Health and Disasters and the NSF-funded earthquake engineering research centers. A pattern is also emerging in which social scientists are included on teams focusing on multidisciplinary research related to terrorism. Relevant examples here are the Homeland Security Center for Risk and Economic Analysis of Terrorism Events headquartered at the University of Southern California and the Homeland Security Center of Excellence for Behavioral and Social Research on Terrorism and Counter-Terrorism headquartered at the University of Maryland. Once again, the research activity carried out at these and other centers is combined with important educational experience for the emerging generation of social science hazards and disaster researchers. The more students serve as research assistants while they work on graduate degrees, the larger is the pool of scholars that may become situational, periodic, or core researchers in the future. And where the centers have multidisciplinary and interdisciplinary thrusts, as defined in Chapter 5, students have a tremendous opportunity to become more adept at team building.

## DESIGNING A WORKFORCE TO MEET FUTURE CHALLENGES

As documented throughout this report, the social science hazards and disaster research workforce has accomplished much over the years, but maintaining and expanding a talented workforce will be even more important in the future. The field will need core researchers throughout the social sciences who can conduct research using the most advanced technologies and methods (see Chapter 7), who are committed to training subsequent generations of hazards and disaster researchers and practitioners, who will promote the field proactively to wider audiences, and who are committed to mainstream disciplinary theories and willing to work on multidisciplinary and interdisciplinary teams (see Chapter 5). The field will need the continuing presence of research centers that combine both research and graduate education. The field will need the increasing representation of women and minorities. And the field will need a continuing pool of periodic and situational researchers to break new ground and challenge existing paradigms. It will also need researchers who work in a variety of other research settings besides academia, such as government agencies, consulting firms, and nonprofit organizations. The desired outcomes of such a workforce will be significant advances in knowledge, junior researchers who build on and extend the work of previous generations, and policy makers and practitioners who make knowledge-driven decisions (see Chapter 8). The challenge of maintaining and expanding a research workforce capable of realizing such accomplishments is essential to meet. Existing and emerging societal risks and knowledge gaps related to them demand a vigorous and sustained response from the social sciences.

As highlighted in Chapter 2, there are any number of societal changes that will put added pressure on the hazards and disaster research workforce. The emergence of a more diverse society, with racial and ethnic minorities comprising an ever-increasing percentage of the population, is an important demographic shift that will require greater attention to hazard vulnerability and mitigation. Changing settlement patterns will continue to have direct implications for emergency preparedness and response. Well-being and quality-of-life issues will continue to have important implications for disaster recovery. The growing threat of terrorism is a key element in a changing risk environment, calling for new core researchers who will develop and apply knowledge in coping with multidimensional threats. Environmental alterations such as global climate change will also create new demands on the hazards and disaster workforce, forcing a reconsideration of its size and composition as the frequency and severity of such disasters as floods and droughts increase as a result of changing climatic conditions.

Notwithstanding the above and other societal trends and changes highlighted in Chapter 2, the supply of core, periodic, and situational hazards and disaster researchers in the social sciences has been too low historically, and the current composition of the research workforce clearly is inadequate. A historically larger workforce would have deepened our understanding of the five core topics of hazards and disaster research identified in Figures 1.1 and 1.2, providing the basis for stemming the rising costs of hazards and disasters. The aging of the research workforce is a serious issue that must be addressed as soon as possible. An increase in the number of hazards and disaster researchers from underrepresented social science disciplines such as political science, economics, and anthropology will be required. Also required will be greater representation of racial and ethnic minorities who can provide better entrée to minority communities and greater cultural sensitivity when dealing with minority populations. The involvement of women in the field should continue to be encouraged and monitored carefully to avoid any possible slippage in their degree of participation in research and related professional activities. Increasing gender parity should remain an important goal for the field.

## Human Resource Development

Various institutional sectors—including government, academia, and professional associations—should have a major role in shaping the social science hazards and disaster research workforce to meet future needs. Building and sustaining a viable workforce is a competitive process wherein potential recruits have various career options from which to choose. Developing the needed workforce of the future can be characterized as involving four elements: recruitment, education, retention, and reward. Systems of reward underlie recruitment, education, and retention efforts.

Whether they are extrinsic or intrinsic in nature, rewards provide the basis for success in recruiting and educating new workforce members and obtaining long-term commitments from them. Various stakeholders—in government, academia, professional organizations, or the private sector—are part of any reward system that aims to achieve recruitment, education, and retention goals. The more collaborative the efforts of stakeholders are, the more successful they are likely to be overall. Such collaboration may involve the establishment of partnerships and the leveraging of vital resources. Thus, a holistic strategy would best further the development of human resources in the hazards and disaster research community.

One of the basic extrinsic rewards of any workforce, of course, is that it provides a means of earning a living. But aside from this purely financial outcome, there are intrinsic rewards that come to those involved in a research workforce such as creating and applying new knowledge. And because hazards and disasters are global risks, having the opportunity to conduct research in an international as well as domestic context is an important inducement for entering the field. Such extrinsic and intrinsic rewards are only important when they become specific incentives, proactively marketed to potential recruits by multiple stakeholders. Experience suggests that once recruited, the complexities and importance of the subject matter will be more than sufficient to keep researchers involved on at least a periodic, and hopefully a sustained, basis. The admixture of situational, periodic, and core researchers is healthy and beneficial for the field. Perhaps the ideal blending would be like a pyramid, with a large number of situational researchers at the base, a significant number who become periodic researchers, leading to a vital number of core researchers.

## RECOMMENDATIONS

Recommendation 9.1: *Relevant stakeholders should develop an integrated strategy to enhance the capacity of the social science hazards and disaster research community to respond to societal needs, which are expected to grow, for knowledge creation and application. A workshop should be organized to serve as a launching pad for facilitating communication, coordination, and planning among stakeholders from government, academia, professional associations, and the private sector. Representatives from the NSF and DHS should play key roles in the workshop because of their historical (NSF) and more recent (DHS) shared commitment to foster the next generation of hazards and disaster researchers.*

As noted above, the social science hazards and disaster research workforce is estimated by this committee to be about 200, which is very small in comparison to other relevant disciplines such as earthquake engineering and earthquake science. For example, the vast majority of EERI's approximately 2,400 members come from these two disciplines. Among other benefits, a larger social science hazards and disaster research workforce would contribute to meeting both the substantial disciplinary research needs and opportunities identified in Chapters 3 and 4 and the interdisciplinary needs and opportunities associated with them that are identified in Chapter 5 (see Figure 5.2).

The future development of the hazards and disaster research workforce within the social sciences requires an integrated strategy. Key components of that strategy should be stakeholder policies and initiatives that include specific incentives for the recruitment, education, and retention of new researchers. The strategy should necessarily build on the many strengths of the existing workforce, as noted above and highlighted throughout this report. The strategy should synergistically leverage the resources of government, academia, professional associations, and the private sector. The initial planning workshop would launch the effort, and the involvement of both NSF and DHS is essential. Working collaboratively with other stakeholders, NSF and DHS should ultimately develop complementary or joint programs that strengthen social science research in this field. These programs should include educational opportunities for undergraduate and graduate students, opportunities for junior faculty members, and financial incentives for becoming engaged in the field. New programs should be developed where there are unmet needs, and these programs should complement successful existing ones. Existing programs that have proved their effectiveness should be continued whenever possible. Underpinning an integrative approach should be a system of data collection that provides real-time information on the status of the workforce and the outcomes of various programs. The resulting database would enable multiple stakeholders to make knowledge-based decisions about future commitments and investments.

> Recommendation 9.2: *NSF should expand its investments in both undergraduate and graduate education to increase the size of the social science hazards and disaster research workforce and its capacity to conduct needed disciplinary, multidisciplinary, and interdisciplinary research on the core topics discussed in this report. NSF should also give special consideration to investing in innovative ways to further workforce development, especially when they involve partnerships such as NSF's recent joint initiative with the Public Entity Research Institute (PERI) and the Natural Hazards Research and Applications Information Center at the University of Colorado. This initiative, discussed below, exempli-*

*fies the collaboration needed across government, academia, professional associations, and the private sector.*

Recommendation 9.3: *In parallel fashion, DHS should make a conscious effort to increase significantly the number of awards its makes to social science students through its scholarship and fellowship program. Because much that must be investigated about the terrorist threat is related to social and institutional forces, more social scientists need to be recruited to adequately study them. With its broader cross-hazards congressional mandate, DHS should contribute to a larger social science hazards and disaster research workforce, one that complements research in other science and engineering disciplines.*

Recommendation 9.4: *NSF and DHS should consider ways that they can cooperate programmatically to enhance the social science hazards and disaster research workforce. Jointly sponsored university research and education programs by the two agencies would be of major benefit to the nation.*

Recommendations 9.2 through 9.4 should be viewed in tandem. Because its mission is to nurture science in the United States, since its inception NSF has supported science and engineering education at both the undergraduate and the graduate levels. Currently, NSF grantees may receive support enabling them to include both undergraduate and graduate students on the same projects. Indeed, NSF's Research Experience for Undergraduate (REU) program is aimed specifically at the inclusion of undergraduates on NSF-funded research projects. It also provides the opportunity to establish what are termed REU sites. These sites provide summer research opportunities for 15 to 20 undergraduate students from different colleges and universities, most often between their junior and senior years. A recently awarded REU site program to the Disaster Research Center at the University of Delaware is the first of its kind for this field and is a very positive development because it enables a sizable group of undergraduate students to undergo intense summer research training at a common location. NSF funding of such social science REU sites, as well as REU funding provided for individual research projects through supplemental grants, is essential for building the next generation of hazards and disaster researchers. Well before NSF established its REU program, some of today's core hazards and disaster researchers were first introduced to the field as undergraduate students and subsequently retained their interest as they moved on to graduate school and then into the research profession. It is important that this vital NSF support for REUs continues to be available, allowing core researchers to seek new recruits without having to wait until students reach the graduate level. Undergraduate students comprise an important talent pool that can help meet future human resource needs in the field.

In 2003, and with NSF support, the Public Entity Risk Institute (PERI) and the Natural Hazards Center at the University of Colorado established a joint program of dissertation fellowships for graduate students in science and engineering disciplines to work on topics related to natural, technological, and human-induced hazards. Initially established as a two-year pilot effort, the aim of the program was to attract new researchers to the field, including social scientists, to meet the growing need for multidisciplinary and interdisciplinary research. Ten fellows were selected each year during the pilot phase of the program. This innovative program complements other graduate and undergraduate student enabling efforts made possible through NSF funding as well as its enabling projects for junior faculty. The committee recommends that NSF increase its support for this type of program, particularly in terms of the number of dissertation awards given to social science graduate students in order to create a more balanced multidisciplinary workforce.

NSF has been the major supporter of undergraduate and graduate education for the social science hazards and disaster research field at least since the creation of the National Earthquake Hazards Reduction Program (NEHRP) more than 25 years ago. This has enabled the workforce to remain at a fairly steady state. Given the expected new demands on the social science hazards and disaster research community, it is an opportune time for NSF to leverage its efforts with those of the recently established DHS. DHS has launched its own programs to support higher education, including its undergraduate scholarship and graduate fellowship program. This program is intended for undergraduate and graduate students interested in research careers that help advance knowledge about societal prevention of and response to terrorism. The program, which is open to students in a variety of science and technology fields, provides multiple-year support. Some 58 awards were made to undergraduate students and 48 to graduate students in 2004, the second year of the program. Notably, however, only 13 scholarships were awarded to social science undergraduate majors and only 5 fellowships were awarded to social science graduate students (Petonito, 2004). The social science proportion of awards should be increased substantially. Additionally, opportunities should be provided for such DHS awardees to work with core hazards and disaster researchers within the social sciences. There would be no better mentors for preparing new initiates to hazards and disaster research.

Thus far, DHS has also sponsored, on a competitive basis, four university-based Homeland Security Centers for research on terrorism, and recently announced plans to fund a fifth one. The first center, established at the University of Southern California, has a major social science component. The most recently established center at the University of Maryland focuses almost exclusively on social science and criminal justice issues. The fifth center will also have a major social science emphasis. As noted earlier, all DHS centers of excellence are mandated to support both research and education. By supporting both graduate and undergraduate students, these DHS-sponsored centers, and perhaps others in the planning stage, will play a major role in developing the hazards and disaster workforce of the future.

Finally, the committee sees direct collaboration among stakeholders as a vital strategy, particularly between such key entities as NSF and DHS. For many years, NSF has been working with the university community and supporting the integration of research and higher education, multidisciplinary hazards and disaster research, and center-based research and education. Thus NSF has significant experience to share with DHS, which is now moving into these areas in significant ways. As noted above, through its current and planned university investments, DHS has in a very short time become a key agency in developing the future hazards and disaster research workforce. Direct collaboration between NSF and DHS would therefore result in greater efficiency through the leveraging of scarce resources for higher education. It would also further the integration of this field, which—without due diligence—might grow apart: DHS-funded researchers who are focused primarily on terrorism; and NSF-sponsored researchers who give attention to other types of societal risks. The two groups will learn much from each other if meaningful and sustainable connections are fostered. To this end, the committee believes that a cross-hazards research and education agenda is the preferred strategy. Through their respective policies and programs, NSF and DHS should collaborate directly in making this perspective salient to academic and other stakeholders.

**Recommendation 9.5:** *As the leader in furthering U.S. science through research and workforce development, NSF should make greater use of its enabling mechanisms, including standard research grants, center grants, grant supplements, and REU programs to attract more minorities to the social science hazards and disaster research workforce.*

As noted above, minorities interested in the social sciences have not been attracted to the hazards and disaster research field. It is possible that many minority persons do not consider hazards and disaster research because of greater opportunities to study chronic social problems in their communities such as poverty and crime. The fact that there are so few minority role models in the field is another major barrier to the greater participation of minorities. Thus, more aggressive minority recruitment efforts are essential to make the workforce more diverse. A more diverse workforce would result in needed new research perspectives, increased linkages to the growing minority communities in the United States, a greater understanding of the vulnerabilities of these communities, and new opportunities to significantly reduce these vulnerabilites.

Major recruiting efforts should focus on institutions with large numbers of underrepresented minorities such as Hispanics and African-Americans. Public community colleges are potential venues because they tend to have higher percentages of minority students than traditional four-year colleges or research universities. A possible model for social science was a program in the geosciences established with NSF support in 2001 for local community colleges and high schools in the Long Beach, California, region. This program involved giving minority participants an intensive summer research experience. Some of the resulting research projects dealt specifically with hazards. Additionally, minority-serving institutions in general are fertile recruiting grounds for the hazards and disaster research pipeline. These institutions include both historically black colleges and universities (HBCUs) and many other institutions that have large enrollments of Hispanics and Native Americans. NSF should encourage majority academic institutions with social science hazards and disaster research programs to seek opportunities to diversify the workforce by establishing collaborations and direct partnerships with students and faculty of minority-serving institutions.

The committee also strongly encourages NSF to further the establishment of hazards and disaster research programs at minority-serving institutions. Such an NSF initiative would be a major step at a time when not only a larger social science hazards and disaster research workforce is needed, but a much more diversified one as well. This would attract new faculty and student talent from the minority community to the field. And in keeping with the spirit of Recommendation 9.4, NSF and DHS should consider jointly sponsoring such programs. In addition to leveraging agency resources, this kind of cooperation would serve as a means to further cross-hazards dialogue and research.

Assigning individuals with specific responsibilities for furthering diversity in majority institutions might also provide needed breakthroughs in diversity recruitment. The Multidisciplinary Center for Earthquake Engineering Research (MCEER), for example, has appointed a diversity program director with the responsibility for reaching out to underrepresented groups in earthquake engineering and assisting them in pursuing engineering degrees. Such an outreach specialist is needed at all three NSF-funded earthquake engineering research centers, but the assigned role should be more broadly defined. Because the centers have a multidisciplinary research agenda, but no underrepresented minority social scientists involved in their research programs, outreach specialists should have the added responsibility of increasing the participation of minority social scientists. Such diversity program directors should also give attention to the continuing recruitment of women social scientists. The committee suggests a similar approach to furthering diversity within the new or planned DHS-sponsored centers. Given their extensive resources and networks, research centers of various sizes and missions can do much more to enhance the diversity of the social science hazards and disaster research workforce.

**Recommendation 9.6:** *The NSF Enabling Project for junior faculty development (discussed below) should be continued if the second pilot proves to be a success.*

Recognizing that the social science hazards and disaster research community needed to supplement traditional means of recruiting and training new talent, in 1996 NSF funded what became known as the Enabling Project. The Enabling Project began as an innovative experiment under the auspices of Texas A&M University. It involved the mentoring of 13 junior faculty members, chosen nationally on a competitive basis, by 6 core social science hazards and disaster researchers from several different universities. Many of the 13 assistant professors had no previous involvement with the hazards and disaster research field, but were judged as having significant potential for eventually contributing to its advancement. The Enabling Project was designed to realize this potential. Its two-year program included workshops, mentoring sessions by senior faculty mentors, and the preparation of project proposal drafts by the Enabling Fellows. Overall the program was a notable success. Several of the fellows received funding for proposals they submitted to NSF, and most remain part of the workforce.

The success of the first two-year Enabling Project led to NSF funding the second Enabling Project, administered through the University of North Carolina at Chapel Hill. This two-year follow-up project was initiated in 2003 with a slightly expanded pool of 15 fellows working with 8 mentors. The fellows were selected competitively as before and again mentored by core faculty members in the field. While it is still too early for a final assessment of the second Enabling Project, early indications suggest that it was a success as well. In fact, the quality of this class of fellows was so high that three received competitive awards from NSF and other sources before the mentoring on proposal development was completed. Also, the second cohort demonstrated a significant interest in multidisciplinary research and, like the first Enabling Project, had a solid representation of participants from various social science disciplines. One major shortcoming of the two projects is that neither has had much success in attracting minority participation. Only one person from an underrepresented minority group was in the first cohort and none was in the second cohort. Women fared better, with three participating in the first Enabling Project and five in the second.

The mentoring the fellows receive from core hazards and disaster researchers puts them rather quickly on the track to join the ranks of the social science hazards and disaster research workforce. If the program is continued by NSF, a more focused strategy should be developed to overcome constraints on the inclusion of promising minority scholars. At a minimum, collaboration with the few minority researchers in the field and minority-serving institutions should be part of the enabling strategy. Additionally, focused efforts should be made to boost the level of participation of junior women faculty in the program, increasing their numbers above those seen in the first two Enabling Projects.

> **Recommendation 9.7:** *Stakeholders in government, academia, professional societies, and the private sector should be open to exploring a variety of innovative approaches for developing the future social science hazards and disaster research workforce.*

NSF's Enabling Projects offer the lesson that alternative educational paradigms provide novel opportunities for developing the social science hazards and disaster research workforce. All learning in preparation for joining the field and shoring up existing skills does not take place in the classroom. Student chapters for aspiring practitioners in professional associations and continuing education activities come immediately to mind. Both of these tools have been used by the Earthquake Engineering Research Institute (EERI)—in the first instance, to recruit the next generation of earthquake engineers and, in the second, to increase the expertise of the existing generation and related professionals. Various types of internships in hazards-related government agencies and professional associations also warrant consideration as further means of providing valuable experience for future workforce members. Internships provide benefits to both the interns and their host organizations. Finally, social science professional organizations should give increased emphasis to workforce development at their workshops and conferences. By being "student friendly," such meetings can become valuable recruitment opportunities. More meeting organizers should follow the lead of the Natural Hazards Center and the Hazards Specialty Group, discussed above, both of which have workshop orientations for students and organize special sessions at which students can present their work and forge networks with each other as well as with senior professionals in the field.

## CONCLUSION

The above and potentially other workforce development strategies cannot replace more traditional approaches. Instead, they should be seen as parts of a more holistic strategy. The strategy must be geared to expanding the pools of core, periodic, and situational hazards and disaster researchers. The strategy must evolve from collaborative efforts by stakeholders in government, academia, professional associations, and the private sector. It must approach hazards and disasters inclusively rather than separately as societal risks. Sensitive to what is known about societal response to hazards and disasters, the strategy must emphasize the need for a larger, more skilled, and more diverse social science workforce to address what is not known. While disasters remain nonroutine events in societies or their larger subsystems, the actual and potential impacts of these events equate with their increasing prominence as public policy issues. Addressing these issues will require the best efforts of social science researchers and also their willingness to collaborate with each other and their counterparts in the natural sciences and engineering.

# References

Abkowitz, M., and E. Meyer. 1996. Technological advancements in hazardous materials evacuation planning. Transportation Research Record 1522:116-121.

Adams, B.J., C.K. Huyck, M.Z. Mio, S. Cho, S. Ghosh, S, H. Chung, R.T. Eguchi, B. Houshmand, M. Shinozuka, and B. Mansouri. 2004a. The Bam (Iran) Earthquake of December 26, 2003: Preliminary Reconnaissance Using Remotely Sensed Data and the VIEWS System; available at http://mceer.buffalo.edu/research/Bam/bam.pdf.

Adams, B.J., C.K. Huyck, M.Z. Mio, S. Cho, R.T. Eguchi, J.A. Womble, and K. Mehta, K. 2004b. Streamlining post-disaster data collection and damage assessment, using VIEWS (Visualizing Impacts of Earthquakes with Satellites) and VRS (Virtual Reconnaissance System). Proceedings of 2nd International Workshop on Remote Sensing for Post-Disaster Response, Newport Beach, CA, October 7-8.

Adams, B.J., M.Z. Mio, K. Mehta, J.A. Womble, and S. Ghosh. 2004c. Collection of Satellite Referenced Building Damage Information in the Aftermath of Hurricane Charley. MCEER Quick Response Report Series, Multidisciplinary Center for Earthquake Engineering Research: Buffalo.

Addison, T. 2000. Aid in Conflict. In Rap Finn (ed.) Foreign Aid and Development: Lessons Learnt and Direction for the Future. London: Routledge.

Adger, W.N., T.P. Hughes, C. Folke, S. Carpenter, and J. Rockstrom. 2005. Social-ecological resilience to coastal disasters. Science 309:1036-1039.

Aguirre, B.E. 2002. Can sustainable development sustain us? International Journal of Mass Emergencies and Disasters 20(2):111-126.

Aguirre, B.E. 2004. Homeland Security Warnings: Lessons Learned and Unlearned. University of Delaware Disaster Research Center Preliminary Paper 342.

Aguirre, B.E., W. Anderson, S. Balandran, B.E. Peters, and H.M. White. 1991. Saragosa, Texas, Tornado, May 22, 1987: An Evaluation of the Warning System. Washington, DC: National Academy Press.

Aguirre, B.E., D. Wenger, T.A. Glass, M. Diaz-Murillo, and G.Vigo. 1995. The social organization of search and rescue: Evidence from the Guadalajara gasoline explosion. International Journal of Mass Emergencies and Disasters 13:93-106.

Aguirre, B.E., D. Wenger, and G. Vigo. 1998. A test of the emergent norm theory of collective behavior. Sociological Forum 13:301-320.

Akaka, Daniel. 2001. FEMA's Project Impact. Congressional Record, March 1, S1742.

Albala-Bertrand, J.M. 1993. The Political Economy of Large Natural Disasters. Oxford: Clarendon Press.

Alesch, D.J., and J.N. Holly. 1998. Small business failure, survival, and recovery: Lessons from the January 1994 Northridge earthquake. NEHRP Conference and Workshop on Research on the Northridge, California Earthquake of January 17, 1994.

Alesch, D.J. and W.J. Petak. 1986. The Politics and Economics of Earthquake Hazard Mitigation: Unreinforced Masonry Buildings in Southern California. Boulder, CO: Natural Hazards Research and Applications Information Center, Institute of Behavioral Science, University of Colorado.

Alesch, D.J., C. Taylor, S. Ghanty, and R.A. Nagy. 1993. Earthquake risk reduction and small business. Pp. 133-160 in Committee on Socioeconomic Impacts (eds.) 1993 National Earthquake Conference Monograph 5: Socioeconomic Impacts. Memphis, TN: Central United States Earthquake Consortium.

Alesch, D.J., J.N. Holly, E. Mittler, and R. Nagy. 2001. Organizations at Risk: What Happens When Small Businesses and Not-for-Profits Encounter Natural Disasters. Fairfax, VA: Public Entity Risk Institute.

Alesch, D.J., P. May, R. Olshansky, W. Petak, and K. Tierney. 2004. Promoting Seismic Safety: Guidance for Advocates. Joint publication of the Mid-America Earthquake Center, the Multidisciplinary Center for Earthquake Engineering Research, and the Pacific Earthquake Engineering Research Center. Buffalo, NY: State University of New York at Buffalo, Multidisciplinary Center for Earthquake Engineering Research.

Alexander, D.A. 1993. Natural Disasters. New York: Chapman and Hall.

Alexander, D.A. 2000. Confronting Catastrophe. New York: Oxford University Press.

Anderson, J.E. 1994. Public Policy making: An Introduction. Boston, MA: Houghton Mifflin Company.

Anderson, W. 1998. A history of social science earthquake research: From Alaska to Kobe. In The EERI Golden Anniversary Volume 1948-1998. Oakland, CA, Earthquake Engineering Research Institute. Pp. 29-33.

Anderson, W. 2000. The Natural Hazards Research Community: Comments on the 25th Anniversary of the Annual Hazards Research and Applications Workshop, Natural Hazards Workshop, Boulder, CO, July.

Anderson, W. 2005. Bringing Children into Focus on the Social Science Disaster Research Agenda. International Journal of Mass Emergencies and Disasters 23:159-175.

Anderson, W., and S. Mattingly. 1991. Future directions. Pp. 311-335 in T. Drabek and G. Hoetmer (eds.) Emergency Management: Principles and Practice for Local Government. Washington, DC: International City Management Association.

Applegate, D. 2004. Personal communication, U.S. Geological Survey, Reston, VA, July.

Arlikatti, S., M.K. Lindell, and C.S. Prater. 2004. Perceived Stakeholder Role Relationships and Adoption of Seismic Hazard Adjustments. College Station, TX: Texas A&M University Hazard Reduction and Recovery Center.

Arlikatti, S., M.K. Lindell, C.S. Prater, and Y. Zhang. In press. Risk area accuracy and hurricane evacuation expectations of coastal residents. Environment and Behavior.

Arnold, C. 1998. Earthquake as Opportunity: Downtown Santa Cruz: The Tenth Year After the Earthquake. Palo Alto, CA: Building Systems Development Inc.

Arthur, W., Jr., B.D. Edwards, S.T. Bell, A.J. Villado, and W. Bennett, Jr. In press. Team task analysis: Identifying tasks and jobs that are team based. Human Factors.

Asada, J., T. Katada, D. Okajima, and S. Kobatake. 2001. A study of providing the information to make inhabitants understand the refuge information at the time of a flood. Annual Journal of Hydraulic Engineering, 45:37-42 (in Japanese).

Association of American Geographers Global Change in Local Places (GCLP) Research Group. 2003. Global Change and Local Places: Estimating, Understanding, and Reducing Greenhouse Gases. Cambridge, UK: Cambridge University Press.

ATLAS.ti Scientific Software Development GmbH. 2002. Available at http://www.atlasti.com/index.php (Accessed July 21, 2005).

Atwood, L.E. and A.M. Major. 1998. Exploring the "Cry Wolf" Hypothesis. International Journal of Mass Emergencies and Disasters 16:279-302.

Backer, Thomas E., Susan L. David, and Gerald Saucy. 1995. Reviewing the Behavioral Science Knowledge Base on Technology Transfer. NIDA Research Monograph 155. Rockville, MD: U.S. Department Of Health And Human Services, Public Health Service, National Institutes of Health National Institute on Drug Abuse.

Baker, E.J. 1991. Hurricane evacuation behavior. International Journal of Mass Emergencies and Disasters 9:287-310.

Bankoff, G. 2001. Rendering the world unsafe: "Vulnerability" as Western discourse. Disasters 25(1):19-35.

Bankoff, G. 2004. Time is of the essence: Disasters, vulnerability, and history. International Journal of Mass Emergencies and Disasters 22:23-42.

Barrett, B., B. Ran, and R. Pillai. 2000. Developing a dynamic traffic management modeling framework for hurricane evacuation. Transportation Research Record 1733:115-121.

Barrows, H.H. 1923. Geography as human ecology. Annals of the Association of American Geographers 13(1):1-14.

Barton, A.H. 1969. Communities in Disaster: A Sociological Analysis of Collective Stress Situations. Garden City, NY: Anchor, Doubleday.

Barton, A.H. 1989. Taxonomies of disaster and macrosocial theory. Pp. 346-351 in G.A. Kreps (ed.) Social Structure and Disaster. Newark, DE: University of Delaware and Associated University Presses.

Barton, A.H. 2005. Disaster and collective stress. Pp. 100-130 in E.L. Quarantelli and R.W. Perry (eds.) What Is a Disaster? New Answers to Old Questions. Philadelphia: Xlibris.

Bates, F.L. (ed.) 1982. Recovery, Change and Development: A Longitudinal Study of the 1976 Guatemalan Earthquake. Athens, GA: Department of Sociology, University of Georgia.

Bates, F.L., and W.G. Peacock. 1987. Disasters and social change. Pp. 291-330 in R.R. Dynes, B. DeMarchi, and C. Pelanda (eds.) Sociology of Disasters. Milan: Franco Angeli.

Bates, F.L., and W.G. Peacock. 1992. Measuring disaster impacts on household living conditions: The domestic assets approach. International Journal of Mass Emergencies and Disasters 10(1):133-160.

Bates, F.L. and W.G. Peacock. 1993. Living Conditions, Disasters, and Development: An Approach to Cross-Cultural Comparisons. Athens, GA: University of Georgia Press.

Bates, F.L., W. Timothy Farrell, and J.K. Glittenberg. 1979. Some changes in housing characteristics in Guatemala following the February 1976 earthquake and their implications for future earthquake vulnerability. International Journal of Mass Emergencies and Disasters 4:121-133.

Beady, C.H. Jr., and R.C. Bolin. 1986. The Role of the Black Media in Disaster Reporting to the Black Community. Natural Hazards Research Working Paper 56. Boulder: Natural Hazard Research and Applications Information Center, University of Colorado.

Been, V. and F. Gupta. 1997. Coming to the nuisan,ce of going to the barrios? A longitudinal analysis of environmental justice claims. Ecology Law Quarterly 24(1):1-56.

Belardo, S., H.L. Pazer, W.A. Wallace, and W.D. Danko. 1983. Simulation of a crisis management information network: A serendipitous evaluation. Decision Sciences 14 (4):588-606.

Berke, P.R. 1998. Reducing natural hazard risks through state growth management. Journal of the American Planning Association 64(1):76-88.

Berke, P.R., and T. Beatley. 1992. Planning for Earthquakes: Risk, Politics and Policy. Baltimore, MD: Johns Hopkins University Press.

Berke, P.R., and T. Beatley. 1997. After the Hurricane: Linking Recovery to Sustainable Development in the Caribbean. Baltimore, MD: Johns Hopkins University Press.

Berke, P.R., J. Kartez, and D. Wenger. 1993. Recovery after disaster: Achieving sustainable development, mitigation and equity. Disasters 17 (2):93-109.

Berke, P.R., N.E., J. Crawford, and J. Dixon. 2002. Planning and indigenous people: Human rights and environmental protection in New Zealand. Journal of Planning Education and Research 22 (2):115-134.

Berry, B.J.L. (ed.) 1977. The Social Burdens of Environmental Pollution: A Comparative Metropolitan Data Source. Cambridge, MA: Ballinger.

Berry, B.J.L., E.C. Conkling, and D.M. Ray. 1996. The Global Economy in Transition (2nd Edition). Englewood Cliffs, NJ: Prentice-Hall.

Beyer, J.M., and H.M. Trice. 1982. The utilization process: A conceptual framework and synthesis of empirical findings. Administrative Science Quarterly 27:591-622.

Bhaduri, B.L., E.A. Bright, P.R. Coleman, and J.E. Dobson. 2002. LandScan: Locating people is what matters, GeoInformatics, 5:34-37.

Bingham, R.D., B.W. Hawkins, J.P. Frendreis, and M.P. LeBlanc. 1981. Professional Associations and Municipal Innovation. Madison: University of Wisconsin Press.

Birkland, T.A. 1997. After Disaster: Agenda Setting, Public Policy, and Focusing Events. Washington, DC: Georgetown University Press.

Blaikie, P., T. Cannon, I. Davis, and B. Wisner. 1994. At Risk: Natural Hazards, People's Vulnerability, and Disasters. London: Routledge.

Bolin, R.C. 1976. Family recovery from natural disaster: A preliminary model. Mass Emergencies 1:267-277.

Bolin, R.C. 1982. Long-Term Family Recovery from Disaster. Monograph 36. Boulder, CO: Natural Hazards Research and Applications Information Center, Institute of Behavioral Science, University of Colorado.

Bolin, R.C. 1985. Disaster characteristics and psychosocial impacts. Pp. 3-28 in B.J. Sowder (ed.) Disasters and Mental Health: Selected Contemporary Perspectives. Rockville, MD: National Institute of Mental Health.

Bolin, R.C. 1993a. Household and Community Recovery After Earthquakes. Monograph 56. Boulder, CO: Natural Hazards Research and Applications Information Center, Institute of Behavioral Science, University of Colorado.

Bolin, R.C. 1993b. Post-earthquake shelter and housing: Research findings and policy implications. Pp. 107-131 in Committee on Socioeconomic Impacts (ed.) 1993 National Earthquake Conference Monograph 5: Socioeconomic Impacts. Memphis, TN: Central United States Earthquake Consortium.

Bolin, R., and P. Bolton. 1986. Race, Religion, and Ethnicity in Disaster Recovery. Monograph No. 42. Boulder, CO: University of Colorado, Institute for Behavioral Science, Natural Hazard Research and Applications Information Center, Environment and Behavior.

Bolin, R., and D.J. Klenow. 1988. Older people in disasters: A comparison of black and white victims. International Journal of Aging and Human Development 26(1):29-43.

Bolin, R., and L. Stanford. 1998. The Northridge Earthquake: Vulnerability and Disaster. New York: Routledge.

Bollens, S. 2000. On Narrow Ground: Urban Policy and Conflict in Jerusalem and Belfast. Albany, NY: State University of New York Press.

Bollens, S. 2002. Urban planning and intergroup conflict: Confronting a fractured public interest, Journal of the American Planning Association 68(1):79-91.

Bond, A., and L. Gasser (eds.) 1988. Readings in Distributed Artificial Intelligence. San Mateo, CA: Kaufmann.

Bosworth, S.L. and G.A. Kreps. 1986. Structure as process: Organization and role. American Sociological Review 51:699-716.

Bourque, L.B., K.I. Shoaf, and L.H. Nguyen. 2002. Pp. 157-194. Survey research. In R.A. Stallings (ed.) Methods of Disaster Research. Newark, Delaware: International Research Committee on Disasters.

Bowen, W.M. 2001. Environmental Justice Through Research-Based Decision making. New York: Garland Publishing.

Bowen, W.M. 2002. An analytic review of environmental justice research: What do we really know? Environmental Management 29:3-15.

Bowen, W.M., and M.V. Wells. 2002. The politics and reality of environmental justice: A history and consideration for public administrators and policy makers. Public Administration Review 62(6):688-698.

Bram, J., J. Orr, and C. Rapaport. 2002. Measuring the effects of the September 11 attack on New York City. FRBNY Economic Policy Review 8(2):5-20.

Breslau, N., R.C. Kessler, H.D. Chilcoat, L.R. Schultz, G.C. Davis, and P. Andreski. 1998. Trauma and posttraumatic stress disorder in the community: The 1996 Detroit area survey of trauma. Archives of General Psychiatry 55:626-632.

Brewer, C.A., and T.A. Suchan. 2001. Mapping Census 2000: The Geography of U.S. Diversity. Washington, DC: U.S. Department of Commerce, Census Bureau.

Briechle, K.J. 1999. Natural hazard mitigation and local government decision making. Pp. 3-9 in The Municipal Yearbook 1999. Washington, DC: International City/County Management Association.

Briggs, X. 1998. Brown kids in white suburbs: Housing mobility and the many faces of social capital, Housing Policy Debate 9(1):177-221.

Briggs, X. 2004. Social capital: Easy beauty or meaningful resource? Journal of the American Planning Association 70(2):151-158.

Britton, N. 2004. Higher education in emergency management: What is happening elsewhere. 7th Annual By-Invitation Emergency Management Higher Education Conference, Emmitsburg, MD, June.

Bruneau, M. 2004. Invited presentation to the committee, August 4.

Bruneau, M., S.E. Chang, R.T. Eguchi, G.C. Lee, T.D. O'Rourke, A.M. Reinhorn, M. Shinozuka, K.J. Tierney, W.A. Wallace, and D. von Winterfeldt. 2003. A framework to quantitatively assess and enhance the seismic resilience of communities. Earthquake Spectra 19(4):733-752.

Bruzewicz, A.J. 2003. Remote sensing imagery for emergency management. Pp. 87-97 in S.L. Cutter, D.B. Richardson, and T.J. Wilbanks (eds.) Geographical Dimensions of Terrorism. New York: Routledge.

Buckle, P. (ed.) 2004. Special issue: New perspectives on vulnerability. International Journal of Mass Emergencies and Disasters 22:5-109.

Buika, J. and L. Comfort. 2004. Building researcher and practitioner coalitions: Safeguarding our future against disasters. Natural Hazard Observer 29(2):3-4.

Bullard, R. 1990. Dumping in Dixie: Race, Class, and Environmental Quality. Boulder, CO: Westview Press.

Burby, R.J. 1994. Floodplain planning and management: Research needed for the 21st century. Water Resources Update (97):44-47.

Burby, R.J. (ed.) 1998. Cooperating with Nature: Confronting Natural Hazards with Land-use Planning for Sustainable Communities. Washington, DC: Joseph Henry Press.

Burby, R.J., P.J. May, P.R. Berke, L.C. Dalton, S.P. French, and E.J. Kaiser. 1997. Making Governments Plan: State Experiments in Managing Land-use. Baltimore, MD: Johns Hopkins University Press.

Burby, R.J., S.P. French, and A.C. Nelson. 1998. Plans, Code Enforcement, and Damage Reduction: Evidence from the Northridge earthquake. Earthquake Spectra 14:59-74.

Burby, R.J., T. Beatley, P.R. Berke, R.E. Deyle, S.P. French, D. Godschalk, E.J. Kaiser, J.D. Kartez, P.J. May, R.Olshansky, R.G. Paterson, and R.H. Platt. 1999. Unleashing the power of planning to create disaster-resistant communities. Journal of the American Planning Association 65(3):247-258.

Burton, I.R., R.W. Kates, and G.F. White. 1993. The Environment as Hazard (2nd Edition) New York: Oxford University Press.

Calthorpe, P., and W. Fulton. 2001. The Regional City. Washington, DC: Island Press.

Campbell, J.P., and N.R. Kuncel. 2002. Individual and team training. Pp. 278-312 in N. Anderson, D.S. Ones, H.K. Sinangil, and C. Viswesvaran (eds.) Handbook of Industrial, Work and Organizational Psychology. Thousand Oaks, CA: Sage.

Cannon, T. 2002. Gender and climate hazards in Bangladesh. Gender and Development 10(2):45-50.

Carley, K.M., and W.A. Wallace. 2001. Computational Organization Theory. In Encyclopedia of Operations Research and Management Science (2nd Edition) Norwell, MA: Kluwer Academic Publishers.

Caro-Coops. 2005. Carolinas Coastal Oceans Observing and Prediction System. Available at http://nautilus.baruch.sc.edu/carocoops_website/general_info.htm (Accessed May 9, 2005).

Carrara, A., and F. Guzzetti (eds.) 1995. Geographical Information Systems in Assessing Natural Hazards. Dordrecht: Kluwer Academic Publishers.

Cash, D.W., W.C. Clark, F. Alcock, N.M. Dickson, N. Eckley, D.H. Guston, J. Jager, and R.B. Mitchell. 2003. Knowledge systems for sustainable development. Proceedings of the National Academy of Sciences, 100(14):8086-8091.

Cederman, L.E. 2005. Computational models and social forms: Advancing generative process theory. American Journal of Sociology 110:864-894.

Chakraborty, J. 2001. Acute exposure to extremely hazardous substances: An analysis of environmental equity risk analysis 21(5):883-894.

Chakraborty, J., and M. Armstrong. 1997. Exploring the use of buffer analysis for the identification of impacted areas in environmental equity assessment. Cartography and Geographic Information Systems 24(3):145-157.

Chang, S.E. 2000. Disasters and transport systems: Loss, recovery and competition at the Port of Kobe after the 1995 earthquake. Journal of Transport Geography 8:53-65.

Chang, S.E. 2001. Structural change in urban economies: Recovery and long-term impacts in the 1995 Kobe earthquake, Kokumin Keizai Zasshi (Journal of Economics and Business Administration) 183(1):47-66.

Chang, S.E. 2003. Evaluating disaster mitigations: a methodology for infrastructure systems. Natural Hazards Review 4(4):186-196.

Chang, S.E. 2005. Modeling How Cities Recover from Disasters. Paper presented at the International Conference on Urban Disaster Reuction, Kobe, Japan, January 18-20.

Chang, S.E., and A. Falit-Baiamonte. 2002. Disaster vulnerability of businesses in the 2001 Nisqually earthquake. Environmental Hazards 4:59-71.

Chang, S.E., and M. Shinozuka. 2004. Measuring improvements in the disaster resilience of communities. Earthquake Spectra 20(3):739-755.

Chang, S.E., W.D. Svekla, and M.Shinozuka. 2002. Linking infrastructure and urban economy: Simulation of water disruption impacts in earthquakes. Environment and Planning B 29(2):281-301.

Charvériat, C. 2000. Natural Disasters in Latin America and the Caribbean: An Overview of risk. Working Paper 434. Washington, DC: Inter-American Development Bank.

Christopolis, I., J. Mitchell, and A. Liljelund. 2001. Re-framing risk: The changing context of disaster mitigation and preparedness. Disasters 25(3):185-198.

Clark, G.L., M.P. Feldman, and M.S. Gertler. 2000. The Oxford Handbook of Economic Geography. New York and Oxford: Oxford University Press.

Clark, W.C., and N.M. Dickson. 2003. Sustainability science: The emerging research program. Proceedings of the National Academy of Sciences 100(14):8059-8061.

Clark, W.C., J. Jager, R. Corell, R. Kasperson, J.J. McCarthy, D. Cash, S.J. Cohen, P. Desanker, N.M. Dickson, P. Epstein, D.H. Guston, J.M. Hall, C. Jaeger, A. Janetos, N. Leary, M.A. Levy, A. Luers, M. MacCracken, J. Melillo, R. Moss, J.M. Nigg, M.L. Parry, E.A. Parson, J.C. Ribot, D.P. Schrag, G.A. Seiselstad, E. Shea, C. Vogel, and T.J. Wilbanks. 2000. Assessing Vulnerability to Global Environmental Risks. Belfer Center for Science and International Affairs (BCSIA) Discussion Paper 2000-12. Cambridge, MA: Environment and Natural Resources Program, John F. Kennedy School of Government, Harvard University. Available at http://ksgnotes1.harvard.edu/BCSIA/sust.nsf/pubs/pub1.

Clarke, L. 1989. Acceptable Risk: Making Decisions in a Toxic Environment. Berkeley, CA: University of California Press.

Clarke, L. 2002. Panic: Myth or reality? Contexts (Fall):21-26.

Clarke, L. 2003. Conceptualizing responses to extreme events: The problem of panic and failing gracefully. Pp. 123-141 in L.B. Clarke (ed.) Terrorism and Disaster: New Threats, New Ideas. Research in Social Problems and Public Policy (11). Amsterdam: Elsevier.

Cochrane, H.C. 1974. Predicting the Economic Impact of Earthquakes. Working Paper 15. Boulder, CO: University of Colorado Institute of Behavioral Science.

Cochrane, H.C. 1975. Natural Hazards and Their Distributive Effects. Boulder, CO. Natural Hazards Research and Applications Information Center, Institute of Behavioral Science, University of Colorado.

Cohen, J. 1977. Statistical Power Analysis for the Behavioral Sciences (Revised Edition). New York: Academic Press.

Cohen, M.D. 1986. Artificial intelligence and the dynamic performance of organizational designs. In Ambiguity and Command: Organizational Perpsectives on Military Decision Making. J.G. March and R. Weissinger-Baylon (eds.) Marshfield, MA: Pitman.

Coleman, J. 1988. Social Capital in the Creation of Human Capital, American Journal of Sociology (Supplement) 94:S95-S120.

Comerio, M.C. 1997. Housing issues after disasters. Journal of Contingencies and Crisis Management 3:166-178.

Comerio, M.C. 1998. Disaster Hits Home: New Policy for Urban Housing Recovery. Berkeley, CA: University of California Press.

Comfort, L.K. 1999. Shared Risk: Complex Systems in Seismic Response. Amsterdam: Pergamon.

Congress of New Urbanism. 2004. New Urban Projects on a Neighborhood Scale in the United States. Ithaca, NY: New Urban News.

Congressional Research Service. 2006. FEMA Reorganization Legislation in the 109th Congress. Available at http://www.openers.com/document/RL33522.

Converse, P.E. 1964. The nature of belief systems in mass publics. Pp. 206-261 in D.E. Apter (ed.) Ideology and Discontent. New York: Free Press.

Cova, T.J., and R.L. Church. 1997. Modeling community evacuation vulnerability using GIS. International Journal of Geographical Information Science 11:763-784.

Cova, T.J., and J.P. Johnson. 2002. Microsimulation of neighborhood evacuations in the urban-wildland interface. Environment and Planning A 34(12):2211-2229.

Crapo, A., L.B. Waisel, W.A. Wallace and T.R. Willemain. 2000. Visualization and the process of modeling: A cognitive-theoretic view. Pp. 218-226 in Proceedings KDD-2000: The Sixth ACM SIGDD, International Conference on Knowledge Discovery and Data Mining, Boston, MA.

Cuny, Fred. 1983. Disasters and Development. London: Oxford University Press.

Cutter, S.L. 1993. Living with Risk. London: Edward Arnold.

Cutter, S.L. 1995. The forgotten casualties: Women, children, and environmental change. Global Environmental Change 5(3):181-194.

Cutter, S.L. 1996. Vulnerability to environmental hazards. Hazards Progress in Human Geography 20:529-539.

Cutter, S.L. (ed.) 2001. American Hazardscapes: The Regionalization of Hazards and Disasters. Washington, DC: Joseph Henry Press.

Cutter, S.L. 2003a. The science of vulnerability and the vulnerability of science. Annals of the Association of American Geographers 93(1):1-12.

Cutter, S.L. 2003b. GI science, disasters, and emergency management. Transactions in GIS 7(4):439-445.

Cutter, S.L., and C. Emrich, 2005. Are natural hazards and disaster losses in the U.S. increasing? EOS, Transactions, American Geophysical Union 86(41):381-389.

Cutter, S.L., and W.R. Renwick. 2004. Exploitation, Conservation Preservation: A Geographic Perspective on Natural Resource Use (4th Edition). Hoboken, NJ: John Wiley and Sons.

Cutter, S.L., D. Holm, and L. Clark. 1996. The role of geographic scale in monitoring environmental justice. Risk Analysis 16:517-526.

Cutter, S.L., J.T. Mitchell, and M.S. Scott. 2000. Revealing the vulnerability of people and places: A case study of Georgetown County, South Carolina. Annals of the Association of American Geographers 90(4):713-737.

Cutter, S.L., M.E. Hodgson, and K. Dow. 2001. Subsidized inequities: The spatial patterning of environmental risks and federally assisted housing. Urban Geography 22(1):29-53.

Cutter, S.L., B.J. Boruff, and W.L. Shirley. 2003. Social Vulnerability to Environmental Hazards. Sociological Quarterly 84:242-261.

Cvetkovich, G., and Earle, T. C. (1985). Classifying hazardous events. Journal of Environmental Psychology 5:5-35.

Dacy, D.C., and H. Kunreuther. 1969. The Economics of Natural Disasters. New York: Free Press.

Dahlhamer, J.M. 1998. Rebounding from Environmental Jolts: Organizational and Ecological Factors Affecting Business Disaster Recovery. Doctoral dissertation. Newark, DE: Department of Sociology and Criminal Justice, University of Delaware.

Dahlhamer, J.M., and M.J. D'Sousa. 1997. Determinants of business-disaster preparedness in two U.S. metropolitan areas. International Journal of Mass Emergencies and Disasters 15:265-281.

Dahlhamer, J.M., and L.M. Reshaur. 1996. Businesses and the 1994 Northridge earthquake: An analysis of pre- and post-disaster preparedness. Newark, DE: University of Delaware Disaster Research Center.

Daines, G.E. 1991. Planning, training, and exercising. Pp. 161-200 in T.S. Drabek and G.J. Hoetmer (eds.) Emergency Management: Principles and Practice for Local Government. Washington, DC: International City/County Management Association.

Dash, N., W.G. Peacock, and B.H. Morrow. 1997. And the poor get poorer: A neglected black community. Pp. 206-225 in W.G. Peacock, B.H. Morrow, and H. Gladwin (eds.) Hurricane Andrew: Ethnicity, Gender and the Sociology of Disasters. London: Routledge.

Davis, M. 1998. Ecology of Fear: Los Angeles and the Imagination of Disaster. New York: Metropolitan Books.

Davis, M., and S.T. Seitz. 1982. Disasters and governments. Journal of Conflict Resolution 26:547-568.

Department of the Interior. 2003. U.S. Geological Survey Circular 1242. A Plan to Coordinate NEHRP Post-earthquake Investigations. Reston, VA.

DeVoe, R.F., Jr. 1997. The natural disaster boom theory; Or window-breaking our way to prosperity. Pp. 181-188 in B.G. Jones (ed.) Economic Consequences of Earthquakes: Preparing for the Unexpected. Report No. NCEER-SP-0001. Buffalo, NY: State University of New York at Buffalo, Multidisciplinary Center for Earthquake Engineering Research.

Deyle, R.E., S.P. French, R.B. Olshansky, and R.G. Paterson. 1998. Pp. 119-166 in R.J. Burby (ed.) Cooperating with Nature: Confronting Natural Hazards with Land-Use Planning for Sustainable Communities. Washington, DC: Joseph Henry Press.

Dobson, S. 2005. E-mail correspondence with committee, April 12.

Dooley, D., R. Catalano, S. Mishra, and S. Serxner. 1992. Earthquake preparedness: Predictors in a community survey. Journal of Applied Social Psychology 22:451-470.

Dow, K., and S.L. Cutter. 1998. Crying wolf: Repeat responses to hurricane evacuation orders. Coastal Management 26:237-252.

Dow, K., and S.L. Cutter. 2000. Public orders and personal opinions: Household strategies for hurricane risk assessment. Environment 2:143-155.

Dow, K., and S.L. Cutter. 2002. Emerging hurricane evacuation issues: Hurricane Floyd and South Carolina. Natural Hazards Review 3:12-18.

Drabek, T.E. 1983. Shall we leave? A study on family reactions when disaster strikes. Emergency Management Review 1(Fall):25-29.

Drabek, T.E. 1985. Managing the emergency response. Public Administration Review 45(special issue):85-92.

Drabek, T.E. 1986. Human System Responses to Disaster: An Inventory of Sociological Findings. New York: Springer Verlag.

Drabek, T.E. 1987. The professional emergency manager. Boulder, CO: University of Colorado Institute of Behavioral Science.

Drabek, T.E. 1990. Emergency Management: Strategies for Maintaining Organizational Integrity. New York: Springer-Verlag.

Drabek, T.E. 1991a. Anticipating organizational evacuations: Disaster planning by managers of tourist-oriented private firms. International Journal of Mass Emergencies and Disasters 9:219-245.

Drabek, T.E. 1991b. Microcomputers and disaster responses. Disasters 15:186-192.

Drabek, T.E. 1994. Disaster Evacuation and the Tourist Industry. Boulder, CO: Natural Hazards Research and Applications Information Center, Institute of Behavioral Science, University of Colorado. Monograph #57.

Drabek, T.E. 1995. Disaster responses within the tourist industry. International Journal of Mass Emergencies and Disasters 13:7-23.

Drabek, T.E. 2003. Strategies for Coordinating Disaster Responses. Monograph 61. Boulder, CO: Natural Hazards Research and Applications Information Center, Institute of Behavioral Science, University of Colorado.

Drabek, T.E. 2004. Social Dimensions of Disaster (2nd Edition). Instructor Guide. Emmitsburg, MD: Federal Emergency Management Agency.

Drabek, T.E., and J.E. Haas. 1969. Laboratory simulation of organizational stress. American Sociological Review 34(2):223-238.

Drabek, T.E., and G.J. Hoetmer (eds.) 1991. Emergency Management: Principles and Practice for Local Government. Washington, DC: International City Management Association.

Drabek, T.E., and W.H. Key. 1976. The impact of disaster on primary group linkages. Mass Emergencies 1:89-105.

Drabek, T.E., and W.H. Key. 1984. Conquering Disaster: Family Recovery and Long-Term Consequences. New York: Irvington.

Drabek, T.E., and D.A. McEntire. 2002. Emergent phenomena and multiorganizational coordination in disasters: Lessons from the research literature. International Journal of Mass Emergencies and Disasters 20(2):197-224.

Drabek, T.E., W.H. Key, P.E. Erikson, and J.L. Crowe. 1975. The impact of disaster on kin relationships. Journal of Marriage and the Family 37:481-494.

Drabek, T.E., H.L. Tamminga, T.S. Kilijanek, and C.R. Adams. 1981. Managing Multiorganizational Emergency Responses. Boulder, CO: University of Colorado Institute of Behavioral Science.

Drabek, T.E., H.L. Tamminga, T.S. Kilijanek, and C.R. Adams. 1982. Managing Multiorganizational Emergency Responses: Emergent Search and Rescue Networks in Natural Disasters and Remote Area Settings. Boulder, CO: Natural Hazards Research and Applications Information Center, University of Colorado.

Duany, A., and E. Talen. 2002. Transect planning. Journal of the American Planning Association 68(3):245-266.

Dubin, R. 1978. Theory Building. New York: The Free Press.

Durkin, M. 1984. The economic recovery of small businesses after earthquakes: The Coalinga experience. International Conference on Natural Hazards Mitigation Research and Practice, New Delhi, October 6-8.

Dye, T.R. 1992. Understanding Public Policy (7th Edition). Englewood Cliffs, NJ: Prentice Hall.

Dynes, R.R. 1970. Organized Behavior in Disaster. Lexington, MA: Heath-Lexington Books.

Dynes, R.R. 1977. Interorganizational relations in communities under stress. Pp. 49-64 in E.L. Quarantelli (ed.) Disasters: Theory and Research. Beverly Hills, CA: Sage.

Dynes, R.R. 1993. Disaster reduction: The importance of adequate assumptions about social organization. Sociological Spectrum 6:24-25.

Dynes, R.R. 1994. Community emergency planning: False assumptions and inappropriate analogies. International Journal of Mass Emergencies and Disasters 12:141-158.

Dynes, R.R. 1998. Coming to terms with community disaster. Pp. 109-27 in E.L. Quarantelli (ed.) What Is a Disaster: Perspectives on the Question. London and New York: Routledge.

Dynes, R.R. 2002. The Importance of Social Capital in Disaster Research, Preliminary Paper 327. Newark, DE: University of Delaware, Disaster Research Center.

Dynes, R.R. 2004. Expanding the horizons of disaster research. Natural Hazards Observer 27:1-3.

Dynes, R.R., and T.E. Drabek. 1994. The structure of disaster research: Its policy and disciplinary implications. International Journal of Mass Emergencies and Disasters (3rd Edition). 12:5-23.

Dynes, R.R., and E.L. Quarantelli. 1968. What looting in civil disturbances really means. Transaction (May):9-14.

Dynes, R.R., E.L. Quarantelli, and G.A. Kreps. 1981. A Perspective on Disaster Planning (3rd Edition). Newark, DE: University of Delaware, Disaster Research Center.

Eagly, A.H., and S. Chaiken. 1993. The Psychology of Attitudes. Fort Worth, TX: Harcourt, Brace Jovanovich.

Earthquake Spectra. 1997. Theme Issue: Loss Estimation 13(4). Oakland, CA: EERI.

Edson, E. 2001. Bibliographic Essay: History of Cartography 38(11/12):1899-1909; available at http://www.maphistory.info/edson.html.

Edwards, M.L. 1993. Social location and self-protective behavior: Implications for earthquake preparedness. International Journal of Mass Emergencies and Disasters 11:293-304.

EERI (Earthquake Engineering Research Institute). 1995. Northridge Earthquake, January 17, 1994: Reconnaissance Report 1. Earthquake Spectrum, Supplement C to Vol. II. Oakland, CA: EERI.

EERI. 2003. Securing Society Against Catastrophic Earthquake Losses: A Research and Outreach Plan in Earthquake Engineering. Oakland, CA: EERI.

EERI. 2005. Membership Roster, 2004. Oakland, CA: EERI.

Eguchi, R.T., C.K. Huyck, and B.J. Adams. 2005. An urban damage scale based on satellite and airborne imagery. Presented at the First International Conference on Urban Disaster Reduction, Kobe, Japan, January 18-20.

Eguiluz, V.M., M.G. Zimmerman, C.J. Cela-Conde, and M. Miguez. 2005. Cooperation and the emergence of role differentiation in the dynamics of social networks. American Journal of Sociology 110:977-1009.

Ellis, D.G., and B.A. Fisher. 1994. Small group decision making: Communication and the group process. New York: McGraw-Hill.

Enarson, E., and B.H. Morrow. 1998. The Gendered Terrain of Disaster: Through Women's Eyes. Westport, CT: Praeger.

Enarson, E., L. Meyreles, M. González, B.H. Morrow, A. Mullings, and J. Soares. 2003. Working with Women at Risk: Practical Guidelines for Assessing Local Disaster Risk. International Hurricane Center. Miami, FL: Florida International University.

English, M.R. 2004. Environmental risk and justice. Pp. 119-159 in T. McDaniels and M.J. Small (eds.) Risk Analysis and Society: An Interdisciplinary Characterization of the Field, Cambridge, UK: Cambridge University Press.

Ericksen, N., P. Berke, J. Crawford, and J. Dixon. 2004. Plan-making for Sustainability: The New Zealand Experience. London: Ashgate Publishers.

Erikson, K.T. 1994. A New Species of Trouble: Explorations in Disaster, Trauma, and Community. New York: W.W. Norton and Co.

Estrin, D., W. Michener, G. Bonito, and workshop participants. 2003. Environmental Cyberinfrastructure Needs for Distributed Sensor Networks: A Report from a National Science Foundation Sponsored Workshop, August 12-14. La Jolla, CA: Scripps Institution of Oceanography.

Feinberg, W.E., and N.R. Johnson. 2001. The ties that bind: A macro-level approach to panic. International Journal of Mass Emergencies and Disasters 19(3):269-295.

FEMA (Federal Emergency Management Agency). 1996. Guide for All-Hazard Emergency Operations Planning. SLG-101. Washington, DC.

FEMA. 1997. Multi-hazard Identification and Risk Assessment: A Cornerstone of the National Mitigation Strategy. Washington, DC.

FEMA. 2003a. Expanding and Using KnowledgeTo Reduce Earthquake Losses: The National Earthquake Hazards Reduction Program Strategic Plan, 2001-2005. Washington, DC.

FEMA. 2003b. Exercise Design Course, IS 139. Emmitsburg, MD: Emergency Management Institute.

FEMA. 2004. HAZUS-MH. Available at www.fema.gov/hazus (Accessed December, 18 2004).

Fischer, H.W. III. 1998. Response to Disaster: Fact Versus Fiction and Its Perpetuation: The Sociology of Disaster. New York: University Press of America.

Fleiss, J.L. 1971. Measuring minimal scale agreement among many raters. Psychological Bulletin 76:378-382.

Flynn, J., P. Slovic, and H. Kunreuther (eds.) 2001. Risk, Media, and Stigma: Understanding Public Challenges to Modern Science. London: Earthscan.

Ford, J.K., and A.M. Schmidt. 2000. Emergency response training: Strategies for enhancing real-world performance. Journal of Hazardous Materials 75:195-215.

Fordham, M. 1999. The intersection of gender and social class in disasters: Balancing resilience and vulnerability. International Journal of Mass Emergencies and Disasters 17:15-36.

Forrest, T. 1993. A social reconstruction of time. Sociological Inquiry 63:444-457.

Fothergill, A. 1996. Gender, risk and disaster. International Journal of Mass Emergencies and Disasters 14(1):33-56.

Fothergill, A. 1998. The neglect of gender in disaster work: An overview of the literature. Pp. 11-25 in E. Enarson and B.H. Morrow (eds.) The Gendered Terrain of Disaster: Through Women's Eyes. Westport, CT: Praeger.

Fothergill, A. 2000. Knowledge transfer between researchers and practitioners. Natural Hazards Review 2:91-98.

Fothergill, A. 2003. The stigma of charity: Gender, class, and disaster assistance. Sociological Quarterly 44(4):659-680.

Fothergill, A. 2004. Heads Above Water: Gender, Class, and Family in the Grand Forks Flood. Albany, NY: State University of New York Press.

Fothergill, A., E. Maestas, and J.D. Darlington. 1999. Race, ethnicity and disasters in the U.S.: A review of the literature. Disasters 23(2):156-173.

Fradkin, P.L. 2005. The Great Earthquake and Firestorms of 1906: How San Francisco Nearly Destroyed Itself. Berkeley, CA: University of California Press.

Freedy, J., M. Saladin, D. Kilpatrick, H. Resnick, and B. Saunders. 1994. Understanding acute psychological distress following natural disaster. Journal of Traumatic Stress 5:441-454.

French, J.R.P., and B.H. Raven. 1959. The bases of social power. Pp. 150-167 in D. Cartwright (ed.) Studies in Social Power. Ann Arbor, MI: Institute for Social Research.

Freudenburg, W.R., and R. Gramling. 1994. Oil in Troubled Waters: Perceptions, Politics, and the Battle over Offshore Drilling. Albany, NY: State University of New York Press.

Frey, R. 1978. The Club of Mainz for improved worldwide emergency and critical care medicine systems and disaster preparedness. Crit. Care Med. 6(6):389-391.

Friesema. H.P., J. Caparano, G. Goldstein, R. Lineberry, and R. McCleary. 1979. Aftermath: Communities After Natural Disasters. Thousand Oaks, CA: Sage.

Fritz, C.E. 1961. Disasters. Pp. 651-694 in R.K. Merton and R. Nisbet (eds.) Social Problems. New York: Harcourt, Brace and World.

GAO (General Accounting Office). 1983. Siting of Hazardous Waste Landfills and Their Correlation with Racial and Economic Status of Surrounding Communities. Washington, DC: U.S. Government Printing Office.

GAO. 1992. Earthquake Recovery: Staffing and Other Improvements Made Following the Loma Prieta Earthquake. RCED-92-141. Washington, DC: U.S. Government Printing Office.

GAO. 1993a. Disaster Management: Improving the Nation's Response to Catastrophic Disasters, RCED-93-186. Washington, DC: U.S. Government Printing Office.

GAO. 1993b. Disaster Management: Recent Disasters Demonstrate the Need to Improve the Nation's Response Strategy. T-RCED-93-13. Washington, DC: U.S. Government Printing Office.

GAO. 2004. Homeland Security—Communication Protocols and Risk Communication Principles Can Assist in Refining the Advisory System. Washington, DC: U.S. Government Printing Office.

Gasser, L., and M.N. Huhns (eds.) 1989. Distributed Artificial Intelligence 2. New York: Morgan Kaufmann.

Gerrity, E.T., and B.W. Flynn. 1997. Mental health consequences of disasters. Pp. 101-121 in E.K. Noji (ed.) The Public Health Consequences of Disasters. New York: Oxford University Press.

Gilbert, N., and A. Abbot. 2005. Introduction (special issue on computational modeling and visualization) American Journal of Sociology 110:859-864.

Gillespie, D.F. 1991. Coordinating community resources. Pp. 55-78 in T.S. Drabek and G.J. Hoetmer (eds.) Emergency Management: Principles and Practice for Local Government. Washington, DC: International City/County Management Association.

Gillespie, D.F., and C.L. Streeter. 1987. Conceptualizing and measuring disaster preparedness. International Journal of Mass Emergencies and Disasters 5:155-176.

Gillespie, D.F., and R.A. Colignon. 1993. Structural change in disaster preparedness networks. International Journal of Mass Emergencies and Disasters 11:143-162.

Gillespie, D.F., R.A. Colignon, M.M. Banerjee, S.A. Murty, and M. Rogge. 1993. Partnerships for Community Preparedness. Boulder, CO: University of Colorado Natural Hazards Research and Applications Information Center.

Girard, C., and W.G. Peacock. 1997. Ethnicity and segregation: Post-hurricane relocation. Pp. 191-205 in W.G. Peacock, B.H. Morrow, and H. Gladwin (eds.) Hurricane Andrew: Ethnicity, Gender and the Sociology of Disasters. London: Routledge.

Gladwin, H., and W.G. Peacock. 1997. Warning and evacuation: A night for hard houses. Pp. 52-74 in W.G. Peacock, B.H. Morrow, and H. Gladwin (eds.) Hurricane Andrew: Ethnicity, Gender and the Sociology of Disasters. London and New York: Routledge.

Gladwin, C.H., H. Gladwin, and W.G. Peacock. 2001. Modeling hurricane evacuation decisions with ethnographic methods. International Journal of Mass Emergencies and Disasters 19 (2):117-143.

Glass, T.A., and M. Schoch-Spana. 2002. Bioterrorism and the people: How to vaccinate a city against panic. Clinical Infectious Diseases 34:217-223.

Godschalk, D.R. 2000. Smart growth around the nation. Popular Government 66(1):12-20.

Godschalk, D.R. 2003. Urban hazard mitigation: Creating resilient cities. Natural Hazards Review 4(3):136-143.

Godschalk, D.R., E.J. Kaiser, and P.R. Berke. 1998. Pp. 85-118 in R.J. Burby (ed.) Cooperating with Nature: Confronting Natural Hazards with Land-Use Planning for Sustainable Communities. Washington, DC: Joseph Henry Press.

Godschalk, D.R., T. Beatley, P. Berke, D.J. Brower and E.J. Kaiser. 1999. Natural Hazard Mitigation: Recasting Disaster Policy and Planning.Washington, DC: Island Press.

Goggin, M.L., A. O'M Bowman, J.P. Lester, and L.J. O'Toole. 1990. Implementation Theory and Practice: Toward a Third Generation. Glenview, IL: Scott Foresman.

Goldstein, I.L., and J.K. Ford. 2001. Training in Organizations: Needs Assessment, Development and Evaluation (4th edition). Belmont, CA: Wadsworth.

Goodchild, M.F. 2003. Geospatial data in emergencies. Pp. 27-34 in S.L. Cutter, D.B. Richardson, and T.J. Wilbanks (eds.) Geographical Dimensions of Terrorism. New York: Routledge.

Gordon, P., H.W. Richardson, B. Davis, C. Steins, and A. Vasishth. 1995. The Business Interruption Effects of the Northridge Earthquake. Los Angeles, CA: University of Southern California Lusk Center Research Institute.

Grabowski, M. and W.A. Wallace. 1993. An expert system for maritime pilots: Its design and assessment using gaming. Management Science 39(12):1506-1520.

Green, R., and I.A. Ahmed. 1999. Rehabilitation, sustainable peace, and development: Towards reconceptualisation, Third World Quarterly 20(1):189-206.

Greenberg, M.R. 1993. Proving environmental inequity in siting locally unwanted land-uses. Risk: Issues in Health and Safety 4 (summer):235-252.

Greene, R.W. 2002. Confronting Catastrophe: A GIS Handbook. Redlands, CA: ESRI Press.

Grossi, P., H. Kunreuther, and C. Patel. (eds.) 2004. Catastrophe Modeling: A New Approach to Managing Risk. New York: Springer Science+Business Media.

Gruntfest, E., K. Carsell, and T. Plush. 2002. An Evaluation of the Boulder Creek Local Flood Warning System. Colorado Springs, CO: Deparment of Geography and Environmental Studies, University of Colorado.

Gruntfest, E., and C. Huber. 1989. Status Report on Flood Warning Systems in the United States. Environmental Management 13(3):279-286.

Guy Carpenter and Company Inc. 2004. Southern California wildfires October/November 2003. CAT-i Catastrophe Information. Available at http://www.guycarp.com/portal/extranet/pdf/California_Wildfires.pdf;jsessionid=percent407e70ecpercent3a105358b4b8c?vid=1 (Accessed July 20, 2005).

Guzzo, R.A., and M.W. Dickson. 1996. Teams in organizations: Recent research on performance and effectiveness. Annual Review of Psychology 47:307-338.

Haas, J.E., R.W. Kates, and M.J. Bowden (eds.) 1977. Reconstruction Following Disaster. Cambridge, MA: MIT Press.

Hackett, E.J. 2000. Interdisciplinary research initiatives at the U.S. National Science Foundation. Pp. 248-259 in P. Weingart and N. Stehr (eds.) Practising Interdisciplinarity. Toronto: University of Toronto Press.

Hackman, J.R., and R. Wageman. 2005. A theory of team coaching. Academy of Management Review 30:269-287.

Haddow, G.D., and J.A. Bullock. 2003. Introduction to Emergency Management. New York: Butterworth-Heinemann.

Halford, E., and S. Nolan. 2002. Rogue volunteers: Response to the WTC attacks by volunteers who refused to leave. Paper presented at the International Research Committee on Disasters session, annual meeting of the American Sociological Association, Chicago, IL, August 16-19.

Hamilton, R. 2003. Milestones in Earthquake Research. Geotimes (March):5, 38.

Hammond, K. 2000. Judgments Under Stress. Oxford, UK: Oxford University Press.

Harrell-Bond, E. 1986. Imposing Aid. Oxford, UK: Oxford University Press.

Heinz Center. 2002. Human links to coastal disasters. Washington, DC: The H. John Heinz III Center for Science, Economics and the Environment.

Hepner, G.F., and M.V. Finco. 1995. Modeling dense gaseous contaminant pathways over complex terrain using a geographic information system. Journal of Hazardous Materials 42:187-199.

Hewings, G.J.D., and Y. Okuyama, 2003. Economic assessments of unexpected events. Pp. 153-160 in S.L. Cutter, D.B. Richardson, and T.J. Wilbanks (eds.) The Geographical Dimensions of Terrorism. New York and London: Routledge.

Hewitt, K., and I. Burton, 1971. The Hazardousness of a Place: A Regional Ecology of Damaging Events. Toronto: Department of Geography, University of Toronto.

Hiroyuki, K. 2003. Overview to the 6th EQTAP Multilateral Workshop," Kashikojima, Japan, December 1-2.

Hobeika, A.G., and C. Kim. 1998. Comparison of traffic assignments in evacuation modeling. IEEE Transactions on Engineering Management 45:192-198.

Hodgson, M.E., and S.L. Cutter. 2001. Mapping and the spatial analysis of Hazardscapes. Pp. in S.L. Cutter (ed.) American Hazardscapes: The Regionalization of Hazards and Disasters. Washington, DC: Joseph Henry Press.

Hollenbeck, J.R., D.R. Ilgen, D.J. LePine, J.A. Colquitt, and J. Hedlund. 1998. Extending the multilevel theory of team decision making: Effects of feedback and experience in hierarchical teams. Academy of Management Journal 41:269-282.

Holling, C.S. 1973. Resilience and stability of ecological systems. Annual Review of Ecology and Systematics 4:1-23.

Holzer, T., R.D. Borchert, C.D. Comartin, R.D. Hanson, C.R. Scawthorn, K.J. Tierney, and T.L. Youd. 2003. The Plan to Coordinate NEHRP Post-Earthquake Investigations. Washington, DC: U.S. Geological Survey.

Homberger, W.S., J.W. Hall, R.C. Loutzenheiser, and W.R. Reilly. 1996. Fundamentals of Traffic Engineering (14th Edition). Berkeley, CA: University of California Institute of Transportation Studies.

Hoover, C.A., and M.R. Greene. 1996. Construction Quality, Education, and Seismic Safety. Oakland, CA: EERI.

Hoover, G.A., and F.L. Bates. 1985. The impact of a natural disaster on the division of labor in twelve Guatemalan communities: A study of social change in a developing country. International Journal of Mass Emergencies and Disasters 3:7-26.

Houts, P.S., P.D. Cleary, and T.W. Hu. 1988. The Three Mile Island Crisis: Psychological, Social, and Economic Impacts on the Surrounding Population. University Park, PA: The Pennsylvania State University Press.

Hughes, J.W., and M.K. Nelson. 2002. The New York region's post-September 11 economic geography. Transportation Quarterly 56(4):27-42.

Huizer, G. 1997. Participatory Action Research as a Method of Rural Development. In SD Dimensions. Food and Agricultural Organization of the United Nations.

Hwang, H.H.M., H. Lin, and M. Shinozuka. 1998. Seismic Performance Assessment of Water Delivery Systems. Journal of Infrastructure Systems 4:118-125.

Hwang, S.N., W.G. Sanderson, and M.K. Lindell. 2001. Analysis of state emergency management agencies' hazard analysis information on the Internet. International Journal of Mass Emergencies and Disasters 19:85-106.

Intergovernmental Panel on Climate Change (IPCC). 2001. Climate Change 2001: Synthesis Report, Third Assessment Report of the Intergovernmental Panel on Climate Change. Cambridge, UK: Cambridge University Press.

International Red Cross Red Crescent Society (IFRC). 2000. Strategy 2010. Available at www.ifrc.org/publicat/s2010/ (Accessed October 27, 2004).

International Strategy for Disaster Reduction (ISDR). 2004. Living with Risk: A Global Review of Disaster Reduction Initiatives. Volumes I and II. Geneva: United Nations Press.

Inter-University Consortium for Political and Social Research (ICPSR). 2005. Guide to Social Science Data Preparation and Archiving: Best Practice throughout the Data Life Cycle. Ann Arbor, MI.

Jackson, E.L. 1981. Response to earthquake hazard: The West Coast of North America. Environment and Behavior 13:387-416.

Jackson, S.E., K.E. May, and K. Whitney. 1995. Understanding the dynamics of diversity in decision making teams. Pp. in 204-261 in R.A. Guzzo, E. Salas, and Associates (eds.) Team Effectiveness and Decision Making in Organizations. San Francisco, CA: Josey-Bass.

Jensen, J.R. 2000. Remote Sensing of the Environment. Upper Saddle River, NJ: Prentice-Hall.

Johnson, N.R. 1987. Panic at the "Who Concert Stampede": An Empirical Assessment. Social Problems 34:362-373.

Johnson, N.R., W.F. Feinberg, and D.M. Johnston, 1994. MicroStructure and panic: The impact of social bonds on individual action and collective flight from the Beverly Hills Supper Club fire. Pp. 168-189 in R.R. Dynes and K.J. Tierney (eds.) Disasters, Collective Behavior, and Social Organization. Newark, DE: University of Delaware Press.

Jones, B.G., and S.E. Chang. 1995. Economic aspects of urban vulnerability and disaster mitigation. Pp. 311-320 in F.Y. Cheng and M.-S. Sheu (eds.) Urban Disaster Mitigation: The Role of Engineering and Technology. Oxford, UK: Elsevier Science Ltd.

Kahneman, D., P. Slovic, and A. Tversky. 1982. Judgment Under Uncertainty: Heuristics and Biases. New York: Cambridge University Press.

Kaiser Family Foundation. 2002. Sicker and Poorer: The Consequences of Being Uninsured; available at:http://www.kff.org/uninsured/loader.cfm?url=/commonspot/security/getfile.cfmandPageID=13970.

Kang, J.E., M.K. Lindell, and C.S. Prater. 2004. Hurricane evacuation expectations and actual behavior in Hurricane Lili. College Station, TX: Texas A&M University Hazard Reduction and Recovery Center.

Kasperson, J.X., and R.E. Kasperson. 2001. International Workshop on Vulnerability and Global Environmental Change. Stockholm: Stockholm Environment Institute.

Kasperson, R.E., O. Renn, P. Slovic, H.S. Brown, J. Emel, R. Goble, J.X. Kasperson, and S. Ratick. 1988. The social amplification of risk. Risk Analysis 8(2):177-187.

Kasperson, R.E., J. Kasperson, and K. Dow. 2001. Vulnerability, equity, and global environmental change. J. Kasperson and R. Kasperson (eds.) Global Environmental Risk. New York: United Nations University Press.

Katada, T., and M. Kodama. 2001. Study of inhabitant's anxiety and action in a state of Mt. Tokachi volcanic eruption. Journal of Infrastructure Planning and Management 18:239-244 (in Japanese).

Kates, R.W., and D. Pijawka. 1977. From rubble to monument: The pace of reconstruction. Pp. 1-24 in J. Haas, R. Kates, and M. Bowden (eds.) Reconstruction Following Disaster. Cambridge, MA: MIT Press.

Kates, R.W., and T.M. Parris. 2003. Long-term trends and a sustainable transition. Proceedings of the National Academy of Sciences 100(14):8062-8067.

Kates, R.W., C. Hohenemser, and J.X. Kasperson, 1985. Perilous Progress: Managing the Hazards of Technology. Boulder, CO: Westview Press.

Kates, R.W., W.C. Clark, R. Corell, J.M. Hall, C.C. Jaeger, O. Lowe, J. McCarthy, H.J. Schellnhuber, B. Bolin, N.M. Dickson, S. Faucheux, G.C. Gallopin. A. Grubler, B. Huntley, J. Jager, N.S. Jodha, R.E. Kasperson, A. Mabogunje, P. Matson, H. Mooney, B. Moore III, T. O'Riordan, and U. Svedin. 2001. Sustainability science. Science 292:641-642.

Kendra, J.M. and T. Wachtendorf. 2002. Rebel food . . . Renegade supplies: Convergence after the World Trade Center attack. Paper presented at the annual meeting of the American Sociological Association, Chicago, IL, August 16-19.

Kessler, R.C., and S. Zhao. 1999. Overview of descriptive epidemiology of mental disorders. Pp. 127-150 in C.S. Anashensel and J.C. Phelan (eds.) Handbook of the Sociology of Mental Health. New York: Kluwer Academic Publishers.

King, D. 2001. Uses and limitations of socioeconomic indicators of community vulnerability to natural hazards: Data and disasters in Northern Australia. Natural Hazards 24: 147-156.

Kingdon, J.W. 1984. Agendas, Alternatives and Public Policies. Boston, MA: Little, Brown.

Kingdon, J.W. 1995. Agendas, Alternatives and Public Policies. (2nd Edition) New York: Harper Collins.

Klein, J.T. 1990. Interdisciplinarity: History, Theory, and Practice. Detroit, MI: Wayne State University Press.

Kleindorfer, P.R., and H.C. Kunreuther. 1987. Insuring and Managing Hazardous Risks: From Seveso to Bhopal and Beyond. Berlin: Springer-Verlag.

Klinenberg, E. 2002. Heat Wave: A Social Autopsy of Disaster in Chicago. Chicago and London: University of Chicago Press.

Korten, D. 1980. Community organization and rural development: A learning process approach. Public Administration Review September/October.

Korten, D. 1984. People centered development: Toward a framework. In D. Korten and R. Klauss (eds.) People Centered Development: Contributions Toward Theory and Planning Frameworks. West Hartford, CT: Kumarian Press.

Kraiger, K. 2003. Perspectives on training and development. Pp.171-192 in W.C. Borman, D.R. Ilgen, R.J. Klimoski, and I.B. Weiner (eds.) Handbook of Psychology. Hoboken, NJ: John Wiley.

Kreps, G.A. 1978. The organization of disaster response: Some fundamental theoretical issues. Pp. 65-86 in E.L. Quarantelli (ed.) Disasters: Theory and Research. Beverly Hills, CA: Sage.

Kreps, G.A. 1984. Sociological inquiry and disaster research. Annual Review of Sociology 10:309-330.

Kreps, G.A. 1985. Disaster and the social order. Sociological Theory 3:49-64.

Kreps, G.A. 1989a. Future directions in disaster research: The role of taxonomy. International Journal of Mass Emergencies and Disasters 7:215-241.

Kreps, G.A. (ed.) 1989b. Social Structure and Disaster. Newark, DE: University of Delaware Press.

Kreps, G.A. 1990. The Federal Emergency Management System in the United States: Past and Present. International Journal of Mass Emergencies and Disasters 8(3):275-300.

Kreps, G.A. 1991a. Organizing for emergency management. Pp. 30-54 in T.S. Drabek and G.J. Hoetmer (eds.) Emergency Management: Principles and Practice for Local Government. Washington, DC: International City/County Management Association.

Kreps, G.A. 1991b. Answering organizational questions: A brief for structural codes. Pp. 143-177 in G. Miller (ed.) Studies in Organizational Sociology. Greenwich, CT: JAI Press.

Kreps, G.A. 1994. Disaster archives and structural analysis: Uses and limitations. Pp. 45-71 in R.R. Dynes and K.J. Tierney (eds.) Disaster, Collective Behavior, and Social Organization. Newark, DE: University of Delaware and Associated University Presses.

Kreps, G.A. 1998. Disasters as systemic events and social catalysts. Pp. 31-56 in E.L. Quarantelli (ed.) What Is a Disaster: Perspectives on the Question. London and New York: Routledge.

Kreps, G.A. 2001. Disaster, Sociology of. Pp. 3718-3721 in N.J. Smelser and Paul B. Bates (eds.) International Encyclopedia of the Social and Behavioral Sciences. Amsterdam: Elsevier Publishing Company.

Kreps, G.A., and S.L. Bosworth. 1993. Disaster, organizing, and role enactment: A structural approach. American Journal of Sociology 99:428-463.

Kreps, G.A., and S.L. Bosworth. Forthcoming. Organizational adaptation to disaster. In H. Rodriguez, E.L. Quarantelli, and R.R. Dynes (eds.) Handbook of Disaster Research. New York: Springer.

Kreps, G.A., and T.E. Drabek. 1996. Disasters are nonroutine social problems. International Journal of Mass Emergencies and Disasters 14:129-153.

Krishna, A. 2002. Active Social Capital. New York: Columbia University Press.

Krishna, A., and E. Schrader. 1999. The social capital tool assessment. Paper presented at the World Bank Conference on Social Capital and Poverty Reduction, Washington, DC. Available at http://pverty.worldbank.org/library/view/8151/ (Accessed September 12, 2004).

Kroll, C.A., J.D. Landis, Q. Shen, and S. Stryker. 1991. Economic Impacts of the Loma Prieta Earthquake: A Focus on Small Businesses. Working Paper 91-187, Berkeley, CA: U.C. Transportation Center and the Center for Real Estate and Urban Economics, University of California at Berkeley.

Kroll-Smith, J.S., and S.R. Crouch, 1991. What is a disaster? An ecological-symbolic approach to resolving the definitional debate. International Journal of Mass Emergencies and Disasters 9:355-366.

Kruvant, W. 1974. Incidence of Pollution Where People Live in Washington. Washington, DC: Center for Metropolitan Studies.

Kunreuther, H., and A. Rose (ed.) 2004. The Economics of Natural Hazards, Vols. I and II. The International Library of Critical Writings in Economics 178. Cheltenham, UK: Elgar.

Kunreuther, H., and R.R. Roth (eds.) 1998. Paying the Price: The Status and Role of Insurance Against Natural Disasters in the United States.Washington, DC: Joseph Henry Press.

Kunreuther, H., and P. Slovic. 1978. Economics, psychology, and protective behavior. American Economic Review 68(2):64-69.

Kunreuther, H., and P. Slovic. 1996. The process of risk management: science, values, and risk. The Annals of the American Academy of Political and Social Science 545:116-125.

Kunreuther, H., R. Ginsberg, L. Miller, P. Sagi, P. Slovic, B. Borkan, and N. Katz. 1978. Disaster Insurance Protection. New York: John Wiley and Sons.

Kunreuther, H., A. Onculer, and P. Slovic. 1998. Time insensitivity for protective investments. Journal of Risk and Uncertainty 16:279-299.

La Porte, T., and P. Consolini. 1998. Theoretical and operational challenges of "high-reliability organizations: Air traffic control and aircraft carriers. International Journal of Public Administration (Part 2):847-853.

LA RED. 2002. DesInventar Data Bases. Available at www.desinventar.org/en/usuarios/suramerica/co.html (Accessed September 23, 2004).

Lambright, W.H. 1984. The Role of States in Earthquake and Natural Hazard Innovation at the Local Level. Syracuse, NY: Science and Technology Policy Center, Syracuse Research Corporation.

Larson, L., and D. Plascencia. 2001. No adverse impact: A new direction in floodplain management policy. Natural Hazards Review 2(4):167-181.

Laska, S. 2004. What if Hurricane Ivan had not missed New Orleans? Natural Hazards Observer 29(2).

Lazarus, R.S., and S. Folkman. 1984. Stress, Appraisal, and Coping. New York: Springer.

Leaning, J. 1984. The Counterfeit Ark: Crisis Relocation for Nuclear War. Cambridge, MA: Ballinger.

Lecomte, E., and K. Gahagen, 1998. Hurricane insurance protection in Florida. Pp. 97-124 in H. Kunreuther and Richard J. Roth, Sr. (eds.) Paying the Price: The Status and Role of Insurance Against Natural Disasters in the United States. Washington, DC: Joseph Henry Press.

Lester, J.P., D.W. Allen, and K.M. Hill, 2001. Environmental Injustice in the United States: Myths and Realities. Boulder, CO: Westview Press.

Lew, H.S., Richard W. Bukowski, and Nicholas J. Carino. 2005. Design, Construction, and Maintenance of Structural and Life Safety Systems (draft). Gaithersburg, MD: National Institute of Standards and Technology.

Lindell, M.K. 1994. Perceived characteristics of environmental hazards. International Journal of Mass Emergencies and Disasters 12:303-326.

Lindell, M.K., and C.J. Brandt. 2000. Climate quality and climate consensus as mediators of the relationship between organizational antecedents and outcomes. Journal of Applied Psychology 85:331-348.

Lindell, M.K., and M.J. Meier. 1994. Effectiveness of community planning for toxic chemical emergencies. Journal of the American Planning Association 60:222-234.

Lindell, M.K., and R.W. Perry. 1987. Warning mechanisms in emergency response systems. International Journal of Mass Emergencies and Disasters 5:137-153.

Lindell, M.K., and R.W. Perry. 1990. Effects of the Chernobyl accident on public perceptions of nuclear plant accident risks. Risk Analysis 10:393-399.

Lindell, M.K., and R.W. Perry. 1992. Behavioral Foundations of Community Emergency Management. Washington, DC: Hemisphere Publishing Corp.

Lindell, M.K., and R.W. Perry. 1996. Identifying and managing conjoint threats: Earthquake-induced hazardous materials releases in the U.S. Journal of Hazardous Materials 50:31-46.

Lindell, M.K., and R.W. Perry. 1998. Earthquake impacts and hazard adjustment by acutely hazardous materials facilities following the Northridge earthquake. Earthquake Spectra 14:285-299.

Lindell, M.K., and R.W. Perry. 2000. Household adjustment to earthquake hazard: A review of research. Environment and Behavior 32:590-630.

Lindell, M.K., and R.W. Perry. 2001. Community innovation in hazardous materials management: Progress in implementing SARA Title III in the United States. Journal of Hazardous Materials 88:169-194.

Lindell, M.K., and R.W. Perry. 2004. Communicating Environmental Risk in Multiethnic Communities. Thousand Oaks, CA: Sage Publications.

Lindell, M.K., and C.S. Prater. 2000. Household adoption of seismic hazard adjustments: A comparison of residents in two states. International Journal of Mass Emergencies and Disasters 18:317-338.

Lindell, M.K., and C.S. Prater. 2002. Risk area residents' perceptions and adoption of seismic hazard adjustments. Journal of Applied Social Psychology 32:2377-2392.

Lindell, M.K., and C.S. Prater. 2003. Assessing community impacts of natural disasters. Hazards Review 4:176-185.

Lindell, M.K., and D.J. Whitney. 1995. Effects of organizational environment, internal structure and team climate on the effectiveness of local emergency planning committees. Risk Analysis 15:439-447.

Lindell, M.K., and D.J. Whitney. 2000. Correlates of seismic hazard adjustment adoption. Risk Analysis 20:13-25.

Lindell, M.K., C.S. Prater, R.W. Perry, and J.Y. Wu. 2002. EMBLEM: An Empirically-Based Large Scale Evacuation Time Estimate Model. College Station, TX: Texas A&M University Hazard Reduction and Recovery Center.

Lindell, M.K., D.J. Whitney, C.J. Futch, and C.S. Clause. 1996. The local emergency planning committee: A better way to coordinate disaster planning. In R.T. Silves and W.L. Waugh, Jr. (eds.) Disaster Management in the U.S. and Canada: The Politics, Policy making, Administration and Analysis of Emergency Management. Springfield, IL: Charles C. Thomas Publishers.

Lindell, M.K., D. Alesch, P.A. Bolton, M.R. Greene, L.A. Larson, R. Lopes, P.J. May, J-P. Mulilis, S. Nathe, J.M. Nigg, R. Palm, P. Pate, R.W. Perry, J. Pine, S.K. Tubbesing, and D.J. Whitney. 1997. Adoption and implementation of hazard adjustments. International Journal of Mass Emergencies and Disasters Special Issue 15:327-453.

Lindell, M.K., S. Naik, S. Veluswami, and C. Agrawal. 2004. EMDSS: An evacuation management Decision Support System for Hurricane Emergencies. College Station, TX: Texas A&M University Hazard Reduction and Recovery Center.

Litynski, D.M., M.R. Grabowski, and W.A. Wallace. 1997. The relationship between three dimensional imaging and group decision making: An exploratory study. IEEE Transactions on Systems, Man and Cybernetics 27(4): 402-411.

Liu, F. 2001. Environmental Justice Analysis: Theories, Methods, and Practice. New York: Lewis Publishers.

Los Angeles Almanac. 2004. Available at www.losangelesalmanac.com/topics/population/po47a.htm.

Lougeay, R., P. Baumann, and M.D. Nellis, 1994. Two digital approaches for calculating the area of regions affected by the Great American Flood of 1993. Geocarto International 9:53-59.

Lu, J.C., M.K. Lindell, and C.S. Prater, C.S. In press. Household evacuation decision making in response to Hurricane Lili. Natural Hazards Review.

Margai, F.L. 2001. Health risks and environmental inequity: A geographical analysis of accidental releases of hazardous materials. Professional Geographer 53(3):422-434.

Marra, M.A., S.J.M. Jones, C.R. Astell, et al. 2003. The Genome Sequence of the SARS-Associated Coronavirus. Available online at: http://www.sciencemag.org/feature/data/sars/.

May, P.J. 1985. Recovery from Catastrophes: Federal Disaster Relief Policy and Politics. Westport, CT: Greenwood Press.

May, P.J. 1994. Analyzing mandate design: State mandates governing hazard-prone areas. Publius: The Journal of Federalism 24:1-16.

May, P.J. 2005. E-mail correspondence with committee, April 12.

May, P.J., and R.E. Deyle. 1998. Governing land-use in hazardous areas with a patchwork system. Pp. 57-82 in R.J. Burby (ed.) Cooperating with Nature: Confronting Natural Hazards with Land-Use Planning for Sustainable Communities. Washington, DC: Joseph Henry Press.

May, P.J., and W. Williams. 1986. Disaster Policy Implementation: Managing Programs Under Shared Governance. New York: Plenum.

May, P.J., R. Burby, N. Ericksen, J. Handmer, S. Michaels, and D.I. Smith. 1996. Environmental Management and Governance to Hazards and Sustainability. London: Routledge.

Mazmanian, D., and P. Sabatier. 1989. Implementation and public policy. Lanham, MD: University Press of America.

McFarland, D. 2002. Cell phone ownership grows 29 percent from 1999-2001. Scarborough Research, March.

McIntyre, R.M., and E. Salas. 1995. Measuring and managing for team performance: Lessons from complex environments. Pp. 9-45 in R.A. Guzzo, E. Salas and Associates (eds.) Team effectiveness and decision making in organizations. San Francisco, CA: Josey-Bass.

McLuckie, B. 1974. Warning—A call to action: Warning and disaster response, a sociological background. National Weather Service Southern Region.

McMaster, R. B., H. Leitner, and E. Sheppard. 1997. GIS-based environmental equity and risk assessment: Methodological problems and prospects. Cartography and Geographic Information Systems 24:172-189.

Medvedev, Z.A. 1990. The legacy of Chernobyl. New York: Norton.

Mejia-Navarro, M., E.E. Wohl, and S.D. Oaks. 1994. Geological hazards, vulnerability, and risk assessment using GIS: Model of Glenwood Springs, Colorado. Geomorphology 10:31-35.

Mendonca, D., and W.A. Wallace. 2002. Development of a decision logic to support improvisation: An application to emergency response. Hawaii International Conference on System Sciences HICSS-35:220b-221.

Mendonca, D., and W.A. Wallace. 2004. Studying organizationally situated improvisations in response to extreme events. International Journal of Mass Emergencies and Disasters 22:5-31.

Mennis, J. 2002. Using geographic information systems to create and analyze statistical surfaces of population and risk for environmental justice analysis. Social Science Quarterly 83:281-297.

Meo, M., B. Ziebro, and A. Patton. 2004. Tulsa turnaround: From disaster to sustainability. Natural Hazard Review 5(1):1-9.

Messner, S., E. Baumer, and R. Rosenfeld. 2004. Dimensions of social capital and rates of criminal homicide. American Sociological Review 69(6):882-903.

Meszaros, J. and M.K. Fiegener. 2002. Effects of the 2001 Nisqually earthquake on small businesses in Washington State. Draft report for the Economic Development Administration. Seattle, WA: University of Washington.

Mileti, D.S. 1987. Sociological methods and disaster research. Pp. 57-71 in R.R. Dynes and K.J. Tierney (eds.) Disaster, Collective Behavior, and Social Organization. Newark, DE: University of Delaware and Associated University Presses.

Mileti, D.S. 1999a. Design for Future Disasters: A Sustainable Approach for Hazards Research and Application in the United States. Washington, DC: Joseph Henry Press.

Mileti, D.S. 1999b. Disasters by Design: A Reassessment of Natural Hazards in the United States. Washington, DC: Joseph Henry Press.

Mileti, D.S., and J.D. Darlington. 1997. The role of searching in shaping reactions to earthquake risk information. Social Problems 44:89-103.

Mileti, D.S., and C. Fitzpatrick. 1993. The Great Earthquake Experiment: Risk Communication and Public Action. Boulder, CO: Westview Press.

Mileti, D.S., and P. O'Brien. 1992. Warnings during disaster: Normalizing communicated risk. Social Problems 39:40-57.

Mileti, D.S., and L. Peek. 2000. The social psychology of public response to warnings of a nuclear power plant accident. Journal of Hazardous Materials 75:181-194.

Mileti, D.S., and J.H. Sorensen. 1987. Why people take precautions against natural disasters. Pp. 296-320 in N. Weinstein (ed.) Taking Care: Why People Take Precautions. New York: Cambridge University Press.

Mileti, D.S., T.S. Drabek, and J.E. Haas. 1975. Human systems in extreme environments. Boulder, CO: University of Colorado Natural Hazards Research and Applications Information Center.

Mileti, D.S., J.H. Sorensen, and P.W. O'Brien. 1992. Toward an explanation of mass care shelter use in evacuations. International Journal of Mass Emergencies and Disasters 10:25-42.

Mileti D.S., D.M. Cress, and J.D. Darlington. 2002. Earthquake culture and corporate action. Sociological Forum 17:161-180.

Miller, A.M., and M. Heldring. 2004. Mental health and primary care in a time of terrorism: Psychological impact of terrorist attack. Families, Systems and Health 22(1):7-30.

Miranda, M.L., D.C. Dolinoy, and M.A. Overstreet. 2002. Mapping for prevention: GIS models for directing childhood lead poisoning prevention programs. Environmental Heath Perspectives 110(9):947-953.

Mitchell, J. 1989. Hazards Research. Pp. 410-424 in G. Gaile and C. Willmott (eds.) Geography in America. Columbus, OH: Merrill Publishing.

Mitchell, J. 1992. Natural hazards and sustainable development. Paper presented at the Natural Hazards Research and Applications Workshop, Boulder, CO.

Mitchell, J. (ed.) 1999. Crucibles of Hazard: Mega-Cities and Disasters in Transition. New York: United Nations Press.

Mitchell, J. 2004. Personal communication; Bloomsburg University, June.

Mitchell, J., M.S. Scott, D.S.K. Thomas, M. Cutler, P.D. Putnam, R.F. Collins, and S.L. Cutter. 1997. Mitigating against disaster: Assessing hazard vulnerability at the local level. Pp. 563-571 in GIS/LIS '97 Proceedings, Bethesda, MD.

Mitchell, J., D.S.K. Thomas, and S.L. Cutter. 1999. Dumping in Dixie revisited: The evolution of environmental injustices in South Carolina. Social Science Quarterly 80(2):229-243.

Monmonier, M. 1997. Cartographies of Danger: Mapping Hazards in America. Chicago, IL: University of Chicago Press.

Montz, B.E., J. Cross, and S.L. Cutter. 2003a. Hazards. Pp. 479-491 in G. Gaile and C. Willmott (eds.) Geography in America at the Dawn of the 21st Century. Oxford, UK: Oxford University Press.

Montz, B.E., J.A. Cross, and S.L. Cutter. 2003b. Hazards. Pp. 481-91 in G. Gaile and C. Willmott (eds.) Geography in America II. Oxford, UK: Oxford University Press.

Morgan, B.B., Jr. and C.A. Bowers. 1995. Teamwork stress: Implications for team decision making. Pp. 262-290 in R.A. Guzzo, E. Salas and associates (eds.) Team Effectiveness and Decision Making in Organizations. San Francisco, CA: Josey-Bass.

Morrow, B.H. 1997. Stretching the bonds: The families of Andrew. Pp. 141-170 in W.G. Peacock, B.H. Morrow, and H. Gladwin (eds.) Hurricane Andrew: ethnicity, gender and the Sociology of Disasters. London, UK: Routledge.

Morrow, B.H. 1999. Identifying and mapping community vulnerability. Disasters 23:1-18.

Morrow, B.H., and W.G. Peacock. 1997. Disasters and social change: Hurricane Andrew and the reshaping of Miami? Pp. 226-242 in W.G. Peacock, B.H. Morrow, and H. Gladwin (eds.) Hurricane Andrew: Ethnicity, Gender and the Sociology of Disaster. London: Routledge.

Mulilis, J.P., and T.S. Duval. 1995. Negative threat appeals and earthquake preparedness: A person-relative-to-event PrE model of coping with threat. Journal of Applied Social Psychology 25:1319-1339.

Mulilis, J.P., and R.A. Lippa. 1990. Behavioral change in earthquake preparedness due to negative threat appeals: A test of protection motivation theory. Journal of Applied Social Psychology 20:619-638.

Multihazards Mitigation Council. 2005. Natural Hazard Mitigation Saves: An Independent Study to Assess the Future Savings from Mitigation Activities. Washington, DC: National Institute of Building Sciences.

Munich Re. 2002. Topics, Annual Review, Natural Catastrophes. Available at www.munichre.com/default_e.asp (Accessed September 6, 2004).

Murakami, H. 2000. Underground: The Tokyo Gas Attack and the Japanese Psyche. London: The Harvill Press.

Mustafa, D. 2002. To each according to his power: Participation, access, and vulnerability in irrigation and flood management in Pakistan. Environment and Planning D: Society and Space 20:737-752.

NAPA (National Academy of Public Administration). 1993. Coping with Catastrophe: Building an Emergency Management System to Meet People's Needs in Natural and Man-Made Disaster. Washington, DC: NAPA.

NASA. 2005. Remote Sensing in History. NASA Observatorium. Available at http://observe.arc.nasa.gov/nasa/exhibits/history/history_1.html (Accessed May 9, 2005).

Nathe, S. 2000. Public education for earthquake hazards. Natural Hazards Review(4):191-196.

National Commission on Terrorist Attacks upon the United States. 2004. The 9/11 Commission Report. New York: Norton.

National Response Team. 1990. Developing a Hazardous Materials Exercise Program: A Handbook for State and Local Officials. NRT-2. Washington, DC.

National Science Board. 2001. Toward a More Effective Role for the U.S. Government in International Science and Engineering. Arlington, VA: National Science Foundation.

National Science Board. 2004. Science and Engineering Indicators 2004. Arlington, VA: National Science Foundation.

Neal, D.M. 1990. Volunteer organization responses to the earthquake. Pp. 91-98 in R. Bolin (ed.) The Loma Prieta Earthquake: Studies of Short-Term Impacts. Monograph 50. Boulder, CO: Natural Hazards Research and Applications Information Center, Institute of Behavioral Science, University of Colorado.

Neal, D.M. 1997. Reconsidering the phases of disasters. International Journal of Mass Emergencies and Disasters 15(2):239-264.

NEHRP (National Earthquake Hazard Reduction Program). 1992. Indirect Economic Consequences of a Catastrophic Earthquake. Washington, DC: FEMA, July.

Nelson, A.C., and J.B. Duncan. 1995. Growth management principles and practices. Chicago: American Planning Association Planners Press.

Ngo, E.B. 2001. When disasters and age collide: Reviewing vulnerability of the elderly. Natural Hazards Review 2(2):80-89.

NHRAIC (Natural Hazards Research and Applications Information Center). 2003. Beyond September 11th: An account of post-disaster research. Boulder, CO: University of Colorado.

NHRAIC. 2003. Beyond September 11: An Account of Post Disaster Research. Special Publication 39. Boulder, CO: NHRIA Institute of Behavioral Science, University of Colorado.

NIBS (National Institute of Building Sciences). 1998. HAZUS. Washington, DC.

NIBS-FEMA. 1999. HAZUS®99 Earthquake Loss Estimation Methodology, Service Release 1 (SR1). Technical Manual, Developed by the Federal Emergency Management Agency through agreements with the National Institute of Building Sciences, Washington, DC.

Nigg, J.M. 1995. Anticipated Business Disruption Effects Due to Earthquake-Induced Lifeline Interruptions. Pp. 46-58 in W. Yayong and F. Chen (eds.) Proceedings of the Sino-US Symposium on Post-Earthquake Rehabilitation and Reconstruction. Beijing: Ministry of Construction.

Nigg, J.M. 1998. Emergency response following the 1994 Northridge earthquake: Intergovernmental coordination issues. Pp. 245-251 in Proceedings of the NEHRP Conference and Workshop on Research on the Northridge, California Earthquake of January 17, 1994, Vol. IV. Richmond, CA: California Universities for Research in Earthquake Engineering.

NIMS (National Incident Management System). 2004. NIMS and the Incident Command System (ICS). Available at: http://www.nimsonline.com/nims_ics_position_paper.htm (Accessed July 25, 2005).

NIST (National Institute of Standards and Technology). 2005 Reports of the Federal Building and Fire Safety Investigation of the World Trade Center Disaster. Gaithersburg, MD: NIST.

NOAA (National Oceanic and Atmospheric Administration). 1972. A Study of Earthquake Losses in the San Francisco Bay Area: Data and Analysis. Prepared for the Office of Emergency Preparedness by the NOAA, Washington, DC.

NOAA. 1973. A Study of Earthquake Losses in the Los Angeles, California Area, prepared for the Federal Disaster Assistance Administration, Department of Housing and Urban Development by the NOAA, Washington, DC.

Noji, E.K. (ed.) 1997. The Public Health Consequences of Disasters. New York: Oxford University Press.

Noon, J.M. 2001. Revisiting key issues about collective behavior, organizing, and role enactment. Sociological Spectrum 21:479-506.

Norris, F.H., M.J. Friedman, and P.J. Watson. 2002a. 60,000 disaster victims speak: Part II. Summary and implications of disaster mental health research. Psychiatry 65(3):240-260.

Norris, F.H., M.J. Friedman, P.J. Watson, C.M. Byrne, E. Diaz, and K. Kaniasty. 2002b. 60,000 disaster victims speak. Part I. An empirical review of the empirical literature, 1981-2001. Psychiatry 65(3):207-239.

North, C.S., S.J. Nixon, S. Shariat, S. Mallonee, J.C. McMillen, E.L. Spitznagel, and E.M. Smith. 1999. Psychiatric disorders among survivors of the Oklahoma City bombing. Journal of the American Medical Association 282(8):755-762.

NRC (National Research Council). 1970. The Great Alaska Earthquake of 1964. Washington, DC: National Academy of Sciences.

NRC. 1975. Earthquake Prediction and Public Policy. Washington, DC: National Academy Press.

NRC. 1989. Estimating Losses from Future Earthquakes. Washington, DC: National Academy Press.

NRC. 1992. The Economic Consequences of a Catastrophic Earthquake: Proceedings of a Forum, August 1 and 2, 1990. Washington, DC: National Academy Press.

NRC. 1999a. Human Dimensions of Global Environmental Change: Research Pathways for the Next Decade. Washington, D.C.: National Academy Press.

NRC. 1999b. Our Common Journey: A Transition Toward Sustainability. Washington, DC: National Academy Press.

NRC. 1999c. The Impacts of Natural Disasters: A Framework for Loss Estimation. Washington, DC: National Academy Press.

NRC, 1999d. Toward Environmental Justice: Research, Education, and Health Policy Needs. Washington, DC: National Academy Press.

NRC. 2001. Grand Challenges in Environmental Sciences. Washington, DC: -National Academy Press.

NRC. 2002a. Countering Terrorism: Lessons Learned from Natural and Technological Disasters. Washington, DC: The National Academies Press.

NRC. 2002b. Making the Nation Safer: The Role of Science and Technology in Countering Terrorism. Washington, DC: The National Academies Press.

NRC. 2002c. The National Plant Genome Initiative: Objectives for 2003-2008. Washington, DC: The National Academies Press.

NRC. 2003a. Large Scale Biomedical Science: Exploring Strategies for Future Research. Washington, DC: The National Academies Press.

NRC. 2003b. Preventing Earthquake Disasters: The Grand Challenge in Earthquake Engineering. Washington, DC: The National Academies Press.

NRC. 2003c. Protecting Participants and Facilitating Social and Behavioral Sciences Research. Constance F. Citro, Daniel R. Ilgen, and Cora B. Marrett (eds.) Washington, DC: The National Academies Press.

NRC. 2004. Reducing Future Flood Losses: The Role of Human Actions—Summary of a Workshop. Washington, DC: The National Academies Press.

NRC. 2005a. Decision Making for the Environment: Social and Behavioral Science Research Priorities. Washington, DC: The National Academies Press.

NRC. 2005b. Facilitating Interdisciplinary Research. Washington, DC: The National Academies Press.

NSF (National Science Foundation). 2002. Integrated Research in Risk Analysis and Decision Making in a Democratic Society. Workshop report, July 17-18.

NSF and Department of Interior. 1976. Earthquake Prediction and Hazard Mitigation: Options for USGS and NSF Programs. Washington, DC: U.S. Government Printing Office.

Oak Ridge National Laboratory. 2003. Oak Ridge Evacuation Modeling System: OREMS Version 2.60. Oak Ridge, TN.

Oakes, J.M., D.L. Anderton, and A.B. Anderson. 1996. A longitudinal analysis of environmental equity in communities with hazardous waste facilities. Social Science Research 25(2):125-148.

O'Brien, P., and D.S. Mileti. 1992. Citizen participation in emergency response following the Loma Prieta earthquake. International Journal of Mass Emergencies and Disasters 10:71-89.

Ofori-Atta, K., E. Roseman, B. Saha, S. Stuart, M. Lipschultz and J. Smidt. 2004. Profiting from transmission investment: A holistic, new approach to cost/benefit analysis. Public Utilities Fortnightly, (October):72-77.

Okuyama, Y. and S.E. Chang (eds.) 2004. Modeling Spatial and Economic Impacts of Disasters. Berlin: Springer.

Oliver-Smith, A. 1990. Post-disaster housing reconstruction and social inequality: A challenge to policy and practice. Disasters 14.

Oliver-Smith, A. 2001. Displacement, Resistance and the Critique of Development: From the Grass Roots to the Global. Final report prepared for ESCOR R7644 and the Research Programme on Development Induced Displacement and Resettlement, Refugee Study Centre, University of Oxford, Oxford, UK. Available at www.qeh.ac.uk/resp/ (Accessed April 2, 2005).

Oliver-Smith, A., and R. Goldman. 1988. Planning goals and urban realities: Post-disaster Reconstruction in a Third World city. City and Society 2(2):67-79.

Olshansky, R.S., and J. Kartez. 1998. Managing land-use to build resilience. Pp. 167-202 in R.J. Burby (ed.) Cooperating with nature: Confronting Natural Hazards with Land-Use Planning for Sustainable Communities.Washington, DC: Joseph Henry Press.

Olson, R.S. 2000. Toward a politics of disaster: Losses, values, agendas, and blame. International Journal of Mass Emergencies and Disasters 18(2):265-287.

Olson, R.S., and A.C. Drury. 1997. Un-therapeutic communities: A cross-national analysis of post-disaster political unrest. International Journal of Mass Emergencies and Disasters 15:221-238.

Olson, R.S., and V.T. Gawronski. 2003. Disasters as critical junctures? Managua, Nicaragua 1972 and Mexico City 1985. International Journal of Mass Emergencies and Disasters 21(1):5-35.

Olson, R.S., and R.A. Olson. 1993. The rubble's standing up in Oroville, California: The politics of building safety. International Journal of Mass Emergencies and Disasters 11(2):163-188.

Olson, R.S., and R.A. Olson. 1994. Trapped in politics: The life, death, and afterlife of the Utah Seismic Safety Council. International Journal of Mass Emergencies and Disasters 12:77-94.

Olson, R.S., R.A. Olson, and V.T. Gawronski. 1998. Night and day: Mitigation policy making in Oakland, California before and after the Loma Prieta disaster. International Journal of Mass Emergencies and Disasters 16(2):145-179.

Ostrom, E., T. Dietz, N. Dolsak, P.C. Stren, S. Stoninsh, and E.U. Weber (eds.) 2001. The Drama of the Commons. Washington, DC: The National Academies Press.

Palm, R.I. 1998. Demand for disaster insurance: Residential coverage. Pp. 51-66 in H. Kunreuther and Richard J. Roth, Sr. (eds.) Paying the Price: The Status and Role of Insurance Against Natural Disasters in the United States. Washington, DC: Joseph Henry Press.

Palm, R.I., and M.E. Hodgson. 1992. After a California Earthquake: Attitude and Behavior Change. Chicago: University of Chicago Press.

Palm, R.I., M.E. Hodgson, R.D. Blanchard, and D. Lyons. 1990. Earthquake insurance in California. Boulder, CO: Westview Press.

Park, J., P. Gordon, J. Moore, and H. Richardson. 2005. Simulating the state-by-state effects of terrorist attacks on three major U.S. ports: Applying NIEMO. Paper presented at the Second Annual DHS CREATE Center Symposium, University of Southern California, Los Angeles.

Parris, T.M., and R.W. Kates. 2003. Characterizing a sustainability transition: Goals, targets, trends, and driving forces. Proceedings of the National Academy of Sciences 100(14):8068-8073.

Peacock, W.G., and C. Girard. 1997. Ethnic and racial inequalities in hurricane damage and insurance settlements. Pp. 171-190 in W.G. Peacock, B.H. Morrow, and H. Gladwin (eds.) Hurricane Andrew: Ethnicity, Gender and the Sociology of Disasters. London: Routledge.

Peacock, W.G., C.D. Killian, and F.L. Bates. 1987. The effect of disaster damage and housing aid on household recovery following the 1976 Guatemalan earthquake. International Journal of Mass Emergencies and Disasters 5:63-88.

Peacock, W.G., B.H. Morrow, and H. Gladwin. 1997. Hurricane Andrew: Ethnicity, Gender, and the Sociology of Disaster. New York: Routledge.

Peacock, W., B.H. Morrow, and H. Gladwin. 2000. Hurricane Andrew and the Reshaping of Miami: Ethnicity, Gender, and the Socio-political Ecology of Disasters. Miami: Florida International University, International Hurricane Center.

Peek-Asa, C., M. Ramirez, and K.I. Shoaf. 2001. Population-based case-control study of earthquake-related deaths and hospitalized admissions sustained during the 1994 Northridge, CA Earthquake. Proceedings 1st Workshop for Comparative Study on Urban Earthquake Disaster Management. Kobe, Japan, January 18-19.

Peek-Asa, C., M. Ramirez, H.A. Seligson, and K.I. Shoaf. 2003. Seismic, structural, and individual factors associated with earthquake-related injury. Injury Prevention 9:62-66.

Pelling, Mark. 2003. The Vulnerability of Cities: Natural Disasters and Social Resilience. London: Earthscan.

Pellmar, T.C., and L. Eisenberg, (eds.) 2000. Bridging Disciplines in the Brain, Behavioral, and Clinical Sciences. Washington, DC: National Academy Press.

Perilla, J.L., F.H. Norris, and E.A. Lavizzo. 2002. Ethnicity, culture, and disaster response: Identifying and explaining ethnic differences in PTSD six months after Hurricane Andrew. Journal of Social and Clinical Psychology 21(1):20-45.

Perrow, C. 1984. Normal Accidents: Living with High Risk Technologies. New York: Basic Books.

Perry, R.W. 2005. Disasters, definitions, and theory building. Pp. 282-296 in R.W. Perry and E.L. Quarantelli (eds.) What Is a Disaster? New Answers to Old Questions. Philadelphia: Xlibris.

Perry, R.W., and M.K. Lindell. 1978. The psychological consequences of natural disaster: A review of research on American communities. Mass Emergencies 3:105-115.

Perry, R.W., and M.K. Lindell. 1990. Living with Mt. St. Helens: Human Adjustment to Volcano Hazards. Pullman, WA: Washington State University Press.

Perry, R.W., and M.K. Lindell. 1991. The Effects of ethnicity on evacuation decision making. International Journal of Mass Emergencies and Disasters 9:47-68.

Perry, R.W., and M.K. Lindell. 1997. Earthquake planning for governmental continuity. Environmental Management 21:89-96.

Perry, R.W., M.R. Greene, and M.K. Lindell. 1980. Enhancing evacuation warning compliance: Suggestions for emergency planning Disasters 4(4):433-449.

Perry, R.W., and E.L. Quarantelli (eds.) 2005. What Is a Disaster? New Answers to Old Questions. Philadelphia: XLibris.

Petak, W.J., and A.A. Atkisson. 1982. Natural Hazard Risk Assessment and Public Policy: Anticipating the Unexpected. New York: Springer Verlag.

Petonito, L. 2004. University Programs, Science and Technology Division, Department of Homeland Security. Invited presentation to the committee, August 4.

Pfirman, S., and the AC-ERE. 2003. Complex Environmental Systems, Synthesis for Earth, Life, and Society in the 21st Century. A report summarizing a 10-year outlook in environmental research and education for the National Science Foundation. Washington, DC: NSF.

Phillips, B.D. 1993. Cultural diversity in disasters: Sheltering, housing, and long term recovery. International Journal of Mass Emergencies and Disasters 11:99-110.

Phillips, B.D. 1998. Sheltering and housing of low-income and minority groups in Santa Cruz County after the Loma Prieta earthquake. Pp. 17-28 in J.M. Nigg (ed.) The Loma Prieta, California, Earthquake of October 17, 1989—Recovery, Mitigation, and Reconstruction. Washington, DC: U.S. Geological Survey.

Phillips, B.D. 2004. Personal communication with Brenda Phillips, Oklahoma State University, August.

Pine, J.C., B.D. Marx, and A. Lakshmanan. 2002. An examination of accidental-release scenarios from chemical-processing sites: The relation of race to distance. Social Science Quarterly 83(1):317-331.

Platt, R.H. 1994. Evolution of coastal hazards policies in the United States. Coastal Management 22:265-284.

Platt, R.H. 1998. Planning and land-use adjustments in historical perspective. Pp. 29-56 in R.J. Burby (ed.) Cooperating with Nature: Confronting Natural Hazards with Land-Use Planning for Sustainable Communities. Washington, DC: Joseph Henry Press.

Platt, R.H. 1999. Disasters and Democracy. Washington, DC: Island Press.

Platt, R.H. 2004. Land-Use and Society: Geography, Law and Public Policy (Revised Edition). Washington, DC: Island Press.

Popkins, R., and C. Rubin. 2000. Practitioners' views of the Natural Hazards Center, Natural Hazards Review 1(4):212-221.

Posner, R. 2004. Catastrophe: Risk and Response. New York: Oxford University Press.

Prater, C.S. 2001. The short history of Project Impact: A preliminary evaluation. Presented at the APEC Workshop on Dissemination of Disaster Mitigation Technologies, Taipei, Taiwan, June 18-21.

Prater, C.S., and M.K. Lindell. 2000. The politics of hazard mitigation. Natural Hazards Review 1:73-82.

Prater, C.S., M.K. Lindell, W.G. Peacock, Y. Zhang, and J.C. Lu. 2004. A Social Vulnerability Approach to Estimating Potential Socioeconomic Impacts of Earthquakes. College Station, TX: Texas A&M University Hazard Reduction and Recovery Center.

Prince, S. 1920. Catastrophe and Social Change. New York: Longmans, Green and Co.

Putman, R. 1995. Bowling alone: America's declining social capital. Journal of Democracy 6:65-78.

Putnam, R. 1993. Making Democracies Work: Civic Traditions in Modern Italy. Princeton, NJ: Princeton University Press.

Qualrus Idea Works Inc. 2003. Available at http://www.ideaworks.com/Qualrus.shtml (Accessed July 21, 2005).

Quarantelli, E.L. 1978. Some basic themes in sociological studies of disasters. Pp. 1-14 in E.L. Quarantelli (ed.) Disasters: Theory and Research. Beverly Hills, CA: Sage.

Quarantelli, E.L. 1982. Sheltering and housing after major community disasters: Case studies ad general conclusions. Newark, DE: University of Delaware Disaster Research Center.

Quarantelli, E.L. 1987. Disaster studies: An analysis of the social historical factors affecting the development of research in the area. International Journal of Mass Emergencies and Disasters 5:285-310.

Quarantelli, E.L. 1989. Conceptualizing disasters from a sociological perspective. International Journal of Mass Emergencies and Disasters 7:243-251.

Quarantelli, E.L. 1994. Disaster studies: The consequences of the historical use of a sociological approach in the development of research. International Journal of Mass Emergencies and Disasters 12:25-49.

Quarantelli, E.L. (ed.) 1998. What Is a Disaster? Perspectives on the Question. New York: Routledge.

Quarantelli, E.L. 1999. International Sociological Association Research Committee on the Sociology of Disasters RC 39. Available at http://www.ucm.es/info/isa/rc39orig.htm (Accessed September 9, 2004).

Quarantelli, E.L. 2002. The Disaster Research Center (DRC) field studies of organized behavior in the crisis time period of disasters. Pp. 94-126 in R.A. Stallings (ed.) Methods of Disaster Research. Philadelphia: Xlibris.

Quarantelli, E.L. 2004. Personal communication.

Quarantelli, E.L. 2006. Looting and Other Criminal Behavior in Hurricane Katrina: Atypical and Complex but Seen Before in Other Catastrophes. Newark, DE: University of Delaware Disaster Research Center.

Quarantelli, E.L., and R.R. Dynes. 1970. Property norms and looting: Their patterns in community crises. Phylon 31:168-182.

Quarantelli, E.L., and R.R. Dynes. 1972. When disaster strikes (it isn't much like what you've heard and read about). Psychology Today 5:66-70.

Quarantelli, E.L., and R.R. Dynes. 1977. Response to social crisis and disaster. Annual Review of Sociology 3:23-49.

Radke, J., T. Cova, M.F. Sheridan, A. Troy, L. Mu, and R. Johnson. 2000. Application challenges for geographic information science: Implications for research, education, and policy for emergency preparedness and response. URISA Journal 12(2):15-30.

Ramsey, E.W., M.E. Hodgson, S.K. Sapkota, S.C. Laine, G.A. Nelso, and D.K. Chappell. 2001. Forest impact estimated with NOAA AVHRR and Landsat TM data related to a predicted hurricane windfield distribution. International Journal of Remote Sensing 77:279-292.

Rashed, T., and J. Weeks. 2003. Assessing vulnerability to earthquake hazards through spatial multicriteria analysis of urban areas. International Journal of Geographical Information Science 17:547-576.

Raven, B.H. 1993. The bases of power: Origins and recent developments. Journal of Social Issues 49:227-251.

Raven, P.H. 2002. Science, sustainability and the human prospect. AAAS Presidential Address. Science 297(5583):954-958.

Reddy, S. 2000. Factors influencing the incorporation of hazard mitigation during recovery from disaster. Natural Hazards 22:185-201.

Rhodes, E.L. 2003. Environmental Justice in America: A New Paradigm. Bloomington, IN: Indiana University Press.

Rhoten, D. 2004. Interdisciplinary research: Trend or transition. Items and Issues 5(1/2): 6-11. Available at http://www.ssrc.org/publications/items/items_5.12/interdisciplinary_research.pdf.

Ripley, R., and G. Franklin. 1984. Congress, Bureaucracy, and Public Policy (2nd Edition). Homewood, IL: Dorsey Press.

Roberts, K.H. 1989. New challenges in organizational research: High reliability organizations. Industrial Crisis Quarterly 3:111-125.

Rocha, J.L., and I. Christopolis. 2001. Disaster mitigation and preparedness on the Nicaraguan post-Mitch agenda. Disasters 25(3):24-250.

Rochefort, D.A. and R.W. Cobb. 1994. The Politics of Problem Definition: Shaping the Policy Agenda. Lawrence, KS: University of Kansas Press.

Rodriguez, H., J. Trainor, and E.L. Quarantelli. Forthcoming. Rising to the challenge of a catastrophe: The emergent and pro-social behavior following Hurricane Katrina. Annals of the American Academy of Political and Social Science.

Rogers, G., and J. Sorensen. 1988. Diffusion of emergency warnings. Environmental Professional 10:281-294.

Rohe, W. 2004. Building social capital through community development. Journal of the American Planning Association 70(2):158-164.

Rose, A. 2004. Defining and measuring economic resilience to disasters. Disaster Prevention and Management 13(4):307-314.

Rose, A., and S. Liao. 2005. Modeling regional economic resilience to disasters: A computable general equilibrium analysis of water service disruptions. Journal of Regional Science 45(1)75-112.

Rose, A., J. Benavides, S.E. Chang, P. Szczesniak, and D. Lim. 1997. The regional economic impact of an earthquake: Direct and indirect effects of electricity lifeline disruptions. Journal of Regional Science 37(3):437-458.

Rose, A., G. Oladosu, and D. Salvino. 2004. Regional Economic Impacts of Electricity Outages in Los Angeles: A Computable General Equilibrium Analysis. Pp. 179-210 in Crew, M. and M. Spiegel (eds.) Obtaining the Best from Regulation and Competition. Dordrecht: Kluwer.

Rosenfeld, M.J. 1997. Celebration, politics, selective looting and riots: A micro level study of the Bulls Riot of 1992 in Chicago. Social Problems 44(4):483-502.

Rossi, P.H., J.D. Wright, S.R. Wright, and E. Weber-Burdin. 1978. Are there long term effects of American natural disaster? Estimation of floods, hurricanes, and tornadoes occurring 1960 to 1970 on U.S. Census tracts in 1970. Mass Emergencies 3:117-132.

Rossi, P.H., J.D. Wright, and E. Weber-Burdin. 1982. Natural Hazards and Public Choice: The State and Local Politics of Hazard Mitigation. New York: Academic Press.

Rubin, C.B. 1991. Recovery from disaster. Pp. 224-259 in T.E. Drabek and G.J. Hoetmer (eds.) Emergency Management: Principles and Practice for Local Government. Washington, DC: International City Management Association.

Rubin, C.B. 1999. Disaster Time Line: Selected Milestone Events and U.S. Outcomes (1965-2000). Available at http://www.disaster-timeline.com.

Rubin, C.B., M.D. Saperstein, and D.G. Barbee. 1985. Community Recovery from a Major Natural Disaster. Monograph 41. Boulder, CO: Natural Hazards Research and Applications Information Center, Institute of Behavioral Science, University of Colorado.

Rubin, C.B., W.R. Cumming, and I. Renda-Tanali. 2003. Terrorism Time Line: Major Focusing Events and U.S. Outcomes (1993–2003). Available at http://www.disaster-timeline.com.

Rubonis, A.V., and L. Bickman. 1991. Psychological impairment in the wake of disaster: The disaster-psychopathology relationship. Psychological Bulletin 109(May):384-399.

Russell, L., J.D. Goltz, and L.B. Bourque. 1995. Preparedness and hazard mitigation actions before and after two earthquakes. Environment and Behavior 27:744-770.

Safwat, N., and H.Youssef. 1997. Texas hurricane evacuation time estimates. College Station, TX: Texas A&M University Hazard Reduction and Recovery Center.

Sagan, S.D. 1993. The Limits of Safety: Organizations, Accidents, and Nuclear Weapons. Princeton, NJ: Princeton University Press.

Salas, E., and J.A. Cannon Bowers. 2001. The science of training: A decade of progress. Annual Review of Psychology 52:471-499.

Saunders, S.L., and G.A. Kreps. 1987. The life history of organization in disaster. Journal of Applied Behavioral Science 23:443-462.

Schroeder, R.A. 1987. Gender Vulnerability to Drought: A Case Study of the Hausa Social Environment. Natural Hazards Research Working Paper 58. Boulder, CO: Natural Hazard Research and Applications Information Center, University of Colorado.

Schuman, H., and G. Kalton. 1985. Survey methods. Pp. 635-698 in G. Lindzey and E. Aronson (eds.) The handbook of social psychology (3rd Edition, Vol. 1). Reading, MA: Addison-Wesley.

Schwab, J., K.C. Topping, C.C. Eadie, R.E. Deyle, and R.A. Smith, R.A. 1998. Planning for Post-disaster Recovery and Reconstruction. PAS Report 483/484. Chicago, IL: American Planning Association.

Select Bipartisan Committee to Investigate the Preparation for and Response to Hurricane Katrina. 2006. A Failure of Initiative: Final Report of the Select Bipartisan Committee to Investigate the Preparation for and Response to Hurricane Katrina. Washington, DC: U.S. Government Printing Office.

Seligson, H.A., and K.I. Shoaf. 2002. Human Impacts of Earthquakes. In C. Scawthorn (ed.) Earthquake Engineering Handbook. Boca Raton, FL: CRC Press.

Seligson, H.A., K.I. Shoaf, C. Peek-Asa, and M. Mahue-Giangreco. 2002. Engineering-based earthquake casualty modeling: Past, present and future. Proceedings of the Seventh U.S. National Conference on Earthquake Engineering, Boston, MA, July 21-25.

Sen, A. 1981. Poverty and Famine: An Essay on Entitlement and Deprivation. Oxford: Oxford University Press.

Sexton, K., and J. Adgate. 1999. Looking at environmental justice from an environmental health perspective. Journal of Exposure Analysis and Environmental Epidemiology 9:3-8.

Sexton, K., L. Waller, R.B. McMaster, G. Maldonado, and J. Adgate. 2002. The importance of spatial scale for environmental health policy and research. Human and Ecological Risk Assessment 8(1):109-125.

Shefner, J. 1999. Pre- and post-disaster political instability and contentious supporters: A case study of political ferment. International Journal of Mass Emergencies and Disasters 17:37-160.

Shinozuka, M., S.E. Chang, R.T. Eguchi, D.P. Abrams, H.H.M. Hwang, and A. Rose. 1997. Advances in Earthquake Loss Estimation and Application to Memphis, Tennessee. Earthquake Spectra 13(4):739-758.

Shinozuka, M., A. Rose, and R.T. Eguchi (eds.) 1998. Engineering and Socioeconomic Impacts of Earthquakes: An Analysis of Electricity Lifeline Disruptions in the New Madrid Area. Monograph No.2. Buffalo, NY: Multidisciplinary Center for Earthquake Engineering Research.

Shoaf, K.I., H.S. Sareen, L.H. Nguyen, and L.B. Bourque. 1998. Injuries as a result of California earthquakes in the past decade. Disasters 22(3):218-235.

Shoaf, K.I., H.A. Seligson, C. Peek-Asa, and M. Mahue-Giangreco. 2000. Standardized Injury Categorization Schemes for Earthquake Related Injuries, produced for the National Science Foundation by the UCLA Center for Public Health and Disaster Relief. Available at http://www.ph.ucla.edu/cphdr/scheme.pdf (Accessed July 21, 2005).

Shoaf, K.I., H.A. Seligson, C. Peek-Asa, and M. Mahue-Giangreco. 2001. Enhancement of casualty models for post-earthquake response and mitigation. Proceedings U.S.-Japan Cooperative Research on Urban Earthquake Disaster Mitigation, Third Grantees Meeting, Seattle, WA, August 15-16.

Shoaf, K.I., H.A. Seligson, C. Peek-Asa, and M. Mahue-Giangreco. 2002. Advancing the science of casualty estimation: A standardized classification scheme. Proceedings of the Seventh U.S. National Conference on Earthquake Engineering. Boston, MA, July 21-25.

Shoaf, K., C. Sauter, L.B. Bourque, C. Giangreco, and B. Weiss. 2004. Suicides in Los Angeles County in relation to the Northridge earthquake. Prehospital and Disaster Medicine 19(4):307-310.

Showalter, P.S. 1993. Prognostication of doom: An earthquake prediction's effect on four small communities. International Journal of Mass Emergencies and Disasters 11:279-292.

Shrivastava, P. 1987. Bhopal: Anatomy of a Crisis. Cambridge, MA: Ballinger Publishing Co.

Shrivastava, P., and I. Mitroff. 1984. Enhancing organizational research utilization: The role of decision makers' assumptions. Academy of Management Review 9:18-26.

Siegel, J.M., L.B. Bourque, and K.I. Shoaf. 1999. Victimization after a natural disaster: Social disorganization or community cohesion? International Journal of Mass Emergencies and Disasters 17:265-294.

Siegel, J.M., K.I. Shoaf, and L.B. Bourque. 2000. The C-Mississippi Scale for PTSD in postearthquake communities. International Journal of Mass Emergencies and Disasters 18:339-346.

Sime, J.D. 1999. Crowd facilities, management and communications in disasters. Facilities 17:313-324.

Simile, C. 1995. Disaster Settings and Mobilization for Contentious Collective Action: Case Studies of Hurricane Hugo and the Loma Prieta Earthquake. Doctoral dissertation. Newark, DE: Department of Sociology and Criminal Justice, University of Delaware.

Simpson, D.M. 2001. Community emergency response training (CERTs): A recent history and review. Natural Hazards Review 2:54-63.

Sitarz, D. (ed.) 1993. Agenda 21: The Earth Summit Strategy to Save Our Planet. Covel, CA: Island Press.

Skinner, P., D.F. Borenstein, M. Perreault, M. Rubin, and S. Kobasa. 1991. It Can't Happen Here! Recent Significant Toxic Chemical Incidents: A Compendium and Discussion, New York State Attorney General's Office.

Slovic, P. 1987. Perception of risk. Science 236:280-285.

Slovic, P. 2000. The Perception of Risk. London: Earthscan Publications Ltd.

Slovic, P. 2005. Personal communication, February 11.

Slovic, P., H. Kunreuther, and G.F. White. 1974. Decision processes, rationality, and adjustment to natural Hazards. Pp. 187-205 in G.F. White (ed.) Natural Hazards: Local, National, Global. New York: Oxford University Press.

Slovic, P., M. Funucane, E. Peters, and D.G. MacGregor. 2004. Risk as analysis and risk as feelings: Some thoughts about affect, reason, risk, and rationality. Risk Analysis 24(2):311-322.

Smith, S.K., J. Tayman, D.A. Swanson. 2001. State and Local Population Projections: Methodology and Analysis, New York: Kluwer.

Sorensen, J.H. 1991. When shall we leave? Factors affecting the timing of evacuation departures. International Journal of Mass Emergencies and Disasters 9:153-165.

Sorensen, J.H. 2000. Hazard warning systems: Review of 20 years of progress. Natural Hazards Review 1:119-125.

Sorensen, J.H., and D.S. Mileti. 1987. Programs that encourage the adoption of precautions against natural hazards: Review and evaluation. Pp. 321-339 in N. Weinstein (ed.) Taking Care: Why People Take Precautions. New York: Cambridge University Press.

Sorensen, J.H., and G.O. Rogers. 1988. Local preparedness for chemical accidents: A survey of U.S. communities. Industrial Crisis Quarterly 2:89-108.

Sorensen, J.H., and G.O. Rogers. 1989. Warning and response in two hazardous materials transportation accidents in the U.S. Journal of Hazardous Materials 22:57-74.

Sorensen, J.H., G.O. Rogers and D. Fisher. 1988. Public alert and warning systems for chemical emergencies Proceedings of the AIChE Hazardous Material Spills Conference.

Southwest Educational Development Laboratory. 1996. A Review of the Literature on Dissemination and Knowledge Utilization. Austin, TX.

Spangle, W.E. 1987. Pre-Earthquake Planning for Post-Earthquake Rebuilding (PEPPER). Journal of Environmental Sciences 29:49-54.

Spangle Associates Urban Planning and Research. 1997. Evaluation of use of the Los Angeles recovery and reconstruction plan after the Northridge earthquake. Available at http://spangleassociates.com/.

Srinivasan, D. 2003. Battling hazards with a brand new tool. Planning 69(2):10-13.

Stallings, R.A. 1978. The structural patterns of four types of organizations in disaster. Pp. 87-104 in E.L. Quarantelli (ed.) Disasters: Theory and Research. Beverly Hills, CA: Sage.

Stallings, R.A. 1995. Promoting Risk: Constructing the Earthquake Threat. New York: De Gruyter.

Stallings, R.A. 2002. Methods of disaster research: Unique or not? Pp. 21-47 in R.A. Stallings (ed.) Methods of Disaster Research. Newark, DE: International Research Committee on Disasters.

Stallings, R.A., and E.L. Quarantelli. 1985. Emergent citizen groups and emergency management. Public Administration Review 45:93-100.

Stern, P.C., O.R. Young, and D. Druckman (eds.) 1992. Global Environmental Change: Understanding the Human Dimensions. Washington, DC: National Academy Press.

Stockwell, J.R., J.W. Sorenson, J.W. Eckert Jr., and E.M. Carreras. 1993. The U.S. EPA geographic information system for mapping environmental releases of toxic chemical release inventory (TRI) chemicals. Risk Analysis 13:155-164.

Suparamaniam, N., and S. Dekker. 2003. Paradoxes of power: The separation of knowledge and authority in international disaster relief work. Disaster Prevention and Management 12(4):312-318.

Sustein, C. 2002. Risk and Reason: Safety, Law, and the Environment. New York: Cambridge University Press.

Szasz, A., and M. Meuser. 1997. Environmental inequalities: Literature review and proposals for new directions in research and theory. Current Sociology 45(3):99-120.

Thomas, D.S.K. 2001. Data, data everywhere, but can we really use them? Pp. 61-76 in S.L. Cutter (ed.) American Hazardscapes: The Regionalization of Hazards and Disasters. Washington, DC: Joseph Henry Press.

Thompson, J.D. 1967. Organizations in Action. New York: McGraw-Hill.

Tierney, K.J. 1988. Social aspects of the Whittier Narrows earthquake. Earthquake Spectra 4:11-23.

Tierney, K.J. 1994. Emergency preparedness and Response. Pp. 105-128 in Practical Lessons from the Loma Prieta Earthquake. Washington, DC: National Academy Press.

Tierney, K.J. 1997a. Business impacts of the Northridge earthquake. Journal of Contingencies and Crisis Management 5(2):87-97.

Tierney, K.J. 1997b. Impacts of Recent Disasters on Businesses: The 1993 Midwest Floods and the 1994 Northridge Earthquake. Pp. 189-222 in B.G. Jones (ed.) Economic Consequences of Earthquakes: Preparing for the Unexpected. Buffalo, NY: National Center for Earthquake Engineering Research.

Tierney, K.J. 2000. Controversy and consensus in disaster mental health research. Prehospital and Disaster Medicine 15(4):181-187.

Tierney, K.J. 2002. The field turns fifty: Social change and the practice of disaster fieldwork. Pp. 349-374 in R. Stallings (ed.) Methods of Disaster Research. International Research Committee on Disasters. Philadelphia: Xlibris.

Tierney, K.J. 2003. Disaster beliefs and institutional interests: Recycling disaster myths in the aftermath of 9-11. Pp. 33-51 in L. Clarke (ed.) Terrorism and Disaster: New Threats, New Ideas. Research in Social Problems and Public Policy 11. New York: Elsevier Science Ltd.

Tierney, K.J. Forthcoming. Recent developments in U.S. homeland security policies and their implications for the management of extreme events. In H. Rodriguez, E.L. Quarantelli, and R.R. Dynes (eds.) Handbook of Disaster Research. New York: Springer.

Tierney, K.J., and J.M. Nigg. 1995. Business vulnerability to disaster-related lifeline disruption. Pp. 72-79 in Proceedings of the 4th U.S. Conference on Lifeline Earthquake Engineering. American Society of Civil Engineers.

Tierney, K.J., and J. Trainor. 2004. Networks and Resilience in the World Trade Center Disaster. Pp. 157-172 in Research Progress and Accomplishments 2003-2004 Buffalo, NY: State University of New York at Buffalo, Multidisciplinary Center for Earthquake Engineering Research.

Tierney, K.J., C. Bevc, and E. Kuligowski. Forthcoming. Metaphors matter: Disaster myths, media frames, and their consequences in Hurricane Katrina. Annals of the American Academy of Political and Social Science.

Tierney, K.J., and G.R. Webb. Forthcoming. Business Vulnerability to Earthquakes and Other Disasters. In E. Rovai and C.M. Rodrigue (eds.) Earthquakes. New York: Routledge.

Tierney, K.J., S.E. Chang, R.T. Eguchi, A. Rose, and M. Shinozuka. 1999. Improving earthquake loss estimation: Review, assessment and extension of loss estimation methodologies. Pp. 13-28 in Research Progress and Accomplishments, 1997-1999. Buffalo, NY: State University of New York at Buffalo, Multidisciplinary Center for Earthquake Engineering Research.

Tierney, K.J., M. Lindell, and R. Perry. 2001. Facing the Unexpected: Disaster Preparedness and Response in the United States. Washington, DC: Joseph Henry Press.

Tjosvold, D. 1995. Cooperation theory, constructive controversy, and effectiveness: Learning from crisis. Pp. 79-112 in R. A. Guzzo, E. Salas and Associates (eds.) Team effectiveness and decision making in organizations. San Francisco: Josey-Bass.

Tobin, G.A., and B.E. Montz, 2004. Natural hazards and technology: vulnerability, risk, and community response in hazardous environments. Pp. 547-570 in S.D. Brunn, S.L. Cutter, and J.W. Harrington, Jr. (eds.) Geography and Technology. Dordrecht: Kluwer Academic Publishers.

Tourangeau, R. 2004. Invited presentation to the committee, August 4.

Transportation Research Board. 1998. Highway Capacity Manual. Special Report 209 (3rd Edition). Washington, DC.

Troutman, T.W., R. Smith, and M.A. Rose. 2001. Situation Specific Call-to-Action Statements. NOAA Technical Memorandum NWS SR-202. National Weather Service. Fort Worth, TX: Scientific Services.

Tubbesing, T. 2004. Personal communication. EERI, July.

Turner, B.L. 1978. Man-Made Disasters. London: Wykeham.

Turner, B.L. II. 2005. Personal communication, April 18.

Turner, B.L. II, R.E. Kasperson, P.A. Matson, J.J. McCarthy, R.W. Corell, L. Christensen, N. Eckley, J.X. Kasperson, A. Luers, M.L. Martello, C. Polsky, A. Pulsipher, and A. Schiller. 2003a. A framework for vulnerability analysis in sustainability science. Proceedings of the National Academy of Sciences 100(14): 8074-8079.

Turner, B.L. II, P.A. Matson, J.J. McCarthy, R.W. Corell, L. Christensen, N. Eckley, G.K. Hovelsrud-Broda, J.X. Kasperson, R.E. Kasperson, A. Luers, M.L. Martello, S. Mathiesen, R. Naylor, C. Polsky, A. Pulsipher, A. Schiller, H. Selin, and N. Tyler. 2003b. Illustrating the coupled human-environment system for vulnerability analysis: Three case studies. Proceedings of the National Academy of Sciences 100(14):8080-8085.

Turner, R.H. 1994. Rumor as intensified information seeking: Earthquake rumors in China and the United States. Pp. 244-256 in R.R. Dynes and K.J. Tierney (eds.) Disasters, Collective Behavior, and Social Organization. Newark, DE: University of Delaware Press.

Turner, R.H., and L.M. Killian. 1987. Collective Behavior (3rd Edition). Englewood Cliffs, NJ: Prentice-Hall.

Turner, R.H., J.M. Nigg, and D. Paz. 1986. Waiting for Disaster: Earthquake Watch in California. Berkeley, CA: University of California Press.

Tweedie, S.W., J.R. Rowland, S.J. Walsh, R.P. Rhoten, and P.I. Hagle. 1986. A methodology for estimating emergency evacuation times. The Social Science Journal 23:189-204.

Ulin, David L. 2004. The Myth of Solid Ground. New York: Penguin.

United Church of Christ (UCC), Commission for Racial Justice. 1987. Toxic Wastes and Race: A National Report on the Racial and Socioeconomic Characteristics of Communities with Hazardous Waste Sites. New York.

United Nations Commission on Sustainable Development. 2001. Natural Disasters and Sustainable Development: Understanding the Links Between Development, Environment, and Natural Disasters Background Document for the World Summit on Sustainable Development. Available at http://www.unisdr.org (Accessed September 8, 2004).

United Nations Conference on Environment and Development (UNCED). 1992. Agenda 21: The United Nations Programme of Action from Rio. New York: United Nations.

United Nations Development Programme (UNDP). 2004. Reducing Disaster Risk: A Challenge for Development. Available at http://www.undp.org/bcrp/disred/english/publications/rdr.htm (Accessed September 20, 2004).

University of South Carolina Research. 2005. Bird's-eye View. Breakthrough 2005:14-16.

Uphoff, Norman. 1991. Fitting projects to people. In M. Cernea (ed.) Putting People First: Sociological Variables in Rural Development. Oxford, UK: Oxford University Press.

Urbanik, T. 1994. State of the art in evacuation time estimates for nuclear power plants. International Journal of Mass Emergencies and Disasters 12:327-343.

Urbanik, T. 2000. Evacuation time estimates for nuclear power plants. Journal of Hazardous Materials 75:165-180.

Urbanik, T., M.P. Moeller, and K. Barnes. 1988. Benchmark Study of the I-DYNEV Evacuation Time Estimate Computer Code, NUREG/CR-4873. Washington, DC: U.S. Nuclear Regulatory Commission.

U.S. Census Bureau. 2001. Current Population Survey (CPS)—Home Computers and Internet Use in the United States: August 2000. Available at http://www.census.gov/prod/2001pubs/p23-207.pdf.

U.S. Census Statistical Abstract. 2004. Table 21. Available at http://www.census.gov/prod/www/statistical-abstract-04.html.

U.S. Census Statistical Abstract. 2004. Table 23. Available at http://www.census.gov/prod/www/statistical-abstract-04.html.

U.S. Census Statistical Abstract. 2004. Table 117. (Health). Available at http://www.census.gov/prod/www/statistical-abstract-04.html.

U.S. Census Statistical Abstract. 2004. Table 213. Available at http://www.census.gov/prod/www/statistical-abstract-04.html.

U.S. Census Statistical Abstract. 2004. Table 216. Available at http://www.census.gov/prod/www/statistical-abstract-04.html.

U.S. Census Statistical Abstract. 2004. Table 648. Available at http://www.census.gov/prod/www/statistical-abstract-04.html.

U.S. Geological Survey. 2003. The Plan to Coordinate NEHRP Post-earthquake Investigations, Circular 1242. Reston, VA: U.S. Geological Survey.

U.S. President's Commission on the Accident at Three Mile Island. 1979. Report of the President's Commission on the Accident at Three Mile Island: The Need for Change—The Legacy of TMI. Washington, DC: U.S. Government Printing Office.

Vale, L., and T.J. Campanella (eds.) 2005. A Resilient City. London: Oxford University Press.

Van Meter, D.S., and C.E. Van Horn. 1975. The policy implementation process: A conceptual framework. Administration and Society 6:445-488.

Vaughan, D. 1996. The Challenger Launch Decision: Risky Technology, Culture, and Deviance at NASA. Chicago, IL: University of Chicago Press.

Vaughan, D. 1999. The dark side of organizations: Mistake, misconduct, and disaster. Annual Review of Sociology 25:271-305.

Wachtendorf, T. 2004. Improvising 9-11: Organizational Improvisation Following the World Trade Center Disaster. Newark, DE: doctoral dissertation, Department of Sociology, University of Delaware.

Waisel Laurie B., W.A. Wallace, and T. Willemain. 1998. Using diagrammatic reasoning in mathematical modeling: The sketches of expert modelers. Proceedings of the AAAI 1997 Fall Symposium on Reasoning with Diagrammatic Representation II. Menlo Park, CA: AAAI Press.

Warren, Roland. 1963. The Community in America. Chicago, IL: Rand McNally.

Watts, M.J., and H.G. Bohle. 1993. The space of vulnerability: The causal structure of hunger and famine. Progress in Human Geography 17(1):43-67.

Waugh, W.L., Jr. 2000. Living with Hazards; Dealing with Disasters: An Introduction to Emergency Management. Armonk, NY: M.E. Sharpe.

Waugh, W.L., Jr., and R.T. Sylves. 1996. Intergovernmental relations in emergency management. Pp. 46-68 in R.T. Silves and W.L. Waugh, Jr. (eds.) Disaster Management in the U.S. and Canada: The Politics, Policy making, Administration and Analysis of Emergency Management. Springfield, IL: Charles C. Thomas Publishers.

Waugh, W.L., Jr., and R.T. Sylves. 2002. Organizing the War on Terrorism. Public Administration Review 62:81-89.

Webb, G.R. 2002. Role improvising during crisis situations. International Journal of Emergency Management 2:47-61.

Webb, G.R., K.J. Tierney, and J.M. Dahlhamer. 2000. Businesses and disasters: Empirical patterns and unanswered questions. Natural Hazards Review1(2):83-90.

Webb, G.R., K.J. Tierney, and J.M. Dahlhamer. 2003. Predicting long-term business recovery from disaster: A comparison of the Loma Prieta Earthquake and Hurricane Andrew. Environmental Hazards 4:45-58.

Weick, K.E. 1993. The collapse of sense-making in organizations. Administrative Science Quarterly 38:628-652.

Wenger, D.E. 1978. Community response to disaster: Functional and structural alterations. Pp. in E.L. Quarantelli (ed.) Disasters: Theory and Research. Beverly Hills, CA: Sage.

Wenger, D.E., and T.F. James. 1994. The convergence of volunteers in a consensus crisis: The case of the 1985 Mexico City earthquake. Pp. 229-243 in R.R. Dynes and K.J. Tierney (eds.) Disasters, Collective Behavior, and Social Organization. Newark, DE: University of Delaware Press.

West, C., and D. Lenze. 1994. Modeling the regional impact of natural disaster and recovery. International Regional Science Review 17:121-150.

White, G., and J. Haas. 1975. Assessment of Research on Natural Hazards. Cambridge, MA: MIT Press.

White House. 2006. The Federal Response to Hurricane Katrina: Lessons Learned. Washington, DC: U.S. Government Printing Office.

Whitney, D.J., and M.K. Lindell. 2000. Member commitment and participation in local emergency planning committees. Policy Studies Journal 28:467-484.

Whitney, D.J., A. Dickerson, and M.K. Lindell. 2001. Nonstructural seismic preparedness of Southern California hospitals. Earthquake Spectra 17:153-171.

Whitney, D.J., M.K. Lindell, and D.H. Nguyen. 2004. Earthquake beliefs and adoption of seismic hazard adjustments. Risk Analysis 24:87-102.

Williams, H. 1954. Fewer disasters: Better studied. Journal of Social Issues 10:5-11.

Wisner, B. 2001. Risk and neoliberal state: Why post-Mitch lessons didn't reduce El Salvador's earthquake losses. Disasters 25(3):251-268.

Working Group on Earthquake Hazards Reduction. 1978. Earthquake Hazards Reduction: Issues for an Implementation Plan. Washington, DC: Office of Technology Assessment.

Working Group on "Governance Dilemmas" in Bioterrorism Response. 2004. Leading during bioattacks and epidemics with the public's trust and help. Biosecurity and Bioterrorism: Biodefense Strategy, Practice, and Science 2(1):25-40.

World Bank. 1990. The World Bank and the Environment. Washington, DC: World Bank Press.

World Commission on Environment and Economic Development (WCED). 1987. Our Common Future (also known as the Brundtland Commission report). Oxford, UK: Oxford University Press.

Wright, J.D., P.H. Rossi, S.R. Wright, and E. Weber-Burdin. 1979. After the Clean-Up: Long-Range Effects of Natural Disasters. Beverly Hills, CA: Sage.

Wu, J.Y., and M.K. Lindell. 2004. Housing reconstruction after two major earthquakes: The 1994 Northridge earthquake in the United States and the 1999 Chi-Chi earthquake in Taiwan. Disasters 28:63-81.

Yandle, T., and D. Burton. 1996. Reexamining environmental justice: A statistical analysis of historical hazardous waste landfill siting patterns in metropolitan Texas. Social Science Quarterly 77(3):477-492.

Yezer, A. 2002. The economics of natural disasters. Pp. 213-235 in R. Stallings (ed.) Methods of Disaster Research. International Research Committee on Disasters. Philadelphia: Xlibris.

Yin, R., and G. Andranovich. 1987. Getting Research Used in the Natural Hazards Field: The Role of Professional Associations. Washington, DC: Cosmos Corporation.

Yin, R., and G. Moore. 1985. The Utilization of Research: Lessons Learned from the Natural Hazards Field. Washington, DC: Cosmos Corporation.

Zerubavel, E. 1981. Hidden Rhythms: Schedules and Challenges of Social Life. Chicago, IL: University of Chicago Press.

Zerubavel, E. 1997. Social Mindscapes. Cambridge, MA: Harvard University Press.

Zerubavel, E. 2003. Time Maps: Collective Memory and the Social Shape of the Future. Chicago, IL: University of Chicago Press.

Zhang, Y., C.S. Prater, and M.K. Lindell. 2004a. Risk area accuracy and evacuation from Hurricane Bret. Natural Hazards Review 5:115-120.

Zhang, Y., M.K. Lindell, and C.S. Prater. 2004b. Modeling and managing the vulnerability of community businesses to environmental disasters. Paper presented at the 45th Annual Conference of the Association of Collegiate Schools of Planning, Portland, OR, October 23.

Ziegler, D.J., S. D. Brunn, and J.H. Johnson. 1981. Evacuation from a nuclear technological disaster. Geographical Review 71:1-16.

Zimmerman, R. 1994. Issues of classification in environmental equity: How we manage is how we measure. Fordham Urban Law Journal 21:633-670.

# Appendixes

# Appendix A

# Acronyms

| | |
|---|---|
| AAAS | American Association for the Advancement of Science |
| AAG | Association of American Geographers |
| AAG GCLP | Association of American Geographers Global Change and Local Places |
| ABAG | Association of Bay Area Governments |
| APA | American Planning Association |
| ASCE | American Society of Civil Engineers |
| ASFPM | Association of State Floodplain Managers |
| AVHRR | Advanced Very High Resolution Radiometer |
| | |
| CAT | Computerized Axial Tomography |
| CATI | Computer-Assisted Telephone Interviewing |
| CERT | Community Emergency Response Training |
| CMS | Civil and Mechanical Systems |
| CPE | Complex Political Emergency |
| CSEPP | Chemical Stockpile Emergency Preparedness Program |
| CTA | Call-to-Action |
| | |
| DEM | Division of Emergency Management |
| DHS | Department of Homeland Security |
| DMA 2000 | Disaster Mitigation Act of 2000 |
| DOD | Department of Defense |
| DOI | Department of the Interior |
| DRC | Disaster Research Center |

DRSS            Disaster Research in the Social Sciences

EERC            Earthquake Engineering Research Center
EERI            Earthquake Engineering Research Institute
EMI             Emergency Management Institute
EMON            Emergent Multi-Organizational Network
EOC             Emergency Operations Center
EOP             Executive Office of the President
EPA             Environmental Protection Agency
EqTAP           Development of Earthquake and Tsunami Disaster
                Mitigation Technologies and Their Integration for the
                Asia-Pacific Region
ETE             Evacuation Time Estimate

FDAA            Federal Disaster Assistance Administration
FEMA            Federal Emergency Management Agency
FIRESCOPE       Firefighting Resources of California Organized for Potential
                Emergencies

GAO             Government Accountability Office
GIS             Geographic Information System
GNP             Gross National Product
GPS             Global Positioning Sysetm

HBCU            Historically Black College and University
HMGP            Hazard Mitigation Grant Program
HRL             Hazards Research Lab
HRRC            Hazard Reduction and Recovery Center
HSAS            Homeland Security Advisory System
HSD             Human and Social Dynamics
HSPD            Homeland Security Presidential Directive
HVA             Hazard Vulnerability Analysis

IBHS            Institute for Business and Home Safety
ICS             Incident Command System
IDNDR           International Decade for Natural Disaster Reduction
IDR             Interdisciplinary Research
IFRC            International Federation of Red Cross Red Crescent Societies
IPCC            Intergovernmental Panel on Climate Change
IRB             Institutional Review Board
IRCD            International Research Committee on Disasters
ISA             International Sociological Association
ISDR            International Strategy for Disaster Reduction

LA RED      Network of Social Studies in Disaster Prevention
LAC-DHS      Los Angeles County Department of Health Services
LFE      Learning from Earthquakes Program
LIDAR      Light Detection and Ranging
LTER      Long-Term Ecological Research

MAE      Mid-America Earthquake Center
MCEER      Multidisciplinary Center for Earthquake Engineering Research
MRI      Magnetic Resonance Imaging

NAPA      National Academy of Public Administration
NAS      National Academy of Sciences
NASA      National Aeronautics and Space Administration
NCEER      National Center for Earthquake Engineering Research
NCHM      National Center for Hazards Mitigation
NEES      Network for Earthquake Engineering Simulation
NEHRP      National Earthquake Hazards Reduction Program
NEXRAD      Next Generation Radar
NFIP      National Flood Insurance Program
NGO      Nongovernmental Organization
NHRAIC      Natural Hazards Research and Applications Information Center
NIBS      National Institute of Building Sciences
NIMS      National Incident Management System
NIST      National Institute of Standards and Technology
NOAA      National Oceanic and Atmospheric Administration
NRC      National Research Council
NSC      National Security Council
NSF      National Science Foundation
NSN      Neighborhood Survival Network
NVOAD      National Voluntary Organizations Active in Disaster
NWS      National Weather Service
NWSCA      National Water and Soil Conservation Authority

OES      Office of Emergency Services
OSTP      Office of Science and Technology Policy

PDD      Presidential Disaster Declaration
PEER      Pacific Earthquake Engineering Research Center
PEPPER      Pre-Earthquake Planning for Post-Earthquake Rebuilding
PERI      Public Entity Research Institute
PTSD      Post-Traumatic Stress Disorder

REU          Research Experience for Undergraduates
RSAI         Regional Science Association International

SARS         Sudden Acute Respiratory Syndrome
SCEC         Southern California Earthquake Center
SCEPP        Southern California Earthquake Preparedness Project
SGER         Small Grants for Exploratory Research
SHELDUS      Spatial Hazard Events and Losses Database for the United
             States
SME          Subject Matter Expert
STIM         School of Travel Industry Management

TSA          Transportation Security Administration

UCC          United Church of Christ
UCLA         University of California, Los Angeles
UNCED        UN Conference on Environment and Development
UNDP         United Nations Development Programme
USC          University of South Carolina
USDA         U.S. Department of Agriculture
USGS         U.S. Geological Survey

WCED         World Commission on Economic Development

# Appendix B

# Recommendations

**Recommendation 3.1:** *Research should be conducted to assess the degree to which hazard event characteristics affect physical and social impacts of disasters and, thus, hazard mitigation and preparedness for disaster response and recovery.*

**Recommendation 3.2:** *Research should be conducted to refine the concepts involved in all three components (hazard exposure, physical vulnerability, social vulnerability) of hazard vulnerability analysis (HVA).*

**Recommendation 3.3:** *Research should be conducted to identify better mechanisms for intervening into the dynamics of hazard vulnerability.*

**Recommendation 3.4:** *Research should be conducted to identify the factors that promote the adoption of more effective community-level hazard mitigation measures.*

**Recommendation 3.5:** *Research should be conducted to assess the effectiveness of hazard mitigation programs.*

**Recommendation 3.6:** *Research should be conducted to identify the factors that promote the adoption of more effective emergency response preparedness measures.*

**Recommendation 3.7:** *Research should be conducted to assess the extent to* *which disaster research findings are being implemented in local emergency operations plans, procedures, and training.*

**Recommendation 3.8:** *Research is needed to identify the factors that promote the adoption of more effective disaster recovery preparedness measures.*

**Recommendation 3.9:** *Research should be conducted to develop better models to guide protective action decision making in emergencies.*

**Recommendation 3.10:** *Research is needed on training and exercising for disaster response.*

**Recommendation 3.11:** *Research should be conducted to develop better models of hazard adjustment adoption and implementation by community organizations.*

**Recommendation 3.12:** *There is a continuing need for further research on hazard insurance.*

**Recommendation 4.1:** *Future research should focus on further empirical explorations of societal vulnerability and resilience to natural, technological, and willfully caused hazards and disasters.*

**Recommendation 4.2:** *Future research should focus on the special requirements associated with responding to and recovering from willful attacks and disease outbreaks.*

**Recommendation 4.3:** *Future research should focus on the societal consequences of changes in government organization and in emergency management legislation, authorities, policies and plans that have occurred as a result of the terrorist attacks of September 11, 2001, as well as on changes that will almost certainly occur as a result of Hurricane Katrina.*

**Recommendation 4.4:** *Research is needed to update current theories and findings on disaster response and recovery in light of changing demographic, economic, technological, and social trends such as those highlighted in Chapter 2 and elsewhere in this report.*

**Recommendation 4.5:** *More research is needed on response and recovery for near-catastrophic and catastrophic disaster events.*

**Recommendation 4.6:** *More cross-societal research is needed on natural, technological, and willfully caused hazards and disasters.*

**Recommendation 4.7:** *Taking into account both existing research and future research needs, sustained efforts should be made with respect to data archiving, sharing, and dissemination.*

**Recommendation 5.1:** *As NSF funding for the three earthquake engineering research centers (EERCs) draws to a close, NSF should institute mechanisms to sustain the momentum that has been achieved in interdisciplinary hazards and disaster research.*

**Recommendation 5.2:** *The hazards and disaster research community should take advantage of current, unique opportunities to study the conditions, conduct, and contributions of interdisciplinary research itself.*

**Recommendation 5.3:** *NSF should support the establishment of a National Center for Social Science Research on Hazards and Disasters.*

**Recommendation 6.1:** *Priority should be given to international disaster research that emphasizes multiple case research designs, with each case using the same methods and variables to ensure comparability.*

**Recommendation 6.2:** *Common indicators of disaster risk and development should be constructed.*

**Recommendation 6.3:** *Collaborative international research projects should be the modal form of cross-national research on disasters and development.*

**Recommendation 7.1:** *The National Science Foundation and Department of Homeland Security should jointly support the establishment of a nongovernmental Panel on Hazards and Disaster Informatics. The panel should be interdisciplinary and include social scientists and engineers from hazards and disaster research as well as experts on informatics issues from cognitive science, computational science, and applied science. The panel's mission should be (1) to assess issues of data standardization, data management and archiving, and data sharing as they relate to natural, technological, and willful hazards and disasters, and (2) to develop a formal plan for resolving these issues to every extent possible within the next decade.*

**Recommendation 7.2:** *The National Science Foundation and Department of Homeland Security should fund a collaborative Center for Modeling, Simulation, and Visualization of Hazards and Disasters. The recommended center would be the locus of advanced computing and communications technologies that are used to support a distributed set of research methods and facilities. The center's capabilities would be accessible on a shared-use basis.*

**Recommendation 7.3:** *The hazards and disaster research community should educate university Institutional Review Boards (IRBs) about the unique benefits of, in particular, post-disaster investigations and the unique constraints under which this research community performs research on human subjects.*

**Recommendation 8.1:** *Renewed attention should be given by the social science hazards and disaster research community to the need for formal evaluation research on knowledge utilization in the field. New research should be carried out using all of the relevant methodologies and technologies available to the social sciences today.*

**Recommendation 8.2:** *Building on earlier practice, social scientists should conduct research utilization studies involving knowledge on hazards and disasters produced by other research disciplines.*

**Recommendation 8.3:** *Cross-cultural research utilization studies should be pursued by social scientists. Such research could contribute to global understanding of knowledge dissemination and application.*

**Recommendation 9.1:** *Relevant stakeholders should develop an integrated strategy to enhance the capacity of the social science hazards and disaster research community to respond to societal needs, which are expected to grow, for knowledge creation and application. A workshop should be organized to serve as a launching pad for facilitating communication, coordination, and planning among stakeholders from government, academia, professional associations, and the private sector. Representatives from the NSF and DHS should play key roles in the workshop because of their historical (NSF) and more recent (DHS) shared commitment to foster the next generation of hazards and disaster researchers.*

**Recommendation 9.2:** *NSF should expand its investments in both undergraduate and graduate education to increase the size of the social science hazards and disaster research workforce and its capacity to conduct needed disciplinary, multidisciplinary, and interdisciplinary research on the core topics discussed in this report. NSF should also give special consideration to investing in innovative ways to further workforce development, especially when they involve partnerships such as NSF's recent joint initiative with the Public Entity Research Institute (PERI) and the Natural Hazards Research and Applications Information Center at the University of Colorado. This initiative exemplifies the collaboration needed across government, academia, professional associations, and the private sector.*

**Recommendation 9.3:** *In parallel fashion, DHS should make a conscious effort to increase significantly the number of awards its makes to social science students through its scholarship and fellowship program. Because much that must be investigated about the terrorist threat is related to social and institutional forces, more social scientists need to be recruited to adequately study them. With its broader cross-hazards congressional mandate, DHS should contribute to a larger social science hazards and disaster research workforce, one that complements research in other science and engineering disciplines.*

**Recommendation 9.4:** *NSF and DHS should consider ways in which they can cooperate programmatically to enhance the social science hazards and disaster research workforce. Jointly sponsored university research and education programs by the two agencies would be of major benefit to the nation.*

**Recommendation 9.5:** *As the leader in furthering U.S. science through research and workforce development, NSF should make greater use of its enabling mechanisms, including standard research grants, center grants, grant supplements, and REU (Research Experience for Undergraduate) programs to attract more minorities to the social science hazards and disaster research workforce.*

**Recommendation 9.6:** *The NSF Enabling Project for junior faculty development should be continued if the second pilot proves to be a success.*

**Recommendation 9.7:** *Stakeholders in government, academia, professional societies, and the private sector should be open to exploring a variety of innovative approaches for developing the future social science hazards and disaster research workforce.*

# Appendix C

# Committee Biographies

GARY A. KREPS, *Chair*, is former vice provost and now professor emeritus at the College of William and Mary. Following completion of his Ph.D. in sociology at Ohio State University in 1971, he began and continued his career as a faculty member and administrator at William and Mary until retiring in July 2005. Dr. Kreps has long-standing research interests in organizational and role theories as both relate to structural analyses of community, regional, and societal responses to natural, technological, and willful hazards and disasters. He has served as a staff member, consultant, or member on five National Academies committees: the Committee on the Socioeconomic Effects of Earthquake Prediction (1976-1978), the Committee on U.S. Emergency Preparedness (1979-1981), the Committee on International Disaster Assistance (1978-1980), the Committee on Mass Media Reporting of Disasters (1978-1980), and the Committee on Disaster Research in the Social Sciences (2004-2006). Over the course of the past two decades, Dr. Kreps and his collaborators have developed taxonomies and theories of organizing and role enactment during the emergency periods of disasters. Major findings from his research program have been reported in two books and articles in *Sociological Theory, Annual Review of Sociology, American Sociological Review, American Journal of Sociology, Journal of Applied Behavioral Science, International Journal of Mass Emergencies and Disasters*, and many other basic and applied publications. Dr. Kreps' 2001 entry in the *International Encyclopedia of the Social and Behavioral Sciences* ("Disaster, Sociology of") emphasizes the need to reconcile functionalist and constructivist conceptions of disasters as acute systemic events.

PHILLIP R. BERKE is currently professor of land-use and environmental planning in the Department of City and Regional Planning at the University of North Carolina, Chapel Hill. Dr. Berke is also senior research associate of the New Zealand International Global Change Institute, and a research fellow at the Lincoln Institute of Land Policy. He previously served as associate director of the Hazard Reduction and Recovery Center at Texas A&M University. Dr. Berke's research interests include land-use and environmental planning, state and local development management, sustainable development, and natural hazard mitigation in developed and developing communities. The central focus of his research is to develop a deeper understanding of the connections between human settlements and the natural environment. His research seeks to explore the causes of land-use decisions, how these decisions impact natural environmental systems, and the consequences of these impacts on human settlements. His ultimate goal is to seek solutions to complex urban development problems that help communities live within the limits of natural systems. His current research focuses on a comparative evaluation of the impacts of compact and low-density sprawl development patterns on watersheds in the Eastern United States. He is also studying the influence of New Zealand's national planning mandate that requires local governments to prepare and implement environmental plans, achieve national environmental goals, and advance land-use patterns that support sustainable outcomes. A feature of this mandate that is being investigated involves how well local plans have redressed human rights violations of the indigenous people of New Zealand—the Maori. Dr. Berke received his Ph.D. in urban and regional science from Texas A&M University.

THOMAS A. BIRKLAND is an associate professor of public administration and policy, and political science at the University at Albany, State University of New York, where he also directs the Center for Policy Research. Dr. Birkland received his Ph.D. in political science from the University of Washington. His research interests are concerned with the impact of disasters and crises on media and policy makers' agendas, resulting in a reprioritizing of perceived important problems. Dr. Birkland was a 1993-1994 Earthquake Engineering Research Institute-Federal Emergency Management Agency (EERI-FEMA) fellow, as well as a faculty fellow in social science research applied to hazards and disasters (the first "Enabling Project"). He has written several articles about natural hazards policy and politics. Most recently, Dr. Birkland was a plenary speaker and moderator at the 9/11 Summit on emergency planning and management for the judiciary and is currently a member of the EERI Social Science/Learning from Earthquakes committee.

STEPHANIE E. CHANG is an associate professor at the University of British Columbia and a joint faculty member with the School of Community and Regional Planning (SCARP) and the Institute for Resources, Environment, and Sustainability (IRES). Dr. Chang also holds a Canada Research Chair position in disaster management and urban sustainability. She previously served as a research assistant professor in the Department of Geography at the University of Washington. Dr. Chang received her Ph.D. in regional science from Cornell University. Her work aims to bridge the gap between engineering, natural sciences, and social sciences in addressing the complex issues of natural disasters. Her research has focused on developing integrated regional models for estimating losses from future earthquakes. She has also developed methods for assessing disaster mitigation strategies and inquired into how disasters impact regional economies. Dr. Chang's current research addresses community disaster resilience and sustainability, mitigation of infrastructure system risks (e.g., electric power, water, and transportation), and urban disaster recovery. Dr. Chang was awarded the 2001 Shah Family Innovation Prize by EERI and serves on the editorial board of *Earthquake Spectra*.

SUSAN L. CUTTER is the director of the Hazards Research Lab, a research and training center that integrates geographical information processing techniques with hazards analysis and management, as well as a Carolina Distinguished Professor of Geography at the University of South Carolina. She is the cofounding editor of an interdisciplinary journal, Environmental Hazards, published by Elsevier. She has worked in the risk and hazards fields for more than 25 years and is a nationally recognized scholar in this field. She has authored or edited eight books and more than 50 peer-reviewed articles. In 1999, Dr. Cutter was elected as a fellow of the American Association for the Advancement of Science (AAAS), a testimonial to her research accomplishments in the field. Her stature within the discipline of geography was recognized by her election as president of the Association of American Geographers in 1999-2000. Dr. Cutter received her Ph.D. in geography from the University of Chicago.

MICHAEL K. LINDELL is former director of the Hazard Reduction and Recovery Center at Texas A&M University. He received his Ph.D. in social psychology from the University of Colorado with a specialty in disaster research and has completed hazardous materials emergency responder training through the hazardous materials specialist level. Dr. Lindell has more than 25 years of experience in the field of emergency management, during which time he has conducted a program of research on the processes by which individuals and organizations respond to natural and technological hazards. In addition, he has had extensive experience in providing technical assistance to government agencies, industry groups, and private corporations in development of emergency plans and procedures. Dr. Lindell has written extensively on emergency management and is the author of more than 60 technical reports, 60 journal articles and book chapters, and 5 books/monographs. His research has examined the processes by which affected populations respond to warnings of the imminent threat of a natural or technological hazard. His organizational research has examined the effects of disaster experience and the community planning process upon the development of adaptive strategies for promoting emergency preparedness. Dr. Lindell has served as an adjunct faculty for FEMA's National Emergency Training Center, lecturing on disaster psychology and public response to warning. He also has been an instructor in other workshops federal agencies have sponsored for state and local emergency planners throughout the country, and appeared as a panelist in conferences on protective actions in hazardous materials emergencies. In addition, Dr. Lindell has been a consultant to five of the Department of Energy National Laboratories on a variety of topics in the area of emergency preparedness and response.

ROBERT A. OLSON is president of Robert Olson Associates, Inc., which consults in such areas as vulnerability analysis and loss estimation, hazard mitigation and prevention, and emergency planning and operations. Clients of the 22-year-old firm have included the Kajima Corporation and other private companies as well as numerous public safety organizations. Previously, from 1975-1982, Mr. Olson served as the first executive director of the California Seismic Safety Commission. He has chaired numerous committees, including the Advisory Committee of the National Information Service for Earthquake Engineering, the California Governor's Task Force on Earthquake Preparedness, and the Advisory Group on Disaster Preparedness to the California Joint Legislative Committee on Seismic Safety. He received the 2001 Alfred E. Alquist Award for Achievement in Earthquake Safety from the Earthquake Safety Foundation, and in 2004 was awarded an honorary membership in the Earthquake Engineering Research Institute. Mr. Olson received his bachelor's degree in political science from the University of California at Berkeley and his master's degree from the University of Oregon.

JUAN M. ORTIZ is the emergency management coordinator for the City of Fort Worth Tarrant County Office of Emergency Management. Mr. Ortiz has 10 years experience in the field of emergency management, having also served as emergency management coordinator for the City of Corpus Christi, Texas. He is the former president of the Coastal Bend Emergency Management Association and former chairman of the International Association of Emergency Managers Texas Coastal Advisory Team. He currently serves on the Board of Directors of the Emergency Management Association of Texas. Mr. Ortiz received his B.S. in emergency administration and planning from the University of North Texas.

KIMBERLY I. SHOAF is an adjunct assistant professor in the Department of Community Health Sciences at the University of California, Los Angeles (UCLA) School of Public Health and research director of the UCLA Center for Public Health and Disasters. Dr. Shoaf's expertise is in the combination of qualitative and quantitative methodologies for studying social and health impacts of disasters. In addition to the chapter on "Disaster Public Health" in the *Encyclopedia of Public Health*, she has also recently coauthored a chapter on "Human Impacts of Earthquakes" for the *CRC Handbook of Earthquake Engineering*. Her research interests include disaster impacts on physical injuries, agency utilization in disasters, international health, public health impact of disasters, program planning and evaluation, and health in the Latino community. She has also focused on the role of academia in preparing the U.S. population for bioterrorism; as well as standardizing definitions and procedural protocols to describe structural damage and injury, and refining casualty estimation models. Dr. Shoaf received her Ph.D. in community health sciences from UCLA.

JOHN H. SORENSEN is a distinguished research staff member at Oak Ridge National Laboratory (ORNL). Dr. Sorensen has been involved with research on emergency planning and disaster response for more than 25 years. He has been the principal investigator on over 40 major projects for federal agencies including FEMA, the Department of Energy, the U.S. Environmental Protection Agency (EPA), the Department of Defense, the National Academies, and the U.S. Chemical Safety and Hazard Investigation Board. Dr. Sorensen has participated in research including the Three Mile Island Public Health Fund Emergency Planning Project on Three Mile Island and the Second Assessment of Research on Natural Hazards where he served as the subgroup leader for prediction, forecast warning, and emergency planning. He has also authored more than 140 professional publications including *Impacts of Hazardous Technology: The Psycho-Social Effects of Restarting TMI-1*. He has published extensively on response to emergency warnings, risk communications, organizational effectiveness in disasters, emergency evacuation, and protective actions for chemical emergencies. Sorensen has led the development of emergency management information systems, simulation models, conventional and interactive training courses, and educational videos. He has served on many advisory committees including the Natural Hazard Research and Information Applications Center at the University of Colorado, the Atomic Industrial Forum's National Environmental Studies Task Force on Emergency Evacuation, the International

City Management Association's Emergency Management "Emergency Planning Greenbook" Project, and FEMA's Emergency Management Technology Steering Group. He was a member of the National Research Council, Commission on Physical Sciences, Mathematics, and Resources, Earth Sciences Board, Subcommittee on Earthquake Research. Dr. Sorensen received his Ph.D. in geography from the University of Colorado at Boulder and was an assistant professor at the University of Hawaii.

KATHLEEN J. TIERNEY is currently professor of sociology and director of the Natural Hazards Research and Applications Information Center at the University of Colorado at Boulder. Prior to moving to Boulder in 2003, Dr. Tierney was professor of Sociology and Director of the Disaster Research Center at the University of Delaware. With more than 25 years of experience in the disaster field, she has been involved in research on the social aspects and impacts of major earthquakes in California and Japan, floods in the Midwest, Hurricanes Hugo and Andrew, and many other major natural and technological disaster events. Since September 11, 2001, she has been directing a study on the organizational and community response in New York following the terrorist attack on the World Trade Center. Her other recent research projects include a study on public perceptions of the earthquake threat in the Northern California Bay Area. Dr. Tierney is the author of dozens of articles, book chapters, and technical reports on the social aspects of hazards, disasters, and risk. She is a member of the National Construction Safety Team Advisory Committee, which is overseeing the official federal investigation of the World Trade Center disaster. Dr. Tierney earned a Ph.D. in sociology from the University of Ohio.

WILLIAM A. WALLACE is a professor of decision sciences and engineering systems at Rensselaer Polytechnic Institute and holds joint appointments in cognitive sciences and civil and environmental engineering; he is the research director of Rensselaer's Center for Infrastructure and Transportation Studies. Dr. Wallace has more than 20 years experience in developing, implementing, and evaluating decision support systems. His current research includes decision support for group improvisation, trust and knowledge management, and decision technologies for emergency response and restoration and incident management. He is cofounder and coeditor of *Computational and Mathematical Organization Theory*. Dr. Wallace has authored and edited 6 books and more than 70 articles and papers—out of a total of more than 200 archival publications. Dr. Wallace received his Ph.D. in management science from Rensselaer Polytechnic Institute.

ANTHONY M. YEZER is professor of economics at George Washington University. Dr. Yezer also serves as special consultant to the National Economic Research Associates (N/E/R/A). He previously held positions as economic consultant to the Department on Housing and Urban Development and the World Bank. His primary areas of research are regional and urban economics, the effects of public policy on the location of economic activity, and applied microeconomic theory. Dr. Yezer received his Ph.D. in economics and urban studies from the Massachusetts Institute of Technology.